Geological

and

Soil Evidence

Forensic Applications

Geological
and
Soil Evidence
Forensic Applications

Kenneth Pye

CRC Press
Taylor & Francis Group
Boca Raton London New York

CRC Press is an imprint of the
Taylor & Francis Group, an informa business

CRC Press
Taylor & Francis Group
6000 Broken Sound Parkway NW, Suite 300
Boca Raton, FL 33487-2742

© 2007 by Taylor & Francis Group, LLC
CRC Press is an imprint of Taylor & Francis Group, an Informa business

International Standard Book Number-10: 0-8493-3146-3 (Hardcover)
International Standard Book Number-13: 978-0-8493-3146-6 (Hardcover)

Library of Congress Cataloging-in-Publication Data

Pye, Kenneth.
    Geological and soil evidence : forensic applications / Kenneth Pye.
      p. cm.
    ISBN 978-0-8493-3146-6 (alk. paper)
    1. Forensic geology. 2. Environmental forensics. 3. Soil science. I. Title.

QE38.5.P94 2007
363.25--dc22
                                                  2007060503

Visit the Taylor & Francis Web site at
http://www.taylorandfrancis.com

and the CRC Press Web site at
http://www.crcpress.com

Dedication

To my father, who instilled in me a desire to know the cause of things.

Contents

Preface   xiii

Acknowledgments   xv

Author   xvii

1   Introduction: The Nature and Development of Forensic Geology   1

1.1   The Nature of Forensic Geology and Forensic Soil Science 1
1.2   Development of the Use of Geological and Soil Materials
     as Trace Evidence 3
1.3   Other Types of Geoscience Data Useful in Forensic
     Investigations 8
     1.3.1   Geophysical Survey Data 8
     1.3.2   Meteorological Data 9
     1.3.3   River Flow and Lake Level Data 9
     1.3.4   Tidal Data 10
     1.3.5   Geomorphological Data 10

2   Types of Geological and Soil Materials That May be Useful as Trace Evidence   11

2.1   Overview of Material Types and Available
     Analytical Strategies 11
2.2   Rocks 12
     2.2.1   The General Character of Rocks 12
     2.2.2   Igneous Rocks 13
     2.2.3   Metamorphic Rocks 13
     2.2.4   Sedimentary Rocks 16
2.3   Sediments 17
     2.3.1   Gravel 21
     2.3.2   Sand 22
     2.3.3   Mud 27

2.4   Soils 30
     2.4.1   Soil Properties and Classification 30
     2.4.2   Soil Constituents 31
2.5   Dusts and Particulates 32
     2.5.1   Natural Dusts Derived from Soils and Sediments 32
     2.5.2   Industrial Dusts 33
     2.5.3   Household and Street Dusts 33
     2.5.4   Particulates in Water Bodies 33
2.6   Minerals 34
     2.6.1   The General Nature of Minerals 34
     2.6.2   Light Minerals 35
     2.6.3   Heavy Minerals 35
     2.6.4   Rare Earth Minerals 37
     2.6.5   Clay Minerals 38
2.7   Glasses and Other Amorphous Materials 41
     2.7.1   Natural Glasses 41
     2.7.2   Natural Soaps and Gels 41
2.8   Fossils 42
     2.8.1   General Nature 42
     2.8.2   Macrofossils 42
     2.8.3   Macroscopic Charcoal 44
     2.8.4   Microfossils 46
         2.8.4.1   Pollen and Spores 46
         2.8.4.2   Diatoms and Other Algae 46
         2.8.4.3   Foraminifera 48
         2.8.4.4   Calcareous Nannoplankton:
                  Coccolithophores and Discoasters 49
         2.8.4.5   Testate Amoebae 49
         2.8.4.6   Ostracods 50
         2.8.4.7   Opal Phytoliths 50
         2.8.4.8   Micro-charcoal 51
2.9   Anthropogenic Materials 51
     2.9.1   General 51
     2.9.2   Concrete 51
     2.9.3   Bricks 52
     2.9.4   Tiles, Pipes, Pottery, and Other Ceramics 53
     2.9.5   Slag and Clinker 53
     2.9.6   Metallic Fragments 53
     2.9.7   Man-Made Glass Particles 55
     2.9.8   Paint 55
     2.9.9   Paper 56
     2.9.10   Fibers 56
     2.9.11   Other Unusual Particle Types 57

3   Bulk Properties of Geological and Soil Materials   59

3.1 Physical Characteristics 59
    3.1.1 Color 60
    3.1.2 Density 68
    3.1.3 Hardness 69
    3.1.4 Microfabric 69
    3.1.5 Porosity and Permeability 71
    3.1.6 Specific Surface 72
    3.1.7 Magnetic Characteristics 73
    3.1.8 Particle Size Distributions 75
    3.1.9 Shape Frequency Distributions 82
3.2 Chemical Characteristics 86
    3.2.1 pH 86
    3.2.2 Eh 86
    3.2.3 Electrical Conductivity 87
    3.2.4 Cation Exchange Capacity 87
    3.2.5 Anions 88
    3.2.6 Major and Trace Elements 89
    3.2.7 Rare Earth Elements 93
    3.2.8 Concentrations of Light Elements 96
    3.2.9 Light Element Stable Isotopes 98
    3.2.10 Heavy Element Stable Isotopes 100
    3.2.11 Radioactive Isotopes 102
3.3 Mineralogical Characteristics 104
    3.3.1 Modal Mineralogy 104
    3.3.2 Heavy Mineral Analysis 106
    3.3.3 Automated "Mineralogical" Analysis
           by Computer-Controlled SEM 107
    3.3.4 Clay Mineral Assemblage 109
    3.3.5 Bulk Luminescence Properties 111
3.4 Bulk Organic Matter Characteristics 111
    3.4.1 Organic Matter and Organic Carbon Content 111
    3.4.2 Organic Compounds 111
    3.4.3 Microbial Populations 112
    3.4.4 Pollen Assemblages 113
    3.4.5 Diatom Assemblages 117

4   Properties of Individual Particles   119

4.1 Introduction 119
4.2 Particle Size 119

4.2.1   Aspects of Size.. 119
4.2.2   Weight.. 120
4.2.3   Linear Dimensions... 120
4.2.4   Cross-Sectional Area and Perimeter............................... 122
4.2.5   Volume.. 124
4.2.6   Surface Area.. 124
4.3   Particle Shape 126
4.3.1   Form 127
4.3.2   Sphericity... 133
4.3.3   Roundness 138
4.3.4   Angularity... 141
4.3.5   Irregularity... 145
4.4   Surface Texture... 148
4.5   Characterization of Particle Morphology Using Fourier
      Analysis, Fractal Analysis, and Fourier Descriptors 157
4.6   Three-Dimensional Particle Shape Analysis Using
      X-Ray Tomography and Laser Profilometry......................... 159
4.7   Color 161
4.8   Luminescence Properties 163
4.9   Composite Characterization of Particles and Objects 164
4.9.1   Gravel-Size Particles, Rocks, and Other
        Large Objects... 164
4.9.2   Sand-Size Particles 165
4.10   Elemental Composition.. 167
4.10.1   Alternative Analytical Techniques........................... 167
4.10.2   Electron Microprobe Analysis 167
4.10.3   SEM-EDXRA Analysis 168
4.10.4   Laser Ablation ICP-MS Analysis 169
4.11   Isotopic Composition... 177
4.11.1   Stable Isotope Ratios... 177
4.11.2   Isotopic Dating.. 177
4.12   Mineralogical Identification and Characterization
      of Noncrystalline Materials.. 177
4.13   Micro-Fabric of Rocks and Soils 178
4.14   Identification and Characterization
      of Organic Particles 180

5   Sampling and Sample Handling   183
5.1   The Nature of Samples and Their Limitations...................... 183
5.2   General Sampling Guidance... 184
5.3   Sampling Strategies for Control Samples............................... 186

5.3.1   Crime Scenes.. 186
5.3.2   Sampling Across a Wider Area........................... 201
5.4   Size and Type of Sample 201
5.5   Sampling Tools and Sample Containers..................... 205
5.6   Sample Labeling and Associated Information 208
5.7   Sample Storage 210
5.8   Questioned Soil Samples from Items Submitted
        for Forensic Examination 210

6   Evaluation of the Significance of Geological and Soil Evidence   225

6.1   General Procedures and Principles............................ 225
6.2   Exploratory Data Analysis....................................... 227
6.3   "Classical" Hypothesis Testing 231
6.4   Correlation and Regression Analysis 233
        6.4.1   Correlation 234
        6.4.2   Regression... 235
6.5   Multivariate Analysis.. 242
        6.5.1   Principal Component Analysis 243
        6.5.2   Hierarchical Cluster Analysis 243
6.6   Combined Approaches 245
6.7   Assessment of Coincidence Probabilities
        and Likelihood Ratios.. 246
        6.7.1   Chance Coincidence Probabilities 246
        6.7.2   Bayesian Statistical Approaches 247
6.8   Direct Data Comparison: Deciding if Two Samples
        Are Indistinguishable, Similar, or Different 248
6.9   Use of Multi-Technique Comparison Data
        for Exclusion/Inclusion.. 255
6.10 Geological and Soil Databases and Database
        Interrogation 256
6.11 Evaluation of the Overall Strength of Geological
        and Soil Evidence... 264
6.12 The Future.. 269

References   271

Index   323

Preface

This book aims to provide an authoritative introduction to the nature and properties of geological and soil materials that may be used as trace evidence, the techniques that may be used to analyze them, and the ways in which the 'significance' of the results can be evaluated. It is intended to be accessible to readers who have a relatively limited existing knowledge of this area, including investigating officers, forensic science advisors, scenes of crime officers, lawyers, and students of forensic science, but also to provide an informative reference source for academics and professionals who may have an existing knowledge of earth science but relatively little experience of forensic applications. Discussions of the basic principles are supported by examples drawn from the author's casework experience and from the wider scientific literature.

The forensic potential of geological and soil evidence has been recognized for more than a century, but in the last 15 years these types of evidence have been used much more widely both as an investigative intelligence tool and as evidence in court. However, there is still a widespread poor understanding of the potential value and the limitations of geological and soil evidence among the forensic science and wider legal communities. It is hoped that this book will help to rectify this situation and will act as a catalyst for further basic research in forensic geology and soil science.

Kenneth Pye
London, September 2006

Acknowledgments

I am grateful to a large number of people who have contributed over the last 30 years to the development of my thinking in relation to forensic geology matters, and about sediments and soils more generally. Particular thanks are due to Simon Blott for valuable discussion on certain technical matters and practical assistance with data processing and the preparation of many of the tables and figures that appear in this book, to Debra Croft for assistance in many casework investigations, and to Samantha Witton for office and technical support. I also owe a debt of gratitude to Becky McEldowney of CRC Press for her understanding during the gestation period of the book.

Author

Kenneth Pye is an independent scientific consultant and expert witness with almost 30 years experience in environmental geoscience and forensic geology. He has worked on numerous serious crime cases in the UK and overseas, including many high-profile cases, instructed both by the prosecution and the defense. He has also undertaken investigative work in connection with a large number of civil law cases.

He completed his PhD in geomorphology at Cambridge University in 1980 and was awarded an ScD in earth sciences by the same university in 1992. Previously, he received First Class BA Honours and MA degrees from Oxford University in 1977 and 1981, respectively. His academic appointments have included Natural Environment Research Council and Royal Society 1983 University Research Fellowships at the Department of Earth Sciences, University of Cambridge (1980 to 1988), a Research Fellowship at Girton College, Cambridge, a Lectureship, Readership and Personal Professorship at the Postgraduate Research Institute for Sedimentology, University of Reading (1989 to 1998), and the Established Chair of Environmental Geology at Royal Holloway, University of London (1999 to 2004).

Professor Pye's publications include 12 books and over 200 scientific articles and chapters in edited books and conference proceedings volumes. In addition to casework involvement, he has given numerous invited lectures on forensic geology, and has participated in specialist forensic science workshops in many parts of the world.

Introduction: The Nature and Development of Forensic Geology

1

1.1 The Nature of Forensic Geology and Forensic Soil Science

Forensic geology is concerned with the application of geological information and techniques to problems that may come before a court of law. *Forensic soil science*, sometimes referred to as *forensic pedology* (Brooks and Newton, 1969), involves the application of information relating to the properties of soils in matters that may come before a court of law. No clear distinction can be made between geology and soil science, since there is considerable overlap between the two disciplines in terms of subject matter and methodology, and both can be regarded as branches of environmental earth science. However, geology is often defined as the scientific study of the solid earth and is mainly concerned with rocks, minerals, sediments, fossils, and the processes responsible for their formation, whereas soil science is defined as the scientific study of soils which form in the uppermost layers of Earth's surface through modification of geological parent materials by a combination of physical, chemical, and biological processes.

Members of the forensic and legal communities have often referred to all types of geological and soil evidence as "soil evidence," even if no true soil material is involved (e.g., Camps, 1962). As noted by Peters (1962, p. 27), "there are many definitions for the word soil, but for forensic science purposes we can adopt the civil engineers' conception, that is, all disintegrated material lying on the top of solid rock." This is also essentially a geological description, and for these reasons the general term *geological evidence* is preferred by many, including the author, to *soil evidence* or *mud evidence*.

Although several aspects of geology and soil science may be useful in a forensic context, the most common application involves the use of sediment and soil particles as a type of trace evidence which can be useful in linking a suspect to a crime scene or other location of investigative interest (e.g., Marumo et al., 1999; Hopen, 2004). Other applications include the use of geophysical and other remote sensing techniques to locate clandestine graves, arms caches, or other buried items, investigations of frauds, thefts, wildlife trafficking, adulterations of commercial products, investigation of sources of environmental pollution, and investigation of the causes of accidents and civil engineering failures (e.g., Shuirman and Slosson, 1992; Day, 1998; Rogers, 1999; Morgan et al., 2006). Many of the latter applications

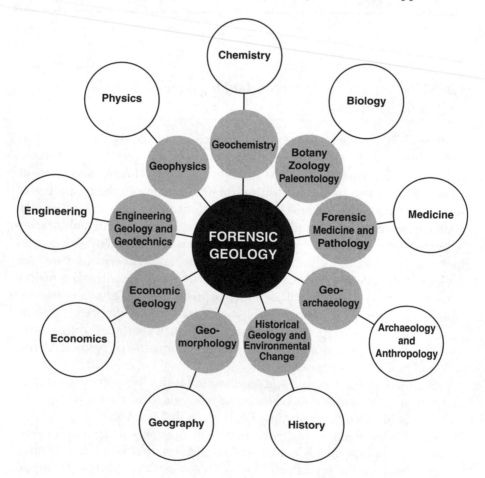

Figure 1.1 The relationship of forensic geology to other disciplines and subdisciplines.

can be classified under the general heading of *environmental forensics* (e.g., Morrison, 2000; Murphy and Morrison, 2002).

Forensic geology is integrative in nature and draws heavily on methods and information from a wide range of other scientific disciplines, including physics, chemistry, biology, and their various sub-disciplines. There are also significant links with the social sciences and other subject areas relating to humans and human activities, including anthropology, archaeology, history, geography, medicine, and engineering (Figure 1.1). Much of the Earth's surface has been heavily modified by human activities, and there are few areas where "soils" can be regarded to be entirely natural. Geological materials have been transported from one area to another for construction and manufacturing purposes for thousands of years, and knowledge of these activities is important in the appraisal of the potential significance of geological evidence. Consequently, forensic geological investigations often involve far more than simple inter-sample comparisons: adequate interpretation often requires detailed background research and interrogation of archival information relating to past land-use, the production and distribution of commercial products, and the disposal of waste materials from mining, quarrying, or similar activities.

1.2 Development of the Use of Geological and Soil Materials as Trace Evidence

The potential value of geological and soil evidence in criminal investigations has been recognized for more than a century, although the early proponents were not, in fact, geologists (for more detailed historical reviews, see Murray and Tedrow, 1975, 1992; Palenik, 1982, Murray, 1988, 2004; Ruffell and McKinley, 2005).

Hans Gross (1847–1915), an Austrian criminal investigator widely credited with being the founder of scientific criminal investigation, drew attention to the value of microscopic study of minerals in dust samples and dirt on shoes in his classic work *Hanbuch fur Untersuchungsrichter* (or *Handbook for Examining Magistrates*), first published in 1893 and as a modified English version in 1962 (Gross, 1962). Also around this time, the English author, Sir Arthur Conan Doyle, creator of the fictional detective Sherlock Holmes, drew attention to the value of trace evidence through his writings, and actually used soil evidence in one of three cases in which he became a real-life investigator. In the early 20th century, Georg Popp, a German chemist and early forensic scientist, studied the composition of dirt on clothing, footwear, and in fingernail scrapings and applied the information in a number of murder investigations.

In France, Edmond Locard (1877–1966), a student of forensic medicine who went on to become Director of the Technical Police Laboratory in Lyon, also developed an early interest in these areas, inspired by the work of Gross and Conan Doyle, among others. Locard founded the Lyon Police Technical Laboratory in 1910 and later became the founder director of the Institute of Criminalistics at the University of Lyon. His work led to the development of what became known as Locard's Exchange Principle, which has frequently been paraphrased as "every contact leaves a trace" (Walls, 1974), or interpreted to imply that "whenever two objects come into contact there is always a transfer of material" (Palenik, 1997). There is no evidence that Locard ever used the expression "every contact leaves a trace," although in a book published in French in 1920, he wrote that:

> Les indices dont je veux montrer ici l'emploi sont de deux ordres: tantôt le malfaiteur a laissé sur les liex les marques de son passage, tantôt, par une action inverse, il a emporté sur son corps ou sur ses vêtements les indices de son séjour ou de son geste. Laissées ou reçues, ces traces sone de sortes extrêmement diverses. (Locard, 1920, p. 139)

This has been translated in English as:

> The material evidence, its exploitation is what I would like to discuss here, is of two orders: on the one hand, the criminal leaves marks at the crime scene of his passage; on the other hand, by an inverse action, he takes with him, on his body or on his clothing, evidence of his stay or of his deed. Left or received, these traces are of extremely varied type. (Horswell and Fowler, 2004, p. 47)

In the late 1920s, Locard published a series of papers in French relating to the analysis of dust traces; these were subsequently translated into German and English (e.g., Locard, 1928, 1930a, 1930b, 1930c). He also published a four-volume treatise that dealt with the entire field of criminalistics (Locard, 1931–1932). However, these publications made no reference to his earlier work, or specifically to the exchange of contact traces. The manner in which Locard's ideas became formalized as a "general principle" is not entirely clear but, as noted by Horswell and Fowler (2004), "there appears to have been some rather crude generalization based on his English translated articles and hence to an over simplification of the concepts involved." In an influential textbook published in 1956, Nickolls stated that "the basis of this reconstruction and of contact traces was laid down by Locard (1928) who stated that when two objects come into contact there is always a transference of material from each object to the other." Similarly, Walls (1974) stated that "Edmund Locard … laid it down as a guiding principle that *every contact leaves a trace*." There is, in fact, no such specific wording in Locard's (1928) paper, although he does describe at length how individuals and their clothing will accumulate "dust" particles from the environment in which they live or work, and gives several examples of how "dust" traces had proved useful in linking suspects to crime scenes. The term dust, as used in the English translations of Locard's work, is somewhat misleading; in fact, it is clear from Locard's papers that he had in mind any kind of powdered, fine-grained debris. The English translation of his 1928 paper states that:

> Dust is an accumulation of *débris* in the form of powder. These *débris* may emanate from all kinds of bodies, organic and inorganic. It is a condition of pulverization which distinguishes dust from mud or grime. Mud is a mixture of dust and liquid in the form of a paste ... Grime is dust mixed with body grease and partly dried.

The concept of "dust" used by Locard is thus rather different from that generally employed in more specific geological contexts, where dust has been defined as "a suspension of solid particles in a gas, or a deposit formed of such particles" (Pye, 1987, p. 1). Airborne transport of the fine particles in suspension, prior to possible deposition, is an essential element of this definition. Fine particles suspended in water, or a deposit of such particles, even after drying, can never be considered as dust; rather, they represent mud. Locard's definition of mud effectively excludes dried residues from which the liquid phase has evaporated. Since such material generally does not occur in a loose powdered form, unless broken up by some agency, it does not sit easily within any of Locard's categories.

The "contact exchange principle" attributed to Locard is best regarded as a theoretical construct rather than a practical reality. There are many instances where contact takes place between two objects, but there is either no transfer of material between the two, or the traces are far too indistinct or short-lived to be of practical use in forensic examination, even using the most sophisticated instrumental techniques available today. The retention, or preservation potential, of any contact traces that may be initially present will be strongly dependent on the nature of the objects that come into contact, the nature of the material exchanged, and the subsequent wear or use history of the objects in question. Conversely, exchange of particles between two objects may take place without requirement for physical contact. This is especially true for dust traces where particles may be transported large distances through the atmosphere before settling or being otherwise deposited on a suitable surface. A whole variety of possible transfer mechanisms need to be considered alongside direct primary contact transfer; in the case of geological and soil evidence, these include secondary contact transfer, or multiple contact transfer, non-contact transfer of individual particles by wind or water, splashing, dripping or dropping of individual particles or "lumps" of material, and explosive transfer.

In the U.S., the potential value of soil, dust, and related mineral matter as trace evidence was also recognized as early as the 1920s by chemists and criminologists such as Edward Oscar Heinrich; by the mid-1930s, soil and mineral analysis was being regularly employed by forensic investigators in the FBI Laboratory at Quantico, Virginia (Schatz et al., 1930; Goin and Kirk, 1947; Murray and Tedrow, 1992; Murray, 2004). Forensic investigations of minerals and other particles found in soils, dirt, and dust samples increased in popularity and sophistication during the 1950s, 1960s, and early 1970s as a result of the enthusiasm of Walter McCrone and his associates, based in Chicago, who applied a variety of innovative optical and other microscopic techniques to their characterization (McCrone, 1982, 1992). McCrone, whose initial work focused on particles in air pollution, was responsible for the development and application of microscopical and microchemical techniques for the study of a wide range of particle types in criminalistic investigations (Weaver, 2003; Moorehead, 2004). This work resulted in the publication of the McCrone Particle Atlas, which appeared in several volumes and editions in the late 1960s and 1970s (McCrone and Delly, 1973a, 1973b, 1973c, 1973d; McCrone et al., 1967, 1979; McCrone, 1980). These approaches were subsequently extended and developed in a forensic context by McCrone's students and professional associates (e.g., Palenik, 1979, 1982, 1988, 1997, 1998, 2000a, 2000b; Petraco and DeForrest, 1993; Petraco, 1994a, 1994b; Hopen, 2004; Petraco and Kubic, 2004).

During the 1970s and 1980s, there was also considerable research interest in the U.S. relating to the use of biochemical and other methods of characterizing bulk soil samples and extracted fractions for forensic applications (Thornton, 1974; Thornton et al., 1975; Thornton and Fitzpatrick, 1978; Thornton and Crim, 1986).

In the U.K., general forensic scientists employed at the Metropolitan Police Laboratory, New Scotland Yard, and later the Home Office Central Research Establishment at Aldermaston, regularly used soil evidence between the 1950s and the late 1970s (Nickolls, 1956, 1962; Peters, 1962; Dudley, 1975, 1976a, 1976b, 1976c, 1977, 1979; Dudley and Smalldon, 1978a, 1978b, 1978c). These studies employed a number of techniques, including microscopic examination, pH determination, color description, sieving, heavy mineral analysis, and visual comparison using density gradient columns. Exploratory investigations were also undertaken using (then) novel techniques, such as cathodoluminescence, thermoluminescence, laser granulometry, elemental compositional analysis, and analysis of organic components such as polysaccharides (Dudley, 1976a, 1976b, 1976c; Chaperlin, 1981; Reuland and Trinler, 1981; Reuland et al., 1992). However, these were not widely used in casework.

During the 1980s and early 1990s, there was something of a hiatus in forensic soil investigations, with relatively few publications of note and declining use in casework. In part, this reflected increasing focus on DNA evidence and a concomitant reduced interest in all forms of "traditional" trace evidence. In the U.K., the closure and reorganization of the former Home Office-funded forensic science laboratories resulted in loss and/or dispersal of experienced staff and a reduction in the general level of employment of staff with "specialist" expertise. Another factor was controversy and disillusionment with soil comparisons based on the density gradient method (Chaperlin and Howarth, 1983). In the U.S., many courts refused to accept the value of soil evidence, based on the widespread belief that "soil is soil, mud is mud," a view which is still erroneously held in some quarters (Hopen, 2004). However, the FBI *Handbook of Forensic Services* (Wade, 2003) does include recommendations for soil sample collection and examinations alongside other forms of trace evidence.

Since the mid-1990s, there has been a revival of interest in "alternative" types of trace evidence, including geological, soil, and botanical evidence. In part, this reflects a realization that DNA analysis does not always provide the answers to crime-related problems, especially in a world where criminals are increasingly forensically aware and often take precautions to avoid leaving blood, fingerprint, or semen evidence at the scene of a crime, and may even "plant" such evidence to mislead the investigation. In part, it also reflects a de-regulation in the field of forensic science, and increasing involvement of expert witnesses drawn from outside the traditional forensic science laboratories. In the past, most forensic evidence was obtained and presented in court almost exclusively by scientists who were trained and employed within the police service or nationally funded support laboratories, such as those of the Forensic Science Service in the U.K. However, since the 1980s, increasing use has been made of scientists drawn from private companies, laboratories, universities, and research institutions, which often have access to a much wider range of scientific facilities and specialist expertise.

Despite the long-standing interest in forensic geology and soils, the first book dealing specifically with this subject area was published only in 1975 (Murray and Tedrow, 1975; reprinted with revisions in 1992). This book remains a valuable introduction to the subject, including historical applications of forensic geological information and techniques. An updated popular introduction to forensic geology was also recently provided by Murray (2004). Useful technical reviews of available methods for the analysis of soil and vegetative matter were provided by Thornton and Crim (1986), Demmelmeyer and Adam (1995), Marumo et al., (1999), and Marumo and Sugita (1998, 2001). There have also been numerous articles in popular magazines, professional journals, and web-based publications, which have illustrated the use of geological and soil evidence in the context of specific cases (Cleveland, 1973; Murray, 1976, 2000, 2005; Murray and Murray, 1980; Rapp, 1987; Lazzarini and Lombardi, 1995; Munroe, 1995; McPhee, 1996; Ovianki, 1996; Lombardi, 1999; Brown, 2000; Lindemann, 2000; Pye, 2004a; Abbott, 2005; Sever, 2005; Ruffell, 2006). However, more recent general reviews of forensic trace evidence have included only very brief reference to soil and other forms of geological evidence (Fisher, 2000; Bisbing, 2001; Saferstein, 2001; Kubic and Petraco, 2002; James and Nordby, 2003; Horswell, 2004; Houck, 2004; Petraco and Kubic, 2004; Townley and Ede, 2004; Gallop and Stockdale, 2005). The *Encyclopedia of Forensic Sciences* (Siegel et al., 2000) contains no entries relating specifically to forensic geology or soil evidence, although there are relevant entries dealing with "dust" and "microscopy" (Palenik, 2000a, 2000b).

There appears still to be a widespread lack of awareness among many forensic scientists, criminal investigators, and lawyers of the ways in which the field of geoscience has developed in the last 30 yrs, of the emergence of a new inter-disciplinary area of "environmental forensics," and how these advances can provide assistance both in terms of investigative tools and forensic evidence. In the specific context of trace evidence, "soil" comparisons no longer need to be based on relatively simple comparisons of color, particle size, and particle types identified principally by optical microscopy; there are now numerous other analytical techniques available and a wealth of contextual information which can assist with interpretation of the significance of the results obtained. Recent developments in instrumentation and methodology have made it possible to obtain several different types of information from very small samples, and in some cases, from single particles. However, this raises a new set of questions relating to the evidential significance of the results obtained (Pye and Croft, 2004a, 2004b).

1.3 Other Types of Geoscience Data Useful in Forensic Investigations

There are a number of other aspects of the earth and environmental sciences that have useful forensic applications. These include geophysical prospecting and remote sensing which are of vital importance in the search for clandestine graves or other buried items. Moreover, information about ground conditions or earth surface processes can often be crucial for correct interpretation of the likely evidential significance of geological and soil trace evidence.

1.3.1 Geophysical Survey Data

Geophysical surveying techniques are most widely used to assist in the search for buried items, including clandestine graves, weapons, and drugs stashes. Several different techniques are available, choice depending on the nature and area of the terrain to be surveyed, the depth of survey required, and the resolution of the data that are required. The principal methods used are shallow seismic reflection profiling, gravity surveys, magnetic anomaly surveys, resistivity surveys, electromagnetic conductivity profiling, metal detector surveys, and ground penetrating radar surveys. The different methods may be used individually but frequently in combination. Reviews of the techniques and their relative merits, including forensic case examples, are given by Milsom (1989), Bevan (1991), Clark (1996), Miller (1996), Nobes (2000), Ruffell (2002, 2005), Bristow and Jol (2003), Buck (2003), Fenning and Donnelly (2004), Scott and Hunter (2004), Watters and Hunter (2004), and Ruffell and McKinley (2005).

1.3.2 Meteorological Data

Data relating to rainfall, temperatures, humidity, and wind speed/direction are important in a variety of forensic contexts which are often case-specific (e.g., as controls on body decomposition rates or blood dispersion patterns), but they are also frequently of assistance when making assessments of the likelihood that soil or mud would be transferred to clothing, footwear, or a suspect vehicle from a ground surface, and retained thereafter. Periods of wet weather generally favor mud/soil transfer from the ground or other sources to items of potential forensic interest, but may also assist rapid loss from exposed surfaces after the initial transfer. Similarly, low temperatures may inhibit soil transfer, especially if the ground is frozen, whereas high temperatures may favor soil retention due to rapid drying.

Wind speed and direction are critical controls on the dispersal pattern of disturbed particles, including soil dust, sand, and pollen or other botanical particles such as seeds. For these reasons, hourly, or at least daily, weather data for the relevant period leading up to and after a crime should be obtained as a matter of routine in all criminal investigations involving outdoor crime scenes. The utility of such information will, however, be dependent on how close geographically the weather station is to the crime scene. Information about the location of weather stations, and the period and frequency of record, can be obtained from national weather organizations such as the U.K. Meteorological Office. Information about methods of recording and interpreting such information is given by Strangeways (2000).

1.3.3 River Flow and Lake Level Data

Where bodies have been disposed of in a river or lake, it is often helpful to obtain information about the water level and current speeds (where applicable) at the time of, and immediately following, the disposal. This can assist both in terms of identifying the method of disposal (e.g., whether a boat was used or the perpetrator walked or waded out), and in predicting the likely movement and location of the body. Information about water levels can often be obtained from regulatory bodies that maintain water level gauges at weirs or landing stages (e.g., The Environment Agency in England and Wales).

1.3.4   Tidal Data

If disposal of a body takes place in the sea, an estuary, or the lower reaches of a river, it may be transported some considerable distance in a relatively short time period, depending on the conditions of wind, tides, and currents. Tidal current velocities are strongly dependent on the tidal range and local seabed/estuarine topography, and can vary considerably over short distances. Several mathematical models are now available to predict both water movements and objects such as bodies that may be transported at different levels in the flow, dependent on their relative buoyancy. Information about tidal levels at the time of an alleged criminal event may also be helpful in checking an alibi and assessing the likelihood that saline mud or seawater would be picked up on clothing or footwear. Guidance on such matters can be obtained from specialist marine observatories, university departments, and research institutes.

1.3.5   Geomorphological Data

Geomorphology is the scientific study of Earth's surface features and the processes responsible for forming them. Knowledge of the nature and significance of landscape features and processes can be of assistance in establishing the locations recorded in photographs or videos seized from a suspect. Knowledge of weathering rates and weathering phenomena can also assist in the interpretation of sub-aerial exposure intervals (e.g., of grave marker stones). An understanding of surface transport processes can also assist in the search for objects, for example, bodies or disarticulated bones dislodged by stream floods and landslips or incorporated in glacier ice (e.g., Ambach et al., 1991). Whatever the issue under investigation, whether relating to serious crime or environmental forensics, it is crucial to understand the environmental setting from which any samples are taken for laboratory analysis. Such contextual information can be obtained from background geological, soil, and topographic maps, and from aerial photography, satellite imagery, or other airborne remote sensing tools (Brillis et al., 2000; Grip et al., 2000).

Types of Geological and Soil Materials That May be Useful as Trace Evidence

2

2.1 Overview of Material Types and Available Analytical Strategies

A basic distinction can be drawn between seven broad groups of geological materials, although the boundaries between them are not rigid: (1) rocks, (2) sediments, (3) soils, (4) dusts, (5) minerals, (6) fossils, and (7) particulates. In addition, we may consider an eighth category, materials of anthropogenic origin, whose properties resemble those of natural rocks or sediments and which show similar behavior in the environment. Sediments, soils, and dusts are mixtures of individual particles that may be either *inorganic* or *organic* in origin. Inorganic particles have formed without any significant involvement of biological processes; they mainly represent rock fragments, individual crystalline mineral grains (e.g., quartz), or fragments of amorphous (noncrystalline) matter such as volcanic glass. Most are naturally formed, although some, such as concrete and brick fragments, may be anthropogenic. Organic particles are those that have formed directly as a result of biological processes. Some types are composed of crystalline mineral material (e.g., shell fragments composed of calcite or aragonite), some are noncrystalline or composed of poorly crystalline quasi-mineral matter (e.g., opal phytoliths and diatoms), and others represent various organic cellular structures or their degradation products composed chiefly of the light elements carbon, hydrogen, and oxygen (e.g., fragments of coal, charcoal, wood, stem material, leaves, and seeds or pollen grains). Assemblages of particles can be examined either in bulk, or individually, usually using different methods. In most forensic geological studies, a combination of bulk material and individual particle examination methods is used, depending on the amount of material available for analysis. Particles in sediments, soils, and dusts can be simple or complex. A simple particle may be compositionally homogeneous and have few, if any, distinguishing morphological features; an example would be a well-rounded quartz sand grain. At the other extreme, complex particles may be composed of more than one mineral type or represent several different particles bound together by some form of cement; such particles often have unusual particle shapes, surface textures, and internal structures. Between these extremes, there is a wide range of possibilities, and most soils and sediments contain a diverse assemblage of several different particle types that show a broad spectrum of morphological and compositional characteristics.

Geological samples and individual particles of potential forensic interest can vary greatly in terms of size. For example, several large stones, each weighing several kilograms, may be used to weigh down a body disposed of at sea or in a lake. At the other extreme, mud or soil smears present on clothing or similar items may consist only of fine silt and clay-size particles. In drowning or asphyxiation cases, particles recovered from the lungs or other internal organs of the victim may also be very small (a few tens of microns to sub-micron in size). The size of the particles in question, and the amount of sample material available, in large measure determine the number and types of analytical techniques that can be employed. In the case of rocks and coarse-grained sediments, useful observations can often be made on the basis of examination in hand-specimen or using a hand lens. Fine-grained sediments, soils, and dust samples, on the other hand, normally require examination using an optical microscope or electron microscope. The particle size distribution of relatively large samples can be determined by dry or wet sieving, but small samples must normally be analyzed using more specialized techniques, such as laser granulometry, or Coulter Counter analysis. Chemical analysis or relatively large samples can often conveniently be undertaken using a technique such as x-ray fluorescence (XRF), but small samples require analysis in solid form by x-ray microanalysis (EDXRA) or using a sensitive solution technique such as inductively coupled plasma spectrometry (ICP). In general, the smaller the size of available sample, the fewer the analytical techniques that can be applied. Since there is usually a requirement to preserve the integrity of forensic samples, nondestructive or minimally destructive techniques are preferred to destructive ones.

2.2 Rocks

2.2.1 The General Character of Rocks

A rock may be defined simply as a mass of solid matter that is composed mainly of mineral crystals or grains that are texturally intergrown or held together by

some form of cement. Pieces of rock have several attributes which can potentially be used as a basis for forensic comparison, including size, shape, surface
texture (including weathering rinds and coatings), density, color, chemical
composition, mineralogical composition, internal textural layering, fabric,
magnetic characteristics, fossil assemblage (both macrofossil and microfossil),
and age. Three main rock groups, *igneous*, *metamorphic*, and *sedimentary*, each
of which can be sub-divided, are traditionally distinguished (Figure 2.1).

2.2.2 Igneous Rocks

Igneous rocks are formed by crystallization from molten magma, either
below Earth's surface (*intrusive*, or *plutonic* rocks) or at the surface (*extrusive*,
or *volcanic* rocks). Several different classification schemes for igneous rocks
have been proposed on the basis of chemical composition, actual or norma-
tive (calculated) mineralogical composition, and/or texture (size and
arrangement of constituent crystals) (e.g., Johannsen, 1931, 1932, 1937, 1938;
Streckeisen, 1976). Johannsen (1931, p. 51) observed that:

> Many and peculiar are the classifications that have
> been proposed for igneous rocks. Their variability
> depends in part upon the purpose for which each was
> intended, and in part from the characters of the rocks
> themselves. The trouble is not with the classifications
> but with nature which did not make things right.

While this might be considered to be an anthropocentric view, the reality
is that natural phenomena do not always fit nicely into regular classification
schemes; overlaps and transitions are always to be expected.

One of the simplest classification schemes for igneous rocks is based on
a combination of silica percentage and texture or crystal size (Table 2.1).
However, this type of classification has fallen out of favor with many igneous
petrologists who prefer more complex schemes based on other mineralogical,
chemical, or textural criteria. Many different textural features can be identified under the microscope in igneous rocks; common examples are illustrated
by MacKenzie et al. (1982), while the relationships between texture, composition, and igneous rock genesis are discussed in more detail by Hall (1996).
However, for forensic purposes, use of the correct rock nomenclature is less
important than having a precise means of inter-sample comparison.

2.2.3 Metamorphic Rocks

A metamorphic rock is one that has been formed from a pre-existing rock
or sediment, involving changes in mineralogy, chemistry, and structure, as a
result of exposure to high temperature, pressure, and/or shearing stress.
Three main types of metamorphism are generally recognized: *burial metamorphism*, which normally occurs at a depth of several kilometers within the

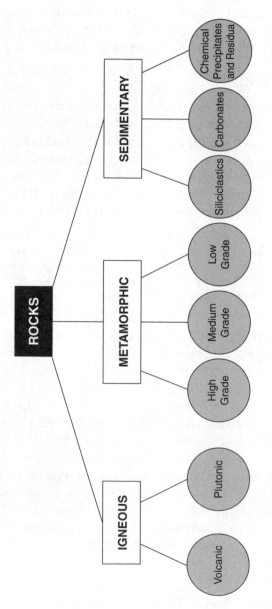

Figure 2.1 A simple classification of rock types.

Table 2.1 A Simple Classification of Igneous Rocks Based on Silica Percentage

%SiO$_2$	Rock Type	Examples
>66	Acidic	Granite, syenite, quartz diorite
52–66	Intermediate	Granodiorite
45–52	Basic	Diorite, gabbro, basalt, andesite
<45	Ultrabasic	Peridotite

Source: Modified after Ehlers, E.G. and Blatt, H., in *Petrology: Igneous, Sedimentary and Metamorphic*, W.H. Freeman and Company, San Francisco, 1982. With permission.

Earth's crust as a result of increasing temperature and pressure; *contact meta-morphism*, which can occur at any depth as a result of contact with an igneous intrusion; and *shock metamorphism,* which results from events such as meteorite impacts or earthquakes. Metamorphism can either be a relatively short-lived event or be progressive over periods of millions of years. During the process of metamorphism, changes in the rock texture and fabric take place, often involving dissolution or recrystallization of some of the original minerals and development of new minerals that are more in equilibrium with the changed environmental conditions. Distinctive textures and mineral assemblages are produced during this process. Different grades of metamorphism can be identified on the basis of the degree to which the parent rock characteristics have been modified by metamorphic processes. The various metamorphic grades are frequently represented by the appearance or disappearance of various "indicator" minerals or mineral assemblages. A distinction between metamorphism and *diagenesis,* which is the process of textural and mineralogical transformation of loose sediment into rock due to compaction, dissolution, and cementation at, or relatively near to, the Earth's surface, is often made by defining the lower limit of metamorphism as the first appearance of a mineral which does not normally form in near-surface environments, such as epidote or muscovite (Ehlers and Blatt, 1982). This is a rather loose definition, since during progressive burial of sediments there is frequently a gradual transformation of relatively poorly crystalline illite into better crystallized forms and then into muscovite. The process is dependent not only on temperature and pressure, but also on pore fluid chemistry. However, in general, metamorphism begins at temperatures of around 150°C and becomes more pronounced with increasing temperature. The upper limit of metamorphism is defined as the onset of melting of the rock, which under most conditions occurs at temperatures of 650 to 800°C. The classification of metamorphic rocks is even more difficult than for igneous and sedimentary rocks; criteria include a combination of textural, structural, mineralogical and genetic features. Rocks that have formed as a result of relatively low-grade metamorphism may show clear evidence of the nature of the parent material, whereas high-grade metamorphic rocks are often entirely

Table 2.2   A Simple Classification of the Main Metamorphic Rock Types

Parent Material	Degree of Metamorphism		
	Low Grade	Medium Grade	High Grade
Igneous Rocks			
Acid	Meta-granite	Hornfels, Gneiss	Gneiss
Intermediate	Meta-granodiorite	Hornfels, Gneiss	Gneiss
Basic	Greenschist, Meta-basite	Amphibolite	Blueschist
Ultrabasic	Greenschist, Meta-basite		Eclogite, Ophiolite
Sedimentary Rocks			
Conglomerate and sandstone	Quartzite	Quartzite	Granulite
Mudrock	Shale	Slate	Schist
Carbonate	Marble	Marble	
Chemical sediments	Banded ironstone, Chert	Banded ironstone, Chert	

re-crystallized and have virtually no textural or mineralogical similarity to the parent rock. Some of the more frequently encountered metamorphic rocks are listed in Table 2.2.

Metamorphic rocks show a wide range of textural features, some of which are visible at the rock outcrop scale and in hand-specimen, whereas others are visible only under a microscope. Common examples of microscopic textures are illustrated in Yardley et al. (1990), and a useful summary of the petrology of metamorphic rocks in relation to their genesis is provided by Yardley (1989).

2.2.4   Sedimentary Rocks

Sedimentary rocks, which comprise c. 70% of the rocks exposed at the Earth's surface (Tucker, 1991), are either composed of individual sediment grains that have been cemented together as a result of diagenesis (detrital sedimentary rocks), or of material formed by chemical precipitation (chemical precipitates). Detrital sedimentary rocks are most commonly classified according to predominant particle size and mineralogy. Several different classification schemes have been proposed, but for forensic purposes a relatively simple system is sufficient (Table 2.3). Descriptions of the properties and genesis of the various rock types are provided by Folk (1974), Blatt et al. (1980), Ehlers and Blatt (1982), Greensmith (1989), and Tucker (1991). The three main groups of sedimentary rocks are: siliciclastic rocks, including sandstones and shales (Pettijohn et al., 1972; Potter et al., 1980); carbonate rocks (Bathurst, 1975; Tucker and Wright, 1990); and chemical sedimentary rocks, including evaporites and surficial weathering crusts (Goudie and Pye, 1983). Specific rock types that have been used as building stones and as other construction raw materials are described in a

Table 2.3   A Simple Classification of the Main Sedimentary Rock Types

Siliciclastic Rocks	Carbonate Rocks	Chemical Precipitates and Residual
Conglomerates	Limestones	Evaporites
Sandstones	Dolomites	Tufas
Mudrocks	Chalks	Ironstones
		Weathering crusts (e.g., bauxites, laterites)
		Cherts
		Coals

number of specialist publications (e.g., Watson, 1911; Leary, 1983, 1986; Mathers et al., 2000). Examples of the textural features found in common sedimentary rocks are illustrated by Scholle (1978, 1979), Adams et al. (1984), Selley (1988), O'Brien and Slatt (1990), Bennett et al. (1991), and Adams and MacKenzie (1998).

Pieces of rock, representing either naturally occurring fragments or blocks of quarried stone, may be encountered in a variety of forensic contexts. They include works of art and archaeological artifacts, which have considerable scientific as well as monetary value (e.g., Kempe and Harvey, 1983; Lazzarini and Lombardi, 1995; Henderson, 2000, 2004). Rocks and stone slabs have also been widely used to conceal drug and arms consignments and to weigh down bodies, bags, and other items in lakes, rivers, and canals (e.g., Figure 2.2).

2.3   Sediments

Sediments are loose mixtures of particles, derived from one or more sources, which have been deposited after being transported by air, water, or ice. The majority of particles in most sediments are of natural origin but, in specific locations (e.g., industrial settling tanks and washing plants), may be man-made.

Sediments can be divided into *detrital* (or *allochthonous*) and *chemical* (or *autochthonous*) types and further sub-divided on the basis of particle size and composition. By convention, the size of sediments is described using the grade scale and terminology first proposed by Udden (1914) and later modified by Wentworth (1922a). The major size class divisions in this scale are *gravel, sand, silt,* and *clay* (Table 2.4). The silt- and clay-size fractions together are often referred to as the *mud fraction*. For convenience, the size of muddy sediments is often quantified in micrometers (μm) or microns (μ) (one micrometer or micron being equivalent to one millionth of a meter, or one thousandth of a millimeter).

Sediments can be classified on the basis of the proportions of gravel, sand, and mud that they contain (Folk, 1954; Figure 2.3a). Sediments that contain no gravel can also be classified in terms of the percentages of sand, silt, and clay (Figure 2.3b).

Figure 2.2 (a) Wooden crates with stacks of hollowed out marble slabs, containing drugs seized from a seagoing container by port customs officers; (b) closer view of one of the stacks of hollowed out slabs, after removal of the drugs.

Table 2.4　The Udden-Wentworth Grain Size Scale, with Class Terminology Modifications Proposed by Friedman and Sanders (1978) and Blott and Pye (2001)

Grain Size		Descriptive Terminology		
mm	phi	Udden(1914) and Wentworth(1922)	Friedman and Sanders (1978)	Blott and Pye (2001)
2048	−11		Very large boulders	
			Large boulders	Very large
1024	−10		Medium boulders	Large
512	−9	Cobbles	Small boulders	Medium
256	−8		Large cobbles	Small
128	−7		Small cobbles	Very small
64	−6		Very coarse pebbles	Very coarse
32	−5		Coarse pebbles	Coarse
16	−4	Pebbles	Medium pebbles	Medium
8	−3		Fine pebbles	Fine
4	−2	Granules	Very fine pebbles	Very fine
2	−1			
1	0	Very coarse sand	Very coarse sand	Very coarse
microns 500	1	Coarse sand	Coarse sand	Coarse
		Medium sand	Medium sand	Medium
250	2	Fine sand	Fine sand	Fine
125	3	Very fine sand	Very fine sand	Very fine
63	4			
31	5		Very coarse silt	Very coarse
			Coarse silt	Coarse
16	6	Silt	Medium silt	Medium
8	7	Mud	Fine silt	Fine
4	8		Very fine silt	Very fine
2	9	Clay	Clay	Clay

Blott and Pye (2001) groupings: Boulders (256 mm and above), Gravel (2–64 mm), Sand (63 microns – 2 mm), Silt (below 63 microns).

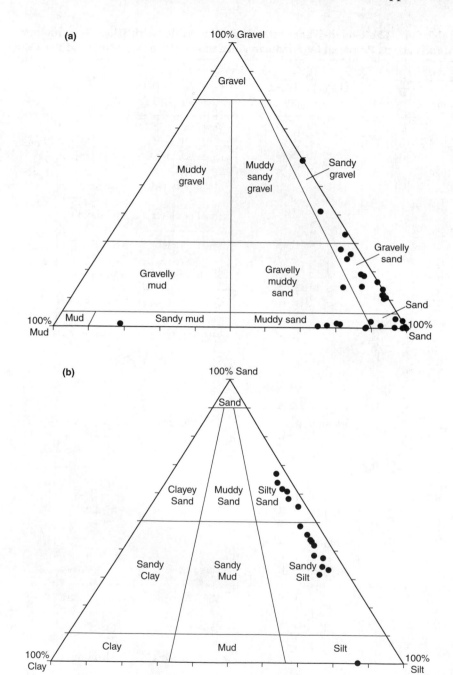

Figure 2.3 (a) Terminology for sediments based on differing percentages of gravel, sand, and mud; (b) sediment terminology based on varying percentages of sand, silt, and clay. Data points relate to a suite of sediment samples collected from Liverpool Bay, northwest England.

Detrital sediments are essentially mixtures of solid particles that have been dislodged from one or more sources by physical processes and transported to a site of accumulation. The distance of transport may be short or long, and transport may involve individual particles, larger masses of intact rock or pre-existing sediment (mass transport), or a combination of the two. Some earth surface processes, like landslides and glacial action, are capable of moving very large masses of material during catastrophic failure. Some glacial sediments contain blocks of ice-rafted rock more than 1 km in maximum dimension, and many other types of sedimentary deposit contain rock blocks up to several meters in size embedded within a matrix of finer sediment. Such sediments are normally very poorly sorted. However, in situations where the sediments are exposed to the action of wind, waves, or other water movements, selective removal of certain sizes (usually the finer ones) results in the formation of better-sorted deposits (e.g., Figure 2.4).

2.3.1 Gravel

Gravel particles are defined as those that are larger than 2 mm but smaller than 64 mm. Larger particles are referred to as *cobbles* (64 to 256 mm) or *boulders* (>256 mm). The term *granule* is sometimes used for particles in the size range 2.0 to 4.0 mm. *Shingle* is a field term which is used in a loose sense to describe particles of fine and medium gravel size which often show a moderate to high degree of rounding, and which occur mainly in marine and lake beach environments. *Pea shingle* is an industry term for fine gravel products widely supplied for paths, drainage blankets, or horticultural purposes.

Figure 2.4 A marine boulder beach at Westward Ho!, Devon, U.K. The lower beach is composed of well-sorted medium sand, whereas the upper beach consists of well-rounded cobbles and boulders up to 1 m in diameter.

Gravel particles may consist of polymineralic rock particles (e.g., granite or limestone) or a single mineral (e.g., vein quartz or calcite). They can be produced either by natural weathering and erosion processes or by artificial means, for example, as a product of rock crushing to produce aggregate (Smith and Collis, 2001) or as a waste product, such as slag, formed by combustion and smelting.

The shapes and surface textures of individual gravel particles can vary greatly, depending partly on the material of which they are composed and partly on the history of weathering, transport, and post-depositional modification experienced by the particles. Gravel particles which are formed by physical weathering processes or mechanisms such as ice action are frequently angular with an irregular surface texture, whereas those which have experienced a long transport history and/or post-depositional modification are frequently well-rounded with relatively smooth surface textures (e.g., Figure 2.5). Aggregate particles formed by rock crushing and used for purposes such as road surface dressing are typically angular (Lees, 1964, 1965).

Compositional analysis of gravel particle assemblages, known as *clast lithological analysis* (Bridgland, 1986), is a technique that has been extensively used to determine the provenance of glacial deposits and river sediments in different parts of the world. Such studies normally involve counting the abundances of different particle types within a specified size range. In many forensic situations, small boulders or coarse gravel particles are often identified initially as rocks or stones, especially if they are essentially free of any finer sediment. In such circumstances, detailed examination of the form of the particle, and of any adhering debris, is required to determine if the source is likely to be a natural rock outcrop or sedimentary body, a quarry or aggregate processing plant, a commercial product, or a secondary source such as a wall or garden.

Rocks, stones, and gravel are frequently encountered in forensic investigations in a variety of contexts. Among the more common instances is their alleged use as weights to sink bodies, bags containing firearms, drugs within a water body (Figure 2.6), as concealment material in drugs or other contraband consignments (Figure 2.7), and as missiles (hand-thrown or catapulted). Smaller gravel particles may be found trapped in the vehicle tires and the sole tread of footwear, within the mouth and upper trachea of individuals who have been drowned or asphyxiated while in contact with a gravelly sediment source, and within items of clothing or on adhesive tape associated with a body (Figure 2.8).

2.3.2 Sand

Sand grains have a size range of 0.063 to 2.00 mm. A sediment is described as *sand* if it contains more than 50% sand-size particles (by mass or volume). A

Figure 2.5 Medium gravel particles from a beach at Towyn, west Wales.

"clean" sand contains less than 10% mud and less than 5% gravel-size particles. Various other adjectives (e.g., muddy sand) can be applied if higher percentages of mud and/or gravel are present (e.g., Figures 2.3a and 2.3b). Relatively clean sands are often found in relatively high energy environments such as beaches (Figure 2.9 and Figure 2.10), whereas muddy sands are typical of lower energy natural environments. Like gravel particles, sand grains may be formed either by natural processes or by human activities such as mining, quarrying, and subsequent crushing to reproduce aggregate or other commercial products such as horticultural sand and equestrian sand (Prentice, 1990; Smith and Collis, 2001). Sand grains may be composed of only one mineral type (i.e., they are *monomineralic*) or of two or more types (*polymineralic*); in the latter case,

Figure 2.6 (a) A rock claimed by the prosecution in the "Lady in the Lake" murder trial to have been taken from a garden wall at the defendant's former home and used to weigh down a package of clothing allegedly belonging to the victim whose body was recovered from Lake Coniston, northwest England, in 1999. At the time of examination, the rock had disintegrated into four pieces, which formed a perfect physical fit. (b) Two parts placed side by side to reveal a distinctive pattern of internal coloration.

the grains are often described as *rock fragments* or *lithic grains*. Monomineralic grains may consist of a single crystal (i.e., they are also *monocrystalline*) or several crystals of the same mineral composition (*polycrystalline*).

  There is a large scientific literature relating to the physical, mineralogical, and chemical properties of natural sand grains, much of it relating to the issue of sand provenance (e.g., Moss, 1966; Blatt, 1967; Potter, 1986; Pettijohn et al, 1987; Pye and Tsoar, 1990; Weltje, 2002, 2004; Weltje and Eynatten, 2004). There is also a substantial literature dealing with industrial sands used in construction and for other industrial purposes such as glass manufacture (e.g., Prentice, 1990). Natural sand deposits are exploited commercially in many locations around the world and exported considerable distances to manufacturing and processing plants.

TOWNDROW A
0054476

NRS/2
0002719

Figure 2.7 (See color insert following page 46.) A comparison of gravel particle types found at two locations of interest in relation to a Customs investigation into cigarette smuggling. The particles represent variants of volcanic rock and tephra which do not occur naturally in the U.K., but which are found in parts of Tenerife, Canary Islands.

Sand grains are often encountered in connection with crimes involving beaches, sand dunes, or construction sites. Sand may be found on the outside and inside of footwear (Figure 2.11), on socks, on clothing (especially in the pockets, turn-ups, and seams), on or inside vehicles, and even inside the bodies of victims and suspects of crime. Sand may be found trapped in washing machine filters and the U-bends of sinks, toilets, etc. in domestic and industrial premises after attempts to clean contaminated hands, clothing, footwear, or other items. Even very small numbers of recovered grains may be of

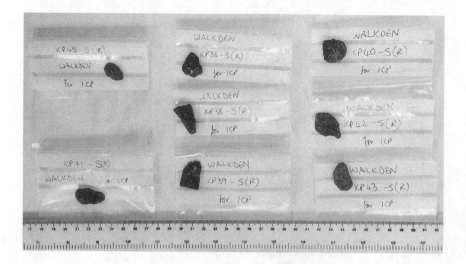

Figure 2.8 Gravel-sized particles of igneous rock recovered from the body of a London businessman whose body was recovered from Poole Bay on the south coast of England several weeks after he was abducted. The gravel had remained trapped on packaging tape wrapped around the victim's head and in parts of his clothing. Gravel particles of the same type were recovered from a drainage trench in Devon, southwest England, where the body had apparently been placed before being exhumed and dumped at sea.

Figure 2.9 A small harbor with a sandy beach in the Netherlands used by small boats to smuggle drugs across the North Sea to southeast England.

20kU X100   100μm 00⊗5 ⊡4⁉9

Figure 2.10 Scanning electron micrograph of sand grains (largely quartz), recovered from the base of a holdall seized in a drugs smuggling investigation, found to have similar size distribution, shape, surface texture, and chemical characteristics to sand samples collected from the Dutch coast close to the harbor shown in Figure 2.9.

considerable forensic significance, for example, if they become jammed in valves or machinery components, leading to fires or explosions (Figure 2.12).

2.3.3 Mud

The term *mud* is often used in a loose sense to describe any sediment which contains a significant proportion of fine particles and which has a "sticky" character when wet. As noted earlier, the *mud fraction* is more specifically defined as material <0.063 mm in size. However, mud in a loose sense usually contains a mixture of silt, clay, and sand-size particles and, in some cases, gravel. Many mud samples also contain a significant proportion of organic matter.

Mud is widespread in the natural environment, but its physical, chemical, and biological properties vary greatly from area to area. Traces of mud are consequently encountered very frequently in forensic casework on a wide range of items, including human skin, fingernails, clothing, footwear, and vehicles. Important environmental sources of mud include coastal and estuarine mudflats, river floodplains, lakeshores, and tracks or paths across fields

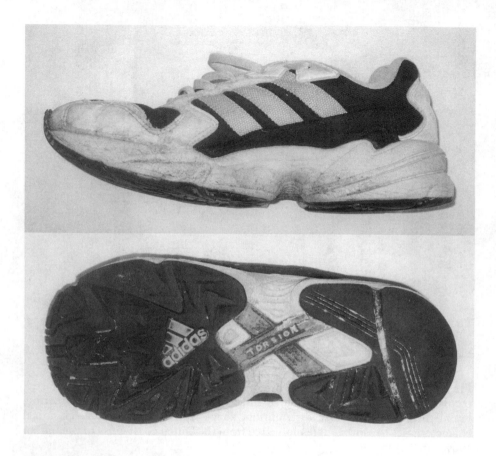

Figure 2.11 Traces of sand in the sole tread and welts of a pair of training shoes seized from a suspect in a murder investigation where the victim's body was found on a marine beach in eastern England. Microscopic examination of the sand indicated it to contain a range of grain types, including marine shell fragments, with adhering marine diatoms and salt crystals similar to those found on control samples taken near the victim's body.

and woodland (Figure 2.13 and Figure 2.14). However, localized occurrences of mud may be found almost anywhere, including urban areas. Clays, silts, and muds are quarried for a variety of industrial uses, including brick making, ceramic manufacture, and production of fillers and coatings (e.g., Keegan, 1999, 2000). Artificial muds with specific physical and chemical properties are also produced for specialized industrial purposes (e.g., drilling muds and cosmetic muds).

Figure 2.12 An SEM micrograph showing fine sand and silt particles around a valve from a gas cylinder. Their presence led to incomplete closure of the valve, contributing to an explosion.

Figure 2.13 Intertidal mudflats and peat beds exposed at low tide on the coast of Merseyside, northwest England, sampled as a location of interest during an investigation.

Figure 2.14 Shoe impressions in soft muddy sediment; samples were taken from plaster casts of the shoe impressions for comparison with mud recovered from a pair of shoes seized from a suspect.

2.4 Soils

2.4.1 Soil Properties and Classification

As noted by Sumner and Wilding (2000), "a concise definition of soil is still elusive bearing in mind the extremely heterogeneous nature of the soil." Much of the scientific work carried out on soils has focused on its properties as a medium for plant growth, or as a foundation for engineering structures, and many of the published definitions and classifications relate to these aspects. However, in simple terms, a soil may be defined as a layer or layers of largely unconsolidated material that formed at the Earth's surface through the combined action of weathering, sediment accumulation, and biological processes. Natural soil properties show great spatial variation as a result of the interaction of five main soil forming factors (Jenny, 1941): (1) geological parent material, (2) climate, (3) topography, (4) biological influences, and (5) time. In addition, the role of a sixth factor, man, is significant. Soils in many areas have been improved for agricultural purposes over many years, involving additions of lime, sand, fertilizers (both natural and artificial), and pesticides. Some soils have also been degraded as a result of human actions, including destruction of natural vegetation, ploughing, and subsequent wind or water erosion of the surface layers.

Any unconsolidated, stable surface material can potentially act as a growing medium for plants, and some "soils" are little more than accumulations of weathered rock or sediment. However, a key feature of more developed soils is the presence of a number of distinct layers, or *horizons*, which show distinctive physical and chemical properties. The development of soil horizons reflects the operation of soil-forming processes over time, including weathering, vertical movement of soluble constituents and finer particles from the surface to deeper levels (and sometimes the reverse), and biological activity that may lead to the accumulation of organic matter within one or more horizons. Soil classification and soil genesis are complex issues, which have been intensively investigated over many years. General reviews of soil properties and processes are provided by Brady and Weil (1999) and Sumner (2000), and detailed descriptions of soil characteristics in particular areas of the world are available in numerous specialized publications (e.g., Mohr et al., 1972; CSIRO, 1983).

2.4.2   Soil Constituents

Soils contain both non-living matter and living matter. The non-living matter includes inorganic mineral grains, chemical precipitates such as calcium carbonate nodules or salt crystals, dead plant matter in various stages of decomposition, and dead animal matter, including the remains of invertebrates and vertebrates. Shell material, insect carapaces, and bones and teeth of small mammals are important constituents of some soils. Traces of past microbial communities may also be preserved as chemically altered organic molecules. The living matter includes bacteria, soil algae, fungi, roots of higher plants, and various species of invertebrate and vertebrate animals. Each of these constituents has been the subject of detailed individual study, and a useful summary is provided by Tan (2000). Detailed accounts of minerals and rock fragments in soil environments are provided in Dixon and Weed (1977), Ugolini et al. (1996), and Dixon and Schulze (2002). A useful overview of soil physical chemistry is provided by Sparks (1998), while Madigan et al. (2003) provide a valuable introduction to the nature of microbes and microbial processes.

The biological, and some chemical, characteristics of soils often show significant seasonal and longer-term change, reflecting fluctuations in ambient environmental conditions (especially temperature, humidity, precipitation, and soil moisture regime). Consequently, in forensic investigations, it is important to collect soil samples as quickly as possible after an incident has taken place. Some soil properties in certain soils are prone to very rapid change after samples are taken from the host environment; for example, reduced (ferrous) iron compounds, which give rise to blue, green, grey, or black coloration, may be rapidly oxidized to ferric compounds which impart brown, yellow, or reddish coloration. These color changes may be accompanied by changes in soil pH and Eh, and some types of soil microorganisms and microfossils may be destroyed as a result of oxidation or change in pH.

In order to minimize the effects of such changes, soil samples should be analyzed as soon as possible after collection or, if this is not possible, stored under conditions that minimize postcollection changes. In situations where there has unavoidably been a long-time period before analysis, attention should be focused on those attributes that are least susceptible to change, for example, the physical and chemical characteristics of the most resistant constituents of the mineral assemblage. If comparisons of potentially volatile properties are made, careful consideration should be given to possible changes during sample storage.

Soils in particular areas often contain assemblages of different particle types that may have a restricted geographical distribution. Sugita and Marumo (2004) refer to such particles as "unique" particles, although in a strict sense very few of the particle types to which they refer (volcanic glass, zeolite minerals, plant fragments, plant opals phytoliths, and microalgae) are truly unique. A better term is *unusual particles* or *unusual matter* (Marumo and Sugita, 2001). Many different types of particle of human origin may be found in soils, especially within urban areas or other places affected by waste tipping. They include cement (calcium aluminosilicate hydrates), plaster (calcium sulfate hydrates), metallic fragments of various kinds, glass spheres, paint flakes, paper fragments, fibers with distinctive color, and plastic fragments. Some of these particles may be regarded as unusual if they show distinctive coloration patterns, shapes, textures, or multiple layering.

2.5 Dusts and Particulates

2.5.1 Natural Dusts Derived from Soils and Sediments

Dust particles transported in suspension in the Earth's atmosphere are generally smaller than 100 μm. Suspended particles larger than about 20 μm tend to fall back to the surface relatively quickly and are not normally transported over large distances unless they are of very low density (e.g., some organic particles), whereas fine particles (<10 μm) can remain in suspension for very long periods and may be transported over distances of thousands of kilometers (Pye, 1987). Any bare, dry surface can act as a local source of dust. Deserts and semiarid regions are major sources at a global scale, but industrial stockpiles, gravel tracks, and car parks and paths can act as important local sources in any part of the world.

Natural dusts normally consist of a mixture of individual silt and clay-size particles, aggregates of such particles and fine organic matter, and some larger particles of low density organic matter. The chemical composition of the dust reflects the "average" composition of the source area from which it is derived and often varies considerably from one area to another. The particle size distribution, mineralogy, chemical composition, and nature of the inorganic and organic particles in a dust sample will often provide a good indication of the type of environment from which it was derived, and in some circumstances define a very specific geographical area. Accumulations of dust may be found on a variety of items of forensic interest, including vehicles, boxes, construction materials, and clothing.

2.5.2 Industrial Dusts

Industrial dusts are derived from sources such as quarries, mines, stockpiles of coal, aggregate, salt or grain, or from particular types of factory or processing plant (e.g., foundry, power station, and chemical works). They often display a particle size distribution, chemical composition, and particle type assemblage that is quite different to that of natural "soil" dusts. Such dusts may contain relatively rare types in considerable abundance, for example, asbestos, glass, metallic alloys, and carbonaceous fly-ash (Harrison, 1986; Watt, 1988; Merefield et al., 2000). The dispersion of particles from industrial sources depends partly on the particle properties, partly on the method of emission, and partly on atmospheric conditions in any given period. Detailed mathematical models are available which can predict distances of travel, airborne concentrations, and deposition rates under different environmental conditions.

2.5.3 Household and Street Dusts

Household dusts are normally a mixture of environmental (street) dust and particles derived from local sources within the property. Such particles include large numbers of skin cells, mites, fibers, and hairs (Benko, 2000; Figure 2.15). The "street dust" component may include particles derived from local factories and activities such as slab-cutting or stone cleaning. While this component is likely to be common to several houses or even streets, the assemblage of fibers and hairs may allow differentiation of one house from another. Street dusts commonly contain relatively high levels of trace metals such as Pb, Cu, Zn, and Cd, although the major elements present are normally Si, Al, and Ca.

2.5.4 Particulates in Water Bodies

Natural water sources such as streams, ponds, and lakes contain variable amounts of suspended sediment particles that are usually both inorganic and organic. Inorganic particles commonly include individual mineral grains and clay–silt aggregates, whereas the organic particle types include diatoms, other algae, fish scales, etc. Domestic tap waters may also contain significant numbers of particles, including calcium precipitates and corrosion products derived

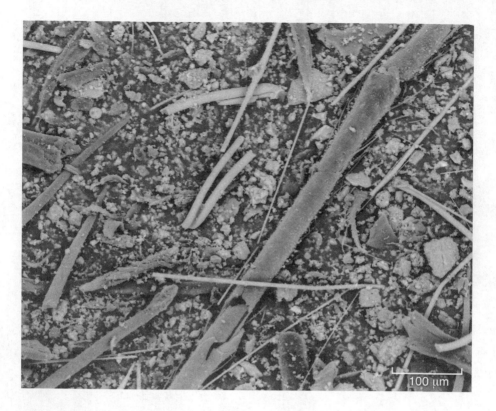

Figure 2.15 An SEM micrograph of a dust sample lifted from the top of a cupboard in a London building. The sample contains numerous fibers and disarticulated insect remains, as well as minerogenic particles derived from "street" sources.

from pipework and other plumbing fitments. Water bodies adjacent to factories, roads, and waste tips often contain both suspended and settled particles which have a relatively high content of heavy metals (e.g., Lee et al., 1997). Analysis of such particles in the airways or gastrointestinal tracts of drowning victims may assist identification of the location where drowning took place.

2.6 Minerals

2.6.1 The General Nature of Minerals

A mineral may be defined as a solid substance with a characteristic chemical composition and a defined crystalline structure. Materials which have a fairly well defined chemical composition but only a poorly defined crystalline structure are sometimes referred to as *mineraloids*, whereas materials with no identifiable structure are referred to as being *amorphous*. There are thousands of named minerals (Frye, 1981), but many are extremely rare in nature.

The most abundant minerals on the Earth's surface belong to the silicate groups, but non-silicates (notably oxides, hydroxides, sulfides, sulfates, carbonates, phosphates, and halides) are also widespread and locally abundant.

The scientific study of minerals (mineralogy) has a long history, and numerous introductory and advanced texts are available (e.g., Deer et al., 1992; Wenk and Bulakh, 2004). Illustrative atlases of minerals and rocks in thin section have been published by MacKenzie and Guilford (1980) and MacKenzie and Adams (1994). Useful reviews of minerals mined commercially for use as raw materials, pigments, fillers, and extenders are provided by Harben and Kuzvart (1996) and Keegan (1999, 2000). More specific information about minerals that pose potential hazards to human health, such as asbestos and silica minerals, is provided by Skinner et al. (1988) and Lewis (2000a, 2000b).

2.6.2   Light Minerals

Minerals vary considerably in density, reflecting variations in their elemental composition and structural arrangement. Minerals that have a density of less than c. 2.90 g cm^{-3} are generally referred to as *light minerals*, whereas those with higher density are referred to as *heavy minerals*. The light and heavy fractions can be separated by gravity flotation using suitable heavy liquids with specified densities. The most frequently used heavy liquids, sometimes diluted with a solvent such as acetone, are bromoform (density 2.90 g cm^{-3}), Thoulet's solution (density 3.19 g cm^{-3}), methylene iodide (density 3.325 g cm^{-3}), Clerici's solution (density 4.4 g m^{-3}), and sodium polytungstate (density 3.1 g cm^{-3}) (Krumbein and Pettijohn, 1938; Tickell, 1965; Carver 1971). The main "light" minerals found in natural rocks, sediments, and soils are quartz, feldspars, and the carbonate minerals calcite, aragonite, and dolomite (Table 2.5).

2.6.3   Heavy Minerals

Heavy minerals are generally far less abundant than light minerals and typically represent up to 1 or 2% of a sediment or soil. Exceptions occur where heavy minerals have been concentrated by natural processes (e.g., by wave and current action on some beaches, within some river channels, and in some aeolian environments). In these cases, the heavy mineral content may reach several tens of percent, forming economically workable ore deposits known as *placers*. The most common heavy mineral groups are oxides such as magnetite, hematite, ilmenite, and rutile, and complex silicates, including pyroxenes, amphiboles, and garnets (Table 2.5). Different types of heavy mineral can be separated either using heavy liquids, as described earlier, or on the basis of differences in magnetic properties using a Frantz isomagnetic separator (e.g., Tickell, 1965, pp. 48–51).

Table 2.5   Some Common Minerals Found in Rocks, Sediments, and Soils showing General Chemical Composition, Density, and Hardness

Mineral		Composition	Density (kg m^{-3})	Hardness
		Light Minerals		
Quartz		SiO_2	2650	7
Albite	Plagioclase	$NaAlSi_3O_8$	2620	6–6 $^1/_2$
Labradorite	feldspar	$(Ca,Na)(Al,Si)AlSi_2O_8$	2700	6–6 $^1/_2$
Anorthite		$CaAl_2Si_2O_8$	2750	6–6 $^1/_2$
Orthoclase	K-feldspar	$KAlSi_3O_8$	2560	6–6 $^1/_2$
Microcline		$KAlSi_3O_8$	2560	6–6 $^1/_2$
Calcite		$CaCO_3$	2710	3
Aragonite		$CaCO_3$	2930	3 $^1/_2$–4
Dolomite		$CaMg(CO_3)_2$	2870	3 $^1/_2$–4
Gypsum		$CaSO_42H_2O$	2320	2
Halite		$NaCl$	2160	2 $^1/_2$
Anhydrite		$CaSO_4$	2890–2980	3–3 $^1/_2$
		Heavy Minerals		
Pyroxenes		$(Ca,Mg,Fe)_2(Si,Al)_2O_6$	3200–3550	5–6
Amphiboles		$NaCa_2(Mg,Fe,Al)_5(Si,Al)_8O_{22}(OH)_2$	3000–3470	5–6
Pyrite		FeS_2	5000–5200	6–6 $^1/_2$
Siderite		$FeCO_3$	3700–3900	4–4 $^1/_2$
Garnet		$(Fe,Al,Mg,Mn,Ca)_5(SiO_4)_3$	3560–4320	6–7 $^1/_2$
Epidote		$Ca_2(Al,Fe)_3O_{12}(OH)$	3250–3500	6–7
Olivine		$(Mg,Fe)_2SiO_4$	3210–4390	6 $^1/_2$–7
Staurolite		$FeAl_4Si_2O_{10}(OH)_2$	3700	7
Kyanite		Al_2SiO_5	3690	5 $^1/_2$–7
Andalusite		Al_2SiO_5	3160–3200	6 $^1/_2$–7 $^1/_2$
Sillimanite		Al_2SiO_5	3230–3270	6 $^1/_2$–7 $^1/_2$
Zircon		$ZrSiO_4$	4670	7 $^1/_2$
Rutile		TiO_2	4250	6–6 $^1/_2$
Anatase		TiO_2	3900	5 $^1/_2$–6
Apatite		$Ca_5(PO_4)_3(F,Cl,OH)$	3100–3250	5
Tourmaline		$Na(Mg,Fe)_3Al_6(BO_3)_3(Si_6O_{18})(OH)_4$	3030–3100	7
Monazite		$(Ce,La,Y,Th)PO_4$	5270	5–6
Ilmenite		$FeTiO_3$	4500–5000	5 $^1/_2$–6
Magnetite		$Fe^{2+}Fe^{3+}_2O_4$	5200	5 $^1/_2$–6 $^1/_2$

Table 2.5 Some Common Minerals Found in Rocks, Sediments, and Soils showing General Chemical Composition, Density, and Hardness (Continued)

Mineral	Composition	Density $(kg\ m^{-3})$	Hardness
	Clay Minerals and Micas		
Muscovite	$KAl_2(AlSi_3O_{10})(OH)_2$	2800–2900	2 $^1/_2$–3
Biotite	$K(Mg,Fe)_3(AlSi_3O_{10})(OH)_2$	2800–3400	2–3
Chlorite	$(Mg,Fe,Al)_6(Al,Si)_4O_{10}(OH)_8$	2600–3300	2–3
Kaolinite	$Al_4Si_2O_5(OH)_4$	2600–2630	2–2 $^1/_2$
Illite	$KAl_2(Al,Si_3O_{10})(OH)_2$	2600–2700	1–2
Palygorskite	$(Mg,Al)_5(Si,Al)_8O_{20}4H_2O(OH)_2$	2200–2360	2–2 $^1/_2$
Smectite	$(Ca,Na)(Al,Mg,Fe)_4(Si,Al)_8O_{20}$ $(OH)_4nH_2O$	1700–2700	1–2
Talc	$Mg_6(Si_8O_{20})(OH)_4$	2580–2830	1
Vermiculite	$(Mg,Fe,Al)_6(Si,Al)_8O_{20}(OH)_4nH_2O$	2400–2700	1 $^1/_2$
Ice	H_2O	920	1 $^1/_2$

[a] According to Moh's scale.

Sediments often contain a considerable number of different heavy mineral types in varying abundances, and different lithostratigraphic units (sediment bodies of the same age and mode of genesis) often have distinguishable heavy mineral assemblages. Heavy mineral assemblages have long been used in sediment provenance studies (Mange and Mauer, 1992) and have been used as a line of evidence in a considerable number of forensic case investigations (e.g., Smale and Trueman, 1969; Isphording, 2004; Stam, 2004). Since heavy minerals form only a very small percentage of most sediments and soils, and in forensic casework only traces of material are available to work with, some investigators have preferred to analyze and compare total mineral assemblages rather than separated light and heavy mineral fractions (e.g., Graves, 1979).

2.6.4 Rare Earth Minerals

Rare earth minerals contain a significant percentage by weight of the rare earth elements, a group of 14 elements in the Periodic Table that show similar chemical behavior. Concentrations of rare earth minerals occur in a number of specific geological settings around the world, but virtually all common rocks, sediments, and soils contain them in trace amounts. There are many different types of rare earth minerals, including carbonates, silicates, oxides, halides, phosphates, and sulfates (Jones et al., 1996). The most frequently encountered rare earth mineral is monazite, a phosphate mineral containing significant amounts of cerium, lanthanum, and thorium, which is a common constituent of detrital sedimentary heavy mineral assemblages.

However, monazite grains may display considerable variation in shape, surface texture, and internal chemical zonation, and can be extremely useful geological provenance indicators.

Isphording (2004) reported a case where detrital heavy mineral assemblages in sediments provided valuable evidence that helped to convict members of the Ku Klux Klan of the murder of a young black male. The body of the victim was recovered from a location in Baldwin County, Alabama, which is underlain by sediments of the Citronelle Formation, a sedimentary unit with a heavy mineral assemblage containing relatively large amounts of ilmenite, kyanite, and zircon, an absence of hornblende and garnet, and a relatively high tourmaline to rutile ratio. Analysis of soil samples taken from items associated with the defendants showed a clear Citronelle signature and showed they had lied about their whereabouts on the night of the murder.

The size, shape, internal compositional zonation, and textural intergrowths of rare earth minerals with other diagenetic minerals in rocks and sediments can often provide useful criteria for forensic comparison (Figure 2.16).

2.6.5 Clay Minerals

On the classical Udden–Wentworth scale, clay-size particles are defined as being smaller than 4 μm in size, but in more classification schemes an upper size limit of 2 μm has generally been preferred (Table 2.4). However, the term *clay mineral* is applied to a group of minerals, the majority of which have a layer silicate (sheet) structure and platy form. Although many clay mineral particles are smaller than 2 μm, some may be 20 μm or larger in maximum dimension. Most clay minerals are hydrous silicates composed principally of the elements silicon, aluminum, oxygen, and hydrogen, with varying amounts of iron, magnesium, potassium, sodium, calcium, and other ions. Depending on their chemical and structural composition, clay minerals display a range of distinctive physical and chemical properties, which make them of considerable interest for use in a wide range of industrial processes (Hall, 1987). Many clay minerals form as alteration products during the processes of weathering, diagenesis, and metamorphism, and their physical and chemical properties often show a close association with conditions of their formation. For these reasons, clay minerals provide useful indicators of environmental conditions and sediment provenance (Grim, 1968; Perrin, 1971; Weaver and Pollard, 1973). Clay mineral analysis is generally performed on standardized size fractions, most commonly the <2 μm fraction but sometimes also the <10, <6, <5, <4, <1, <0.1, or <0.05 μm fractions. The most common method of analysis is by x-ray powder diffraction (Brindley and Brown, 1980; Marumo et al., 1986, 1988; Wilson, 1987; Ruffell and Wiltshire, 2004), although infrared

PARK PDB 5/19 incl2

25 Pa 06-Oct-04   S2600   WD15.7mm 20.0kV x1.2k 25um

BSE2

Figure 2.16 Backscattered electron SEM image of part of a polished section prepared from the rock of interest in the "Lady in the Lake" case, shown in Figure 2.6. The mid-grey "grain" in the upper center is zircon, and the white "grain" adjacent to it is a rare-earth silicate mineral. The surrounding darker grains are quartz, feldspar, and illite/mica. The textural intergrowths, as well as the elemental compositions, of the different minerals were of special interest in this case.

analysis (Farmer, 1974; Russell, 1987) and differential thermal analysis (DTA; Paterson and Swaffield, 1987) are also used (Table 2.6).

Clay mineral assemblages have been examined in many forensic soil and other forensic geological investigations. For example, Isphording (2004) examined the clay mineral assemblages in the <4 μm fractions of soil samples from a body deposition site in Alabama, southern United States, with soil from a defendant's residence and questioned samples taken from the defendant's car, shoes, and clothing. The crime scene was located on terrace sediments belonging to the Citronelle Formation, which had a clay mineral assemblage dominated by kaolinite and illite. In contrast, the clay assemblage of soils at the defendant's home address on Quaternary floodplain deposits in Mobile Bay contained a high proportion of smectite (montmorillonite) in addition to kaolinite and illite. Analysis of the questioned samples indicated that they had a clear "Ciotronelle" clay mineral signature.

Table 2.6   Individual Particle Types and Properties that Can be Used for Forensic Comparison

Particle Type	Main Techniques and Equipment Used
	Inorganic Particles
Quartz grains	Shape and surface textural analysis, cathodoluminescence microscopy, fluid inclusion analysis, isotopic analysis by laser ablation inductively coupled plasma spectrometry, analysis of grain coatings by scanning electron microscopy and energy-dispersive x-ray microanalysis
Other "light" mineral grains "Heavy" mineral grains	Optical and scanning electron microscopy, supplemented by energy-dispersive x-ray microanalysis, cathodoluminescence microscopy, laser Raman spectroscopy, electron-probe analysis, ion-probe analysis, laser ablation inductively coupled plasma spectrometry, microspectrophotometry, dating by Ar-Ar and U-Pb series methods, stable isotope analysis by mass spectrometry
Clay minerals	X-ray diffraction, infrared analysis, differential thermal analysis, electron diffraction, transmission electron microscopy
Gravel particles	Shape and surface textural analysis, internal fabric analysis by measuring and image analysis
	Organic Particles
Opal phytoliths Foraminifera Coccoliths Coralline particles Molluscs Gastropods Ostracods Diatoms Insect remains Pollen and spores Plant seeds Leaf and stem fragments Coal fragments Charcoal fragments Wood fragments	Optical and scanning electron microscopy, supplemented by energy-dispersive x-ray microanalysis

Table 2.6 Individual Particle Types and Properties that Can be Used for Forensic Comparison (Continued)

Particle Type	Main Techniques and Equipment Used
	Anthropogenic Particles
Slag and ash	
Spherules	Optical and scanning electron microscopy, energy-
Brick	dispersive x-ray microanalysis, cathodoluminescence
Concrete	microscopy, laser Raman spectroscopy, electron-probe
Pottery	analysis, ion-probe analysis, laser ablation inductively
Glass	coupled plasma spectrometry, microspectrophotometry
Alloys and pure metals	
Fibers	Optical and scanning electron microscopy, supplemented
Paint	by energy-dispersive x-ray microanalysis, fluorescence
Paper	microscopy, microspectrophotometry, Fourier-transform
	infrared spectroscopy and microscopy, ultraviolet
	spectroscopy

2.7 Glasses and Other Amorphous Materials

2.7.1 Natural Glasses

In geology and mineralogy, the term glass refers to a solid material that has no crystalline structure, that is, is amorphous. Glasses may show compositional variability and inclusions of gas, liquid, or mineral crystals (phenocrysts). They are often transparent or translucent, have a vitreous luster, and show a characteristic conchoidal fracture when broken. Devitrification due to incipient crystal growth or chemical alteration may be evident to varying degrees. The most commonly encountered examples are volcanic glasses, including obsidian and rhyolite glass (Beall, 1981).

2.7.2 Natural Soaps and Gels

Other amorphous or mineraloid (poorly crystalline) substances that are quite frequently found in sediments and soils include gels or soaps. These materials form most commonly as a result of chemical weathering or diagenesis (chemical transformation processes). Many chemical precipitates initially have a poorly defined structure, which progressively becomes more crystalline with time and/or higher temperatures experienced by surface heating or burial. Examples include amorphous or poorly crystalline iron oxy-hydroxides and iron mono-sulfides which form precursors for minerals such as ferrihydrite, goethite, greigite and pyrite, and aluminosilicate gels which later develop into clay minerals such as smectites.

2.8 Fossils

2.8.1 General Nature

Fossils occur in many different shapes and sizes, but a basic distinction can be made between *macrofossils* and *microfossils*. Macrofossils can generally be seen easily in hand specimen, whereas a microscope is normally needed to determine the nature of microfossils. Fossils may represent either the whole or part of a dead organism, an impression or cast of the organism, or a trace left by movement or other activity of the organism (Black, 1989; Benton and Harper, 1997). The organism itself may belong to one of three main groups: vertebrates, invertebrates, and plants. Fossils occur mainly in finer grained sedimentary rocks and sediments, although fossil traces may sometimes be found in metamorphic rocks. Fossils are useful for several reasons: (1) they can assist in determining an age for the rocks or sediments that host them; (2) they provide a means of correlation between different rock outcrops; and (3) they may provide indicators of past environmental conditions.

Fossils vary in their state of preservation, from essentially complete preservation of the original tissue structures of the dead organism, through complete replacement of the primary tissues by secondary mineral phases but retention of many of the original structural details, to preservation only of outlines as three-dimensional "ghosts" or two-dimensional surface "films." In forensic geology, their main value lies in their use as indicators of the age and likely source beds of rocks that may, for example, have been used to weigh down an object in a water body. The list of organisms that may be of value in this regard is long, but the more commonly encountered types include graptolites, brachiopods, bivalves, and gastropods (Table 2.6). The distinction between "modern" and "fossil" plant or animal remains in terms of age is somewhat arbitrary. Although the term "fossil" is often thought of as relating to objects thousands or millions of years old, a case can be made for regarding anything older than a few years as fossil, particularly if the organism in question no longer lives in the area.

2.8.2 Macrofossils

Intact and broken specimens of many different types of shell, bone, tooth, and plant material of varying sizes are commonly found in sediments and soils (Figure 2.17). These may represent reworked material derived from geological formations of considerable age (thousands to tens of millions of years), or represent the remains of organisms that lived in the area only a few years previously. Careful study and identification of the species type, combined with examination of the state of preservation or alteration, can normally allow recognition of modern or "fossil" material. It may also be possible to obtain a direct estimate age using methods such as radiocarbon

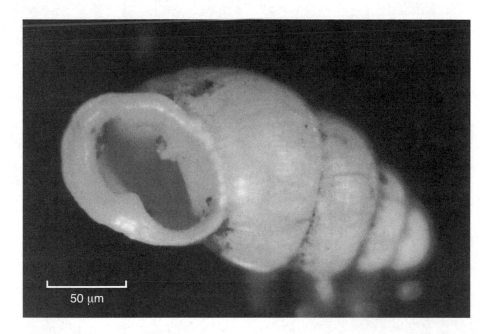

Figure 2.17 Reflected light micrograph showing a small terrestrial gastropod (*Carychium* sp.) recovered from a soil sample taken from scene of murder in a copse near Bristol, West of England. Shells of the same type were present in soil recovered from footwear worn by a suspect later convicted of the murder.

or uranium-series dating. Even if this is not possible, it is often possible to assign a geological age, and to identify possible source areas, by reference to published geological information. For example, Stam (2004) described a case where 3 to 4 mm long shells of the freshwater gastropod *Tryonia protea*, which lived in a now vanished lake, Lake Cahuilla, during the last Ice Age, and which occur in abundance in fossil lake sediments in the southern part of Imperial Valley, southern California, provided one line of evidence that a suspect was present at the scene of an assault in the area.

Fragments of plant material, in the form of roots, bark, leaves, stems, fruits, and seeds (Figure 2.18), are quite frequently found in association with soil and sediment in forensic casework. In most cases, the material is relatively modern (i.e., weeks to a few years old), but exceptions occur in areas where peat and similar fossil organic deposits are found (e.g., Figure 2.13). Identification and interpretation of such material should be undertaken by a suitably qualified forensic botanist or palaeobotanist. Where sample size permits, it may be possible to obtain an absolute age by radiocarbon dating (Pilcher, 1991). There are numerous published reports of the use of botanical evidence in forensic investigations (Bock and Norris, 1997; Hall, 1996; Bates et al., 1997; Coyle, 2004; Ladd and Lee, 2004; Robertson, 2004). In the case

Figure 2.18 (See color insert following page 46.) Reflected light micrograph showing a blackberry (*Rubus fruticosus* agg.) seed and a small spider in the >150 μm fraction of a control soil sample from the copse murder scene near Bristol. The seed contains traces of soil trapped in the surface detail. Blackberry seeds containing soil with an indistinguishable elemental composition were also found in abundance in soil recovered from the suspect's footwear, together with many other particle types found in the crime scene control samples.

of plant material that is modern or relatively young in geological terms, it may be possible to identify specific plant species or sub-species, and potentially individual plants, using DNA techniques (e.g., Korpelainen and Virtanen, 2003; Bever, 2004; Carita, 2004).

2.8.3 Macroscopic Charcoal

Charcoal is often described as an amorphous, usually porous form of carbon that is formed by heating and incomplete combustion (pyrolysis) of organic matter, most commonly wood, in the virtual absence of oxygen at temperatures in the range 200 to 500°C (Teixeira et al., 2002). However, true charcoal usually contains a high proportion of microcrystalline graphite (one of the two forms of crystalline carbon). Charcoal is chemically almost inert and very resistant to weathering. It may persist in soils for centuries or millennia and can be reworked by wind or water to be redeposited a considerable distance away from its place of formation (Clark, 1988a, 1988b; Scott et al., 2000a, 2000b). A broad

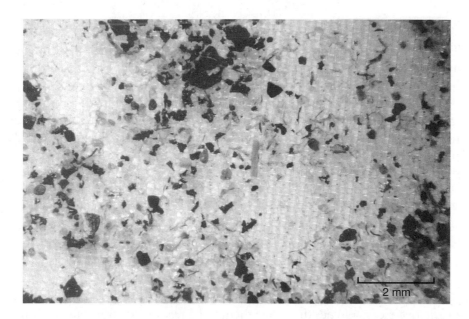

Figure 2.19 (See color insert following page 46.) Reflected light optical micrograph showing abundant fragments of black charcoal in the >150 μm fraction of a soil sample from a waste disposal site near Southport, northwest England, searched in the hunt for a body as part of a murder enquiry.

distinction can be made between *macroscopic charcoal* or *macro-charcoal*, which is >80 μm in size, and *microscopic charcoal* or *micro-charcoal*, which is smaller than that size (e.g., Innes et al., 2004; Figure 2.19). Plant structures are often well preserved in macroscopic-charcoal, allowing the identification of the parent plant material type in larger samples and allowing modern and fossil types to be distinguished using a combination of optical and scanning electron microscopy (Scott and Jones, 1991; Figueral and Mosbrugger, 2000). Detailed microscopic and microchemical examination is also required to distinguish charcoal from fragments of coal and coke (Juckes and Pitt, 1977). Charcoal can be formed either as a result of natural fires or commercial manufacture. In modern sediments, a continuum of particle types is often found between unburnt plant fragments and completely carbonized (burnt) material. However, identification as "charcoal" is normally restricted to black, completely opaque, angular fragments (Clark, 1982). The carbon, oxygen, hydrogen, and sulfur content of charcoal is dependent on the temperature of carbonization; higher temperature forms have a higher relative carbon content.

The age of individual larger charcoal fragments up to several tens of thousands of years old can be determined by radiocarbon accelerator mass spectrometry (^{14}C AMS) dating. The minimum sample mass requirement for this type of analysis is c. 5 mg, which is equivalent to three or four 1–mm-diameter charcoal fragments (Carcaillet, 2001).

2.8.4   Microfossils

There are many different types of microfossil that are found in sediments and soils (and also suspended in air and water, and as deposits on clothing, hair, etc.). The ones of most frequent forensic interest are pollen and spores, diatoms and other algae, foraminifera, opal phytoliths, and charcoal fragments (Table 2.6), although many other types exist and may be of importance in particular contexts (Haslett, 2002; Armstrong and Brasier, 2005). General procedures for the extraction and examination of microfossils are described by Green (2001).

2.8.4.1   Pollen and Spores

Pollen and spores are small "grains" of organic matter produced by plants — pollen by the "higher plants," conifers and angiosperms, and spores by "lower plants" such as bryophytes and ferns. Both types have a cell wall that is very resistant to microbial attack and, therefore, survive prolonged periods of burial and transport by wind and water. Many spores and pollen grains have distinctive morphologies that allow them to be readily identified and counted (e.g., Erdtman, 1986; Moore et al., 1991; Faegri et al., 2000). Although pollen and spores may be dispersed over a considerable distance from the source, assemblages present within soils and sediments can provide a useful reflection of the surrounding vegetation and ambient environmental conditions. The condition of pollen and the nature of the assemblage present on clothing, footwear, or other items may also provide an indication of the time of the year when the item was last exposed. "Exotic" modern pollen types of nonlocal or nonwidespread origin may provide clues to the identity and location of specific sources of exposure. Fossil pollen types may also provide an indication of geographical source if the potential source rocks/sediments have a restricted distribution. Radiocarbon dating of fossil pollen grains by accelerator mass spectrometry has been discussed by Brown et al. (1989). Examples of the methods and applications of forensic palynology are provided by Bryant et al. (1990), Mildenhall (1990, 2004, 2006), Stanley (1992), Bruce and Dettmann (1996), Eyring (1996), Graham (1997), Lewis (1997), Bryant and Mildenhall (1998), Horrocks and Walsh (1998, 1999, 2001), Horrocks et al. (1998, 1999), Brown (2006), Brown et al. (2002), and Milne et al. (2004).

2.8.4.2   Diatoms and Other Algae

Diatoms are a group of single-celled algae that have silicified cell walls comprising a *frustule* that consists of two overlapping halves, or *valves*. There are many different types of diatoms and other algae which live in a variety of aquatic and sub-aerial environments, any of which can provide useful forensic evidence in particular circumstances (Siver et al., 1993; Van den Hoek et al., 1995;

Color Figure 2.7 A comparison of gravel particle types found at two locations of interest in relation to a Customs investigation into cigarette smuggling. The particles represent variants of volcanic rock and tephra which do not occur naturally in the U.K., but which are found in parts of Tenerife, Canary Islands.

Color Figure 2.18 Reflected light micrograph showing a blackberry (*Rubus fruticosus* agg.) seed and a small spider in the >150 µm fraction of a control soil sample from the copse murder scene near Bristol. The seed contains traces of soil trapped in the surface detail. Blackberry seeds containing soil with an indistinguishable elemental composition were also found in abundance in soil recovered from the suspect's footwear, together with many other particle types found in the crime scene control samples.

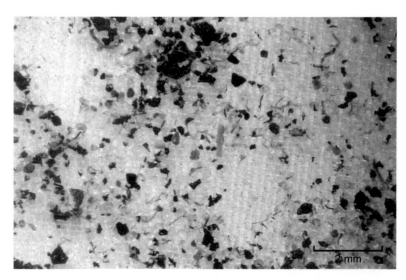

Color Figure 2.19 Reflected light optical micrograph showing abundant fragments of black charcoal in the >150 µm fraction of a soil sample from a waste disposal site near Southport, northwest England, searched in the hunt for a body as part of a murder enquiry.

a.

b.

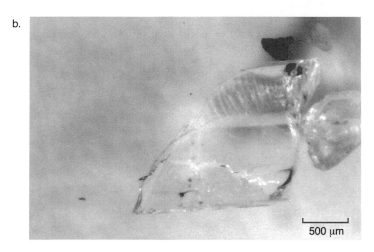

Color Figure 2.23 Reflected light optical micrographs showing examples of anthropogenic particles found in soil samples taken from the scene of a fatal hit and run incident: (a) A "lump" of zinc metal labeled, A, and a fragment of turquoise-colored composite material, labeled B; (b) an angular shard of broken clear glass.

Color Figure 2.24 Reflected light optical micrograph showing two slightly different blue-colored blue paint fragments (A and B) in the >150 μm fraction of a soil sample taken from a garden in South Devon, where a pensioner was bludgeoned to death with a garden edging stone during an attempted robbery. The paint flakes were derived from the drain pipes and soffits at the property. Also visible are several light-colored aggregates of horticultural vermiculite originating from potting compost, which had been emptied onto the garden soil.

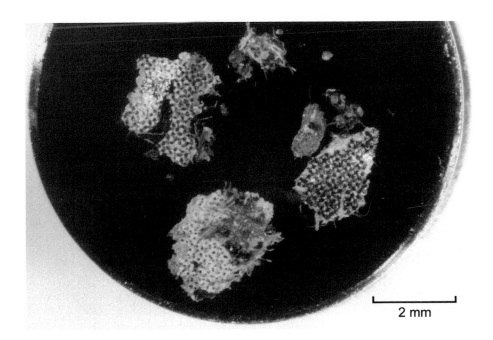

Color Figure 2.25 Fragments of coated magazine paper recovered from the >150 μm fraction of a control soil sample collected at a murder scene near Bristol. There are several composite ink patterns on the same type of backing paper. Similar paper fragments were found in soil recovered from the shoes worn by the murder suspect.

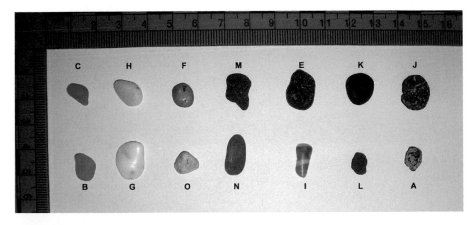

Color Figure 4.27 An assemblage of rounded fine gravel particles from a marine beach in eastern Fuerteventura, Canary Islands.

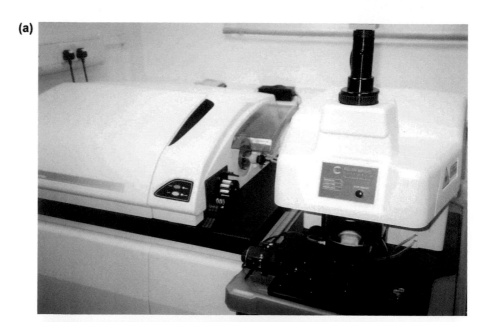

Color Figure 4.35 (a) New Wave laser ablation ICP interface used to obtain *in situ* elemental data from solid samples; (b) closer view of the ablation chamber shown in (a).

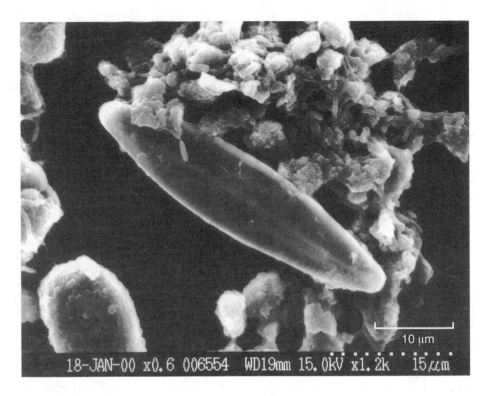

Figure 2.20 An SEM micrograph of a diatom (*Navicula* sp.) in an extract from the lungs of a victim who had inhaled muddy water, sand, and gravel particles, prior to being shot.

Sugita et al., 2004). Diatoms are relatively easy to work with owing to the fact that the frustules are often well preserved and display a considerable variety of forms and textural detail (Figure 2.20). They are very sensitive to environmental conditions, such as temperature, salinity, and nutrient status, and consequently different types of environment are characterized by distinctive diatom assemblages (Round et al., 1990; Stoermer and Smol, 1999). In some lake and marine environments, thick accumulations of sediments composed wholly or largely of diatom frustules may form; these can become lithified to form a lightweight, porous rock known as *diatomite*. Although individual diatoms are small and can be dispersed over a relatively wide area by wind or water, relatively high concentrations can provide a useful locational indicator of potential forensic importance (Peabody, 1977, 1999; Cameron, 2004; Hartley et al., 2006). Diatoms found within the body can also sometimes provide a useful indicator of drowning as a cause of death (Pollanen, 1997; Pollanen et al., 1997; Pollanen 1998a, 1998b; Krstic et al., 2002; Figure 2.21).

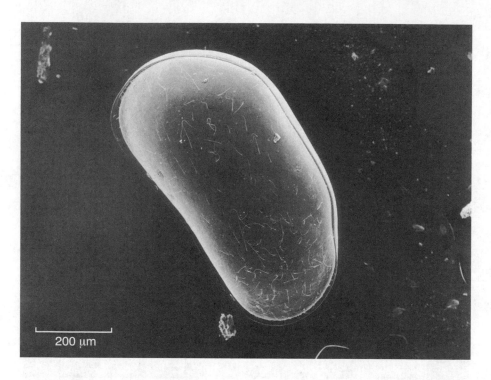

Figure 2.21 An SEM micrograph of an ostracod collected from the clothing of a murder victim whose body was found in a water butt.

2.8.4.3   Foraminifera

Foraminifera are single-celled protozoa that live either on the sea floor (*benthic foraminifera*) or as free-floating components of marine plankton (*planktonic foraminifera*). They possess a "shell" or *test* that consists of a single or multiple chambers composed mainly of calcium carbonate (calcite or aragonite). The largest benthic foraminifera exceed 3 mm in diameter, but most are smaller than 1 mm. Foraminifera have a wide distribution in the modern oceans and also in the geological record, where they form important stratigraphic biomarkers. Some rocks are composed very largely of intact and broken foraminifera. For these reasons, and because changes in ocean water temperature and chemical composition over time are recorded within the tests, foraminifera have been subject to detailed study over many years (e.g., Loeblich and Tappan, 1988; Jenkins and Murray, 1989; Murray, 1991; Sen Gupta, 1999). Forensically, foraminifera have proved useful both in comparison of modern marine sediments and rocks, especially chalks and chalky marls, from the geological record.

2.8.4.4 Calcareous Nannoplankton: Coccolithophores and Discoasters

Calcareous nannoplankton are a diverse group of forms that include *coccoliths*, *discoaters*, and *nannoconids* which are composed mostly of calcium carbonate and range in size from less than 1 μm to c. 30 μm (Winter and Siesser, 1994; Haq and Boersma, 1998). In the geological record they occur widely in deep-water marine sediments and locally become sufficiently abundant to form calcareous rocks such as the Chalk of Upper Cretaceous age.

Coccolithophores are unicellular protozoa that form an important component of oceanic phytoplankton. Small calcareous scales (3 to 15 μm in diameter), known as *coccoliths*, form around these cells and fall off on death of the organism. Some parts of the chalk consist almost entirely of intact and broken coccoliths.

Discoasters are star-shaped calcareous nannofossils which are now extinct but form important biostratigraphic markers in some parts of the geological record, most notably rocks belonging to the Tertiary epoch.

Nannoconids are very small cone-shaped microfossils that consist of closely spaced calcite wedges that together form a spiral. Individual nannoconids are often grouped together to form petal-like structures. They provide a useful biostratigraphic marker and correlation tool in certain rock, especially those of Cretaceous age.

In a forensic context, calcareous nannoplankton have proved most useful in comparing and identifying the potential sources of chalk and chalk-marl type rocks. In a U.K. case example, analysis of calcareous nannoplankton in fragments of chalk rock and deposits of chalky mud provided useful evidence linking a murder suspect's car with a deposition site near Lakenheath, Suffolk, where the bodies of two young girls were discovered in a ditch. The chalk fragments and other components of mud found on the vehicle also showed a high degree of similarity to samples from the track, leading to the body deposition site in terms of several other physical and chemical characteristics.

2.8.4.5 Testate Amoebae

Testate amoebae (*Rhizopoda*) are another type of protozoa that provide sensitive indicators of hydrological conditions, including fluctuations in water table level (Tolonen, 1986; Charman et al., 2000). In a forensic context, they provide a useful means of inter-sample comparison in acidic, organic-rich sediments and soils where mineral matter and fossils composed of calcium carbonate are rare or absent.

2.8.4.6 Ostracods

Ostracods are small, bivalved Crustacea with two valves composed of chitin or calcium carbonate (Figure 2.22). There are more than 33,000 known living and fossil species (Armstrong and Brasier, 2005). Modern types are adapted to live in a range of habitats, including marine, freshwater, and moist terrestrial (damp soil, puddles, and leaf litter). They are frequently found in water butts and troughs, in addition to streams and ponds. They are widely used as environmental/palaeoenvironmental indicators and for purposes of biostratigraphic correlation in the geological record (e.g., Holmes and Chivas, 2002).

2.8.4.7 Opal Phytoliths

Phytoliths are biogenic particles, typically 10 to 200 μm in diameter, which are composed largely of opaline silica and which form around and within certain plant cells, principally those of the stems and leaves in higher plants. Phytoliths show a wide range of sizes and shapes, many of which are specific to individual plant species or groups. Silica is deposited on the surfaces of, between, and within plant cells. Phytoliths are an important component of many arid and semiarid grassland soils, and of dusts derived from such areas (Wilding and Drees, 1971, 1973, 1974; Wilding et al., 1977; Kondo et al., 1994).

Phytoliths have been extensively used in archaeological investigations to provide information about vegetation history at human occupation sites. They have also been shown in forensic evaluation studies to effectively

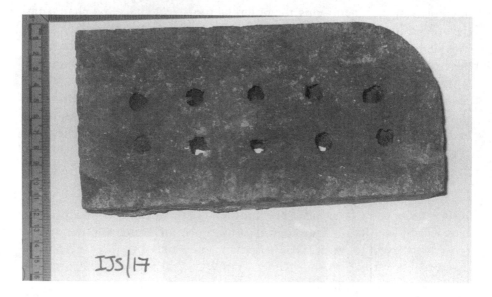

Figure 2.22 A brick used to weigh down a holdall containing body parts found in the Regent's Canal, northwest London.

discriminate between different sites even when the soils were mineralogically similar (Marumo and Yanai, 1986).

2.8.4.8 *Micro-charcoal*

Micro-charcoal abundances in soils have been extensively investigated in palaeoenvironmental studies as an indicator of fire frequency and possible human influence (Clark, 1988a; Scott et al., 2000; Innes et al., 2004). Abundances have often been determined by analysis of the same preparations made for pollen analysis, most frequently by point counting of pollen slides under an optical microscope (Clark, 1982). However, charcoal is relatively brittle and can disintegrate during the pollen preparation process, producing artificially high counts (Clark, 1984). In order to avoid this, there are advantages in counting charcoal in resin-impregnated thin sections under an optical microscope (Clark, 1988b) or in uncovered grain mounts using a scanning electron microscope (SEM). The latter technique offers the added advantage that the identity of the particle as "charcoal" can be confirmed using the x-ray microanalysis attachment on the SEM. Charcoal abundances have been expressed in terms of particle counts, mass, volume, and areas. Statistically representative data normally require counting of at least 200 to 250 fragments per sample or stratigraphic layer (Figueral and Mosbrugger, 2000), and much larger counts are required to reduce relative errors to the 5% level, even in the absence of preparation artifacts (Clark, 1982).

2.9 Anthropogenic Materials

2.9.1 General

There is a considerable variety of man-made products which in some senses can be regarded as artificial "rocks." They include concrete (both poured structural concrete and precast concrete slabs, blocks, pipes, etc.), mortar, render, bricks, and tiles (Doran, 1992). A further type is provided by industrial waste products, such as the slags, formed as a by-product of steel manufacture and similar activities. Some of these man-made materials display similar textural features to natural rocks, and detailed petrographic, mineralogical, and chemical analysis is sometimes necessary to identify their true nature.

2.9.2 Concrete

Concrete consists of aggregate held together by cement. Structural concrete, used for buildings and other construction products, normally contains coarse aggregate (gravel), fine aggregate (sand), and a range of calcium silicate hydrate and other hydration products. Modern concretes also often contain cement replacements such as fly ash, blast furnace slag, and other additives

(Neville and Brooks, 1987; Mehta and Montero, 2005). A range of other products used in construction, including mortars, renders, and screeds, also contain hydraulic cement and some form of aggregate. These products display a great variety of textural, chemical, and mineralogical features, depending on the type of aggregate and cement used, the nature of any additives, and the nature of postplacement environmental exposure.

In forensic work, fragments of concrete are often encountered as a component of samples taken from waste ground, waste tips, and unmade tracks. Larger pieces of concrete are also quite frequently used as weights in holdalls, bags, and packages containing discarded firearms and body parts. On occasions, entire bodies have been encased in blocks of concrete, as in the case of "Baby Lara" found in a garage in the village of Barepot, Cumbria, in 2002. The block, which evidence suggested had been there for several years prior to discovery, contained the skeletal remains of a female baby, approximately 5 to 6 months old, wrapped in polythene. The pathologist's report suggested that the child had experienced a protracted period of neglect and probable physical abuse, prior to death. Examination of the concrete indicated that it was a "homemade" mixture, prepared in two phases. Police investigations led to the identification of the probable parents, who had both died several years previously, but the exact cause of death and the identities of the persons who entombed the baby's body were not determined.

Methods for the petrographic and chemical analysis of concrete and related materials are well established, and a large amount of reference information is available relating to concrete and related materials, including a number of collections made specifically for forensic comparison purposes (e.g., Vermeij, 2003).

2.9.3 Bricks

Bricks have been produced for building purposes since pre-Classical times (Henderson, 2000). Early bricks were made from sun-dried clay, but firing of bricks to produce greater hardness and durability has been undertaken for hundreds of years. More recently, bricks have also been made from mixtures of aggregate and hydraulic cement (silicate bricks) and other materials for specialized purposes (e.g., refractory bricks in furnace linings). However, most common bricks are still made from fired clay and consist chiefly of a mixture of crystalline aluminosilicate phases and noncrystalline glassy material. The color, density, hardness, and porosity of bricks vary greatly, depending on the nature of the raw materials used and the firing conditions (Brick Development Association, 1974).

In forensic casework, whole bricks and half-bricks are frequently encountered as weights, missiles, and murder weapons (Figure 2.21). Small fragments of brick are also a common constituent of soils and sediments, especially in urban areas. In the U.K., brick investigations have played a significant role in many serious crime cases (e.g., Smith, 2003).

2.9.4   Tiles, Pipes, Pottery, and Other Ceramics

Tiles and pipes, which are also commonly made from fired clay or calcium silicate–aggregate mixtures, share many properties in common with bricks. Like bricks, they have been manufactured for thousands of years using a variety of techniques and raw materials (Henderson, 2000, 2004). Roofing tiles usually have relatively coarse textures, containing a significant proportion of sand, whereas pipes are generally made from finer-grained raw material mixtures.

Fragments of tile, pipe, pottery, and other ceramic materials are widely found in soils, especially those of urban areas and in the vicinity of waste tips. Many ceramic types can be dated relatively accurately and the sometimes individual manufacturers identified. Where this is not possible, fragments of glazed ceramics can still be useful since they often show complex surface designs and color patterns that are reflected in small-scale variations in chemical composition. As such they represent one of the "unusual" particle types that can be helpful for inter-sample comparison purposes.

2.9.5   Slag and Clinker

Slag and clinker are by-products of many industrial processes, including metal smelting and burning of low-purity coal. They show a variety of physical and chemical properties, depending on the nature of raw materials, combustion temperatures involved, and conditions of cooling. Slags have been produced since Bronze-Age times and are commonly found in association with archaeological sites (Henderson, 2000). However, they became much more abundant and widespread following the Industrial Revolution. Slag fragments have found their way into many rivers, lakes, and estuaries where they provide "signatures" which reflect the nature and history of local industrial activity. In recent years, slag and related waste products have been sold commercially for a variety of uses, including hard core, road and track surfacing materials and concrete admixtures. Consequently, "slag" can often be found at considerable distances from the centers of production.

2.9.6   Metallic Fragments

Metal fragments occur in many soils in relatively low concentrations. The majority is composed of iron, zinc, copper, nickel, and lead, but a wide variety of alloys and composite particles can be found. The particles may be "fresh" or be coated to varying degrees with secondary alteration products (Figure 2.23a). Metallic particles of human origin can often be distinguished

a.

500 μm

b.

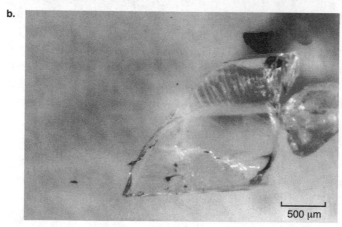

500 μm

Figure 2.23 (See color insert following page 46.) Reflected light optical micrographs showing examples of anthropogenic particles found in soil samples taken from the scene of a fatal hit and run incident. (a) A "lump" of zinc metal, labeled A, and a fragment of turquoise-colored composite material; labeled B (b) an angular shard of broken clear glass.

from naturally occurring particles on the basis of their shape, surface texture, and internal textural characteristics.

Mining, smelting, and manufacture of metal artifacts have taken place for thousands of years (Henderson, 2000), although on a vastly increased scale during the past two centuries. The nature of metal particles in a soil or sediment may provide useful indicators of likely source area; for example, high concentrations of tin, copper, lead, silver, or gold may provide indications of mineralized source terrains (e.g., Thorndycraft et al., 1999). The particles found in soils may have been deposited there directly by human action (e.g., dumping), or indirectly as a result of river-borne or atmospheric transport.

2.9.7   Man-Made Glass Particles

Glass particles are another common constituent of surface soils in populated areas. Two broad groups can be recognized: angular shards which commonly show conchoidally fractured surfaces (e.g., Figure 2.24b), and spheres of sub-spheres that may be either solid or hollow, sometimes containing smaller spheres (*cenospheres*). The former type generally represents fragments of broken glass, whereas the latter type most commonly represents combustion or smelting products that have been dispersed by atmospheric or water-borne transport.

Glass also has a long history of production (Henderson, 2000). Modern, Medieval, or older glass of anthropogenic origin can often be distinguished from each other and from naturally occurring glasses (e.g., obsidian) by detailed analysis of their optical, chemical, and structural properties. The analysis and forensic interpretation of glass fragments is a specialized field, and the reader is referred to sources such as Curran et al. (2000) and Caddy (2001) for further information.

2.9.8   Paint

Dried paint flakes are another particle type commonly found in the soils of urban areas (Figure 2.24). The flakes may consist of a single layer, multiple layers

Figure 2.24 (See color insert following page 46.) Reflected light optical micrograph showing two slightly different blue-colored blue paint fragments (A and B) in the >150 μm fraction of a soil sample taken from a garden in South Devon, where a pensioner was bludgeoned to death with a garden edging stone during an attempted robbery. The paint flakes were derived from the drain pipes and soffits at the property. Also visible are several light-colored aggregates of horticultural vermiculite originating from potting compost, which had been emptied onto the garden soil.

(primer, undercoat, and one or more topcoats), or be attached to a substrate (wood, metal, plastic, etc.). Paint flakes often form part of an assemblage of "unusual" particle types that may be location-specific. Methods for detailed analysis and forensic interpretation of paint are discussed in Caddy (2001).

2.9.9 Paper

Small pieces of paper and cardboard are also frequently found in soils in areas prone to litter accumulation and dumping (e.g., waste tips and road-side locations). Uncoated paper and cardboard tends to break down rapidly when wetted and have a relatively short persistence in soils unless protected from the elements. However, coated papers may survive for several years (Figure 2.25). Paper coatings vary considerably in composition and often include clay minerals, whitening agents, and pigments of diverse types. Printed papers often contain a range of inks in intricate physical patterns that may be specific to particular parts of a printed page. Fragments of printed paper with distinctive patterns therefore provide another "unusual" particle type of potential interest for inter-sample comparison purposes.

2.9.10 Fibers

Fibers of many different types occur in soils, and may be numerous. The fibers can be natural or man-made, representing hairs (animal or human),

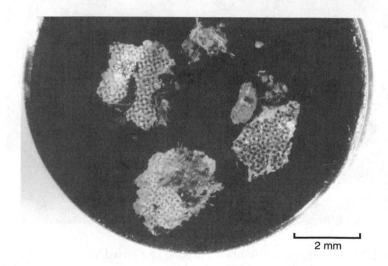

2 mm

Figure 2.25 (See color insert following page 46.) Fragments of coated magazine paper recovered from the >150 μm fraction of a control soil sample collected at a murder scene near Bristol. There are several composite ink patterns on the same type of backing paper. Similar paper fragments were found in soil recovered from the shoes worn by the murder suspect.

textiles, or fragments of decayed rope, sacking, etc. Detailed analysis, comparison, and identification is a specialized operation which is treated in many existing textbooks (e.g., Greaves and Saville, 1995; Ogle and Fox, 1999; Robertson and Grieve, 2001).

2.9.11   Other Unusual Particle Types

Many other anthropogenic particle types are sometimes found in soils, including fragments of plastic and rubber, which may have distinctive shapes, colors, textural patterns, and compositional characteristics. Depending on the nature of the material and the context in which it is found, any such type of material can potentially be useful as an "unusual" particle type forming part of an assemblage.

Bulk Properties of Geological and Soil Materials

3

3.1 Physical Characteristics

Several properties of bulk sediments and soils can be used for the purposes of forensic comparison (Table 3.1). These properties may be physical, chemical, biological, or biochemical in character. The selection of which properties to determine in any given situation will inevitably be governed by a number of factors, including the nature and size of the questioned samples, what is known about the character of the crime scene or other comparison samples, and the availability or otherwise of analytical instrumentation, expertise, time, and budget. Some bulk properties can be determined relatively quickly with simple equipment and at low cost, but others require the use of highly sophisticated equipment that frequently involves longer time periods and significantly higher costs.

Different bulk properties vary in terms of their discriminatory power, and an understanding of this is critical for correct interpretation of the significance of results. In most circumstances, it is desirable to make use of several different techniques that provide information about a range of physical, chemical, mineralogical, and biological characteristics. In previous forensic work, these have commonly included color, particle size distribution, bulk mineralogy, chemical composition, and pollen assemblage (e.g., Bremner 1996; Karathanasis and Hajek (2002); Oldfield 1999; Thornton and Crim, 1986; Demmelmeyer and Adam, 1995; Marumo and Sugita, 1998, 2001; Marumo et al., 1999; Brown, 2000; Brown et al., 2002; Lee et al., 2002; Croft and Pye, 2004a, 2004b, 2004c). Only in very rare situations does a single bulk property or type of data provide sufficient information on which to base conclusions about a possible association.

Table 3.1   Bulk Properties of Sediments and Soils that Can be Used for Forensic Comparison

Bulk Sample Properties	Main Techniques and Equipment Used
Rock/sediment/soil texture	X-radiography, microtomography, optical and scanning electron microscopy, image analysis
Particle size distribution	Dry and wet sieving, laser granulometry
Particle shape properties	Image analysis
Surface area	Nitrogen–gas adsorption
Color	Color charts, spectrophotometry
pH	pH electrode, colorimetry
Water-soluble cations and anions	Atomic absorption, ion chromatography
Enzymes	Enzymatic extraction
Bacteria	Culture experiments, microscopy
Lipid biomarkers	Gas chromatography mass spectrometry
Carbon, nitrogen, and sulfur contents	Wet chemistry, CHNOS elemental analyzer
Bulk organic-matter content	Walkley-Black colorimetric method, Fourier transform infrared spectroscopy, pyrolysis-gas chromatography-mass spectrometry
Polyaromatic hydrocarbons	Gas chromatography-mass spectrometry, high pressure liquid chromatography
Calcium carbonate content	Collins calcimeter, Chittick apparatus
Thermoluminescence characteristics	Heat-induced photon emission
Fluorescence characteristics	Fluorescence microscopy
Major and trace element composition	Atomic absorption, x-ray fluorescence, inductively coupled plasma spectrometry, neutron activation
Bulk mineralogy	Optical microscopy, point counting, automated scanning electron microscopy, x-ray microanalysis, x-ray diffraction
Clay mineralogy	X-ray diffraction, infrared spectroscopy
Mineral magnetics	Magnetic susceptibility, frequency-dependent susceptibility, isothermal remanent magnetization
Light stable isotope ratios (e.g., carbon, oxygen, nitrogen, sulfur)	Continuous flow mass spectrometry, laser fluorination mass spectrometry
Radiogenic isotope ratios (e.g., neodymium, strontium, lead)	Thermal ionization mass spectrometry, inductively coupled plasma spectrometry
Radioactive isotope ratios (e.g., uranium, thorium, lead, cesium)	Alpha-counting, beta-counting, gamma-counting, mass spectrometry

3.1.1   Color

One of the most visually obvious features of a sediment or soil is its color (Bigham and Ciolkosz, 1993). The apparent color of any material, including soil, is governed partly by its structural and optical properties and partly by the lighting conditions under which it is viewed (Thornton, 1997; Tilley 2000). Light can be reflected, scattered, refracted, dispersed, diffracted, or absorbed by a material; the extent to which these different processes occur depends on the nature of the incident light and on the properties of the material, including refractive index, density, crystalline structure, ionic composition, thickness, and surface texture.

The color of most soils and sediments ranges from yellow through various shades of brown to red, although shades of white, grey, green, blue, and black are encountered. For many years, the color of soils has been described by reference to visual color charts, such as the Munsell Chart (Munsell Color, 1994) or, in the case of rocks, to the Rock Color Chart (Goddard et al., 1948). Comparisons can be made in the field or in the laboratory, but in either case it is important to standardize the conditions under which comparisons are made. Antoci and Petraco (1993) described a sample mount apparatus suitable for comparing small forensic soil samples with standard Munsell Soil Color Chart chips. However, visual comparisons are unavoidably subjective and are prone to errors associated with differences in color perception between individuals and the effects of variations in illumination conditions (Thornton, 1997). Post et al. (1993) described an inter-laboratory comparison exercise in which sets of <2 mm soil samples were sent to different soil scientists who were asked to determine the dry and moist color of each soil. The results showed agreement on the same Munsell color chip for a single color component (hue, value, or chroma) 71% of the time, but there was an average of only 52% agreement for all three color components. The standard deviations of the observations ranged from 0.45 for value-moist determinations to 0.68 for chroma-moist determinations.

In view of this variability in color chart-based descriptions, there are advantages in obtaining objective, quantitative descriptions of color using computer-controlled spectrophotometer systems and standardized conditions of illumination and data recording (Melville and Atkinson, 1985; Torrent and Barron, 1993; Croft and Pye, 2004a; Figure 3.1). Such instrumental methods also have an advantage that color can be described in terms of a number of alternative color space and notation systems, such as XYZ tri-stimulus values (Commission Internationale de l'Eclairage, 1931), and L*a*b* values (Commission Internationale de l'Eclairage, 1978), in addition to the Munsell notation.

The Munsell System of color space is conceptually cylindrical, and is based on three attributes: hue (H), denoting red, yellow, green, blue, etc.; value (V), representing the degree of lightness; and chroma (C), indicating degree of difference from black, white, or neutral color. The tri-stimulus XYZ system is based on the quantification of three color stimuli which can be detected by the human eye, where each color stimulus represents radiant energy of a given intensity and spectral composition. The L*a*b* system is a mathematical derivative of the tri-stimulus system which provides a more continuous and uniform color space and a better way of visualizing color. The system can be conceptualized in terms of a spherical color space in which L* forms the vertical axis, representing lightness, and a* and b*

Figure 3.1 Minolta CM2022 spectrophotometer, computer system with Spectramagic software, and set of reference tiles used for quantitative determination of color.

are two orthogonal horizontal axes that represent redness–greenness and yellowness–blueness, respectively.

Color descriptions and comparisons have most commonly been made in the laboratory on air dried, disaggregated bulk samples, but useful additional comparative information and better discrimination between samples can be obtained if comparisons are also made on moist samples, on individual size fractions before and after crushing, after organic matter decomposition, after iron oxide removal, and after heating or ashing (Sugita and Marumo, 1996; Croft and Pye, 2004a). Janssen et al. (1983) advocated the use of clay-size fractions for soil color comparisons since much of the pigmentation in soils is due to components in the finest size fractions. Dudley and Smalldon (1978b) determined the color of the <2 mm fraction of soils which were oven dried at 110°C for 4 h and later ashed at 850°C for 30 min. Color was recorded by reference to Munsell charts. Croft and Pye (2004a) advocated the use of a standardized <150 μm fraction, after oven drying at low temperature (40°C) for at least 4 h, which is analyzed both in an unground state and after grinding to a fine powder using an agate pestle and mortar. Comparison can also be made with color determination on sub-samples of bulk soil material. This procedure has the advantage that it provides information about the effects of both grain surface coatings and mineral composition on resulting color.

It can be undertaken on relatively small forensic samples in a nondestructive manner, prior to other forms of chemical or mineralogical analysis.

Table 3.2 provides an example of L*a*b*, Munsell H, V, and C and percentage reflectance values at different wavelengths for eight bulk samples of beach and dune sand from seven locations in Australia. The average reflectance curves (based on an average of five repeat measurements) for each sample are compared in Figure 3.2a, whereas bivariate plots of the a* vs. b*, a* vs. L*, and b* vs. L* parameters are shown in Figures 3.2b, 3.2c, and 3.2d, respectively. The red sand samples AR1 and AR2 from Ayers Rock are clearly differentiated from the other samples on all three bivariate plots. Although there is a relatively much smaller difference between the other samples in terms of these parameters, the three Queensland samples (TB1, CT1, and FMB2) plot tightly as a clustered group, reflecting a similar mineralogical composition that is dominated by bleached quartz and opaque heavy minerals.

Figure 3.2e shows reflectance curves for five different unground size fractions obtained from sample AR1. Plots of a* vs. b*, a* vs. L*, and b* vs. L* are shown in Figures 3.2b, 3.2c, and 3.2d, and the corresponding data are presented in Table 3.3. The results clearly show that, although all reflectance curves for all of the size fractions show a broadly similar shape, there are observable differences. The greatest degree of redness (highest reflectance at the larger wavelengths) is shown by the <63 μm fraction, followed by the bulk sample, whereas the lowest degree of redness is displayed by the two coarsest sand fractions (250 to 500 μm and 150 to 250 μm). This is consistent with the fact that the red color of these sands is due primarily to the presence of clay–iron oxide coatings on the surfaces of the (largely quartz) sand grains (Folk, 1976), and that the relative importance of the coatings increases on the smaller grain sizes (in proportion to an increasing surface area to volume ratio).

Figure 3.3 illustrates a simple forensic casework example in which a comparison was made between soil on a suspect's trainers, a trowel recovered from his home address, a soil sample taken from an area of woodland where a drugs stash was discovered, and a control sample taken from the garden at the suspect's home address. The reflectance curves for the two soil samples taken from the trowel show a much closer similarity to the soil from the suspect's garden than that from location of the drugs stash, whereas the reflectance curves for the soil taken from the trainers are more similar in shape to that from the location of the drugs stash.

Although color comparisons can undoubtedly be useful in forensic soil work, it is important to recognize that, in some situations, the color of soils, sediments, and rocks can vary significantly over short spatial distances, and also with time, due to fluctuations in field or laboratory conditions. Soils often show marked color differences with depth, notably between different soil horizons, and may show mottling which varies in degree with soil moisture and oxygenation conditions (Hodgson, 1974; Torrent and Barron, 1993).

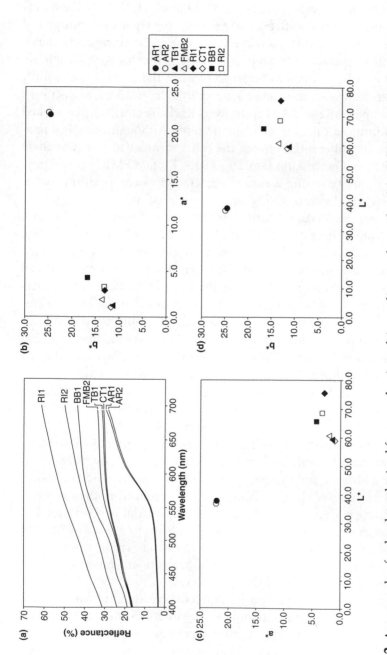

Figure 3.2 An example of color data obtained from a selection of Australian sand samples: (a) to (d) relate to bulk samples, whereas (e) to (h) relate to different size fractions obtained from one of the samples (AR1). AR1 and AR2 are red siliceous dune crest samples from Ayers Rock, Northern Territory; RI1 and RI2 are cream-colored calcareous dune crest samples from Rottnest Island, Western Australia; TB is a white siliceous beach sand sample from Cape Tribulation, North Queensland; CT1 is a white siliceous beach sand sample from Thornton Beach, North Queensland; FMB2 is mixed quartz — heavy mineral beach sand concentrate from Four Mile Beach, Port Douglas, North Queensland; and BB1 is a mixed quartz — carbonate beach sand from Bondi Beach, New South Wales.

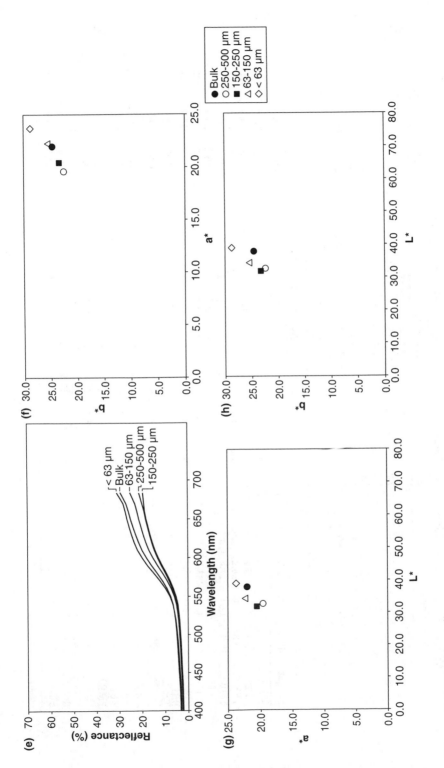

Figure 3.2 (Continued)

Table 3.2 Variation in Color Parameters of Eight Bulk Samples of Australian Sands[a]

Sample	CIE LAB Space Parameters			Munsell Parameters		
	L*	A*	b*	Hue	Value	Chroma
AR1	38.0	22.0	24.6	2.7YR	3.8	5.8
AR2	37.2	22.2	24.9	2.8YR	3.7	5.9
TB1	59.3	1.2	11.3	3.2Y	5.8	1.6
FMB2	60.7	1.9	13.5	2.7Y	5.9	1.9
CT1	58.9	1.1	11.6	3.5Y	5.8	1.6
RI1	75.4	2.9	13.0	0.5Y	7.4	2.0
RI2	68.6	3.3	13.2	0.5Y	6.7	2.0
BB1	65.7	4.3	16.7	0.9Y	6.5	2.6

% Reflectance at Different Wavelengths (nm)

Sample	400	410	420	430	440	450	460	470	480	490	500	510	520	530	540	550
AR1	3.4	3.4	3.5	3.6	3.8	4.0	4.2	4.3	4.5	4.7	4.9	5.2	5.5	6.0	6.5	7.5
AR2	3.1	3.1	3.2	3.3	3.5	3.7	3.9	4.0	4.2	4.3	4.6	4.8	5.1	5.6	6.1	7.1
TB1	16.8	17.4	18.2	19.1	20.2	21.0	21.5	22.1	22.6	23.3	24.0	24.8	25.7	26.6	27.3	28.0
FMB2	16.4	17.1	17.8	18.9	20.1	21.1	21.7	22.2	22.8	23.6	24.4	25.4	26.6	27.7	28.7	29.7
CT1	16.1	16.8	17.6	18.5	19.6	20.4	21.0	21.6	22.1	22.8	23.5	24.4	25.3	26.2	26.9	27.6
RI1	31.2	32.4	33.7	35.0	36.5	37.8	38.9	39.9	41.0	42.0	43.1	44.4	45.6	46.9	48.1	49.3
RI2	23.9	24.8	25.8	26.9	28.1	29.2	30.0	30.8	31.7	32.6	33.6	34.6	35.7	36.9	37.9	39.0
BB1	18.8	19.5	20.4	21.6	23.1	24.2	24.8	25.2	25.8	26.7	27.8	29.2	30.8	32.6	34.1	35.8

Sample	560	570	580	590	600	610	620	630	640	650	660	670	680	690	700
AR1	9.0	10.9	13.1	15.6	17.8	19.7	21.2	22.4	23.6	24.6	25.5	26.6	27.6	28.6	29.5
AR2	8.6	10.5	12.7	15.1	17.2	19.2	20.6	21.8	22.9	23.9	24.7	25.7	26.6	27.6	28.4
TB1	28.6	29.0	29.4	29.7	30.0	30.2	30.3	30.5	30.6	30.7	30.8	30.9	31.1	31.3	31.4
FMB2	30.5	31.1	31.6	31.9	32.2	32.4	32.6	32.7	32.9	33.0	33.1	33.2	33.4	33.6	33.7
CT1	28.2	28.6	29.0	29.2	29.5	29.7	29.8	29.9	30.0	30.0	30.1	30.2	30.4	30.6	30.7
RI1	50.5	51.5	52.4	53.3	54.1	55.0	55.7	56.5	57.2	57.8	58.6	59.4	60.1	60.8	61.3
RI2	40.1	41.0	41.9	42.7	43.4	44.2	44.9	45.5	46.2	46.8	47.4	48.1	48.8	49.4	49.9
BB1	37.3	38.4	39.4	40.0	40.6	41.0	41.3	41.6	41.9	42.2	42.4	42.7	43.0	43.3	43.6

[a] See Figure 3.2 caption for sample identification.

Table 3.3 Variation in Color Parameters of Five Size Fractions of Australian Sand Sample AR1

Size Fraction	CIE LAB Space Parameters			Munsell Parameters		
	L*	a*	b*	Hue	Value	Chroma
Bulk	38.0	22.0	24.6	2.7 YR	3.8	5.8
250–500 μm	32.8	19.6	22.4	3.3 YR	3.3	5.2
150–250 μm	31.9	20.5	23.2	3.1 YR	3.2	5.4
63–150 μm	34.4	22.3	25.3	2.9 YR	3.4	5.9
<63 μm	39.0	23.7	28.7	3.2 YR	3.9	6.5

% Reflectance at Different Wavelengths (nm)

Sample	400	410	420	430	440	450	460	470	480	490	500	510	520	530	540	550
Bulk	3.4	3.4	3.5	3.6	3.8	4.0	4.2	4.3	4.5	4.7	4.9	5.2	5.5	6.0	6.5	7.5
250–500 μm	2.3	2.4	2.5	2.6	2.7	2.9	3.1	3.2	3.3	3.5	3.7	3.9	4.1	4.4	4.8	5.6
150–250 μm	2.1	2.2	2.3	2.3	2.4	2.6	2.7	2.8	3.0	3.1	3.3	3.5	3.7	4.0	4.4	5.1
63–150 μm	2.4	2.4	2.5	2.6	2.7	2.9	3.0	3.2	3.3	3.5	3.7	3.9	4.1	4.5	5.0	5.9
<63 μm	2.9	2.9	3.0	3.1	3.3	3.6	3.8	4.0	4.1	4.4	4.7	5.0	5.3	5.9	6.6	7.8

Sample	560	570	580	590	600	610	620	630	640	650	660	670	680	690	700
Bulk	9.0	10.9	13.1	15.6	17.8	19.7	21.2	22.4	23.6	24.6	25.5	26.6	27.6	28.6	29.5
250–500 μm	6.7	8.1	9.7	11.4	13.1	14.5	15.5	16.4	17.3	18.1	18.8	19.5	20.3	21.1	21.4
150–250 μm	6.2	7.6	9.2	11.0	12.7	14.2	15.3	16.2	17.1	17.9	18.6	19.4	20.2	20.9	21.3
63–150 μm	7.2	8.9	10.9	13.0	15.0	16.8	18.1	19.2	20.1	21.1	21.9	22.8	23.6	24.5	24.9
<63 μm	9.6	11.9	14.5	17.3	19.7	21.7	23.1	24.2	25.2	26.2	27.0	27.9	28.7	29.5	29.9

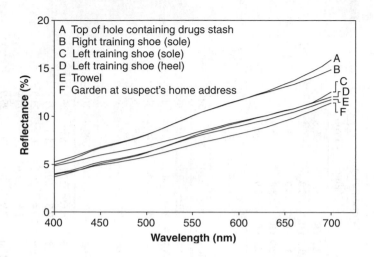

Figure 3.3 Example comparison of reflectance curves for a number of questioned and reference soil samples collected during a drugs investigation.

However, in the case of surface soil sample (0 to 5 cm depth), the small-scale spatial variability is usually less marked and mottling is observed only relatively rarely (e.g., Pye et al., 2006b). In soils that are subject to seasonal wetting and drying, the color may vary systematically over periods of weeks to months. Longer-term changes in soil color may be brought about by alternation of the drainage conditions and/or the supply of organic matter to the soil. Changes in soil color may also occur after collection of soil samples if the storage conditions are significantly different to those in the field. The most marked changes are related to the oxidation of reduced iron compounds, which are often abundant in sub-aqueous and intertidal sediments. This can cause problems if samples are collected and analyzed at different times, especially if precautions are not taken to standardize the storage conditions.

3.1.2 Density

Material density, or *particle density*, which is defined as the mass per unit volume and expressed in g/cm^3 or kg/m^3, varies significantly between different types of mineral or other material. Some examples for common soil minerals are listed in Table 2.5.

The *bulk density* of soil or sediment is also expressed as mass per unit volume. The bulk density is always lower than the particle density owing to the presence of void spaces, or porosity (Flint and Flint, 2002a). The dry mass of a rock or lump of sediment/soil can be determined by using a laboratory balance, whereas the volume can be estimated by measuring the displacement of an equivalent volume of water when the rock is immersed.

Disintegration of a soil lump when it is placed in water can often be prevented by first wrapping it in cling film. Measurement of density provides a quick method of comparing lumps of rock, concrete, soil, or sediment which may appear visually and compositionally similar but which are compacted or cemented to varying degrees. The bulk density is a widely used parameter in geotechnical engineering and soil mechanics (Head, 1984), and to some extent in soil science (Grossman and Reinsch, 2002). In forensic work, it is sometimes determined to assist routine object description.

3.1.3 Hardness

Minerals and rocks also vary considerably in their degree of hardness, reflecting the composition and texture of the material. The relative hardness of many common minerals has long been expressed in terms of Moh's scale of hardness, which ranges from 1 (talc) to 10 (diamond). Values for an unknown mineral can be estimated by scratching with a fingernail, knife, or reference minerals of known hardness. Examples of Moh's hardness values for common soil minerals are shown in Table 2.5. More quantitative estimates of hardness can be obtained using indentation apparatus, such as the Vicker's Hardness tester. The hardness or "soundness" of rocks, bricks and concrete, which is often affected by weathering or other form of chemical alteration, can also be quantified using instruments such as the Schmidt rebound hammer (Sheorei et al., 1984).

3.1.4 Microfabric

Sediments, soils, and rocks often show a characteristic arrangement of different particle constituents, which is referred to as the *microfabric* or, in the case of soils, the *micromorphology*. Many different aspects of micro-fabric can be identified, including particle size lamination, fabric isotropy/anisotropy, and the presence of particular types of structure at different spatial scales. Some larger-scale features of the fabric may be visible to the naked eye, but others require optical microscope (Brewer, 1976; Kemp, 1985) or even scanning electron microscope (SEM) examination (Smart and Tovey, 1981, 1982). The best information about microfabric is obtained by using a combination of examination methods applied both to three-dimensional "fracture surface" specimens and polished thin sections or polished thick sections (Gillott, 1980; Sergeyev et al., 1980; Krinsley et al., 1998). In the case of rocks, additional aspects of fabric include cleavage and textural intergrowths between different crystals, the distribution of cements, fossil remains, and secondary porosity (Figure 3.4, Figure 3.5, and Figure 3.6). Some of these features can be quantified using automated image analysis methods, but others still require visual interpretation (Krinsley et al, 1998). These techniques can be applied to any material, including concrete and brick (Figure 3.7). Procedures for

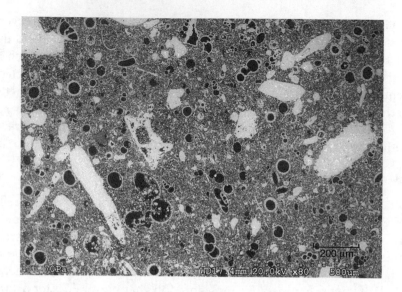

Figure 3.4 Backscattered SEM micrograph of a polished section of chalk collected during the Soham murders investigation. The dark "circles" with pale gray "rims" are resin-filled voids which represent calcareous coccospheres, foraminiferal tests, and small gastropods.

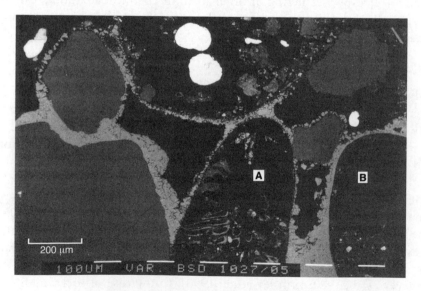

Figure 3.5 Backscattered SEM micrograph of a polished sample of weathered calcareous sandstone. Many of the original high-Mg calcite (calcium carbonate) grains have been largely dissolved, leaving "ghosts" (A and B) whose outlines are preserved by more stable low-Mg calcite cement rims. The white grains that appear to be floating in black void spaces are heavy minerals. Complex textural features of this type are often specific to very restricted amounts of rock (e.g., natural rock outcrops or the surfaces of building stones which have been subject to sub-aerial weathering).

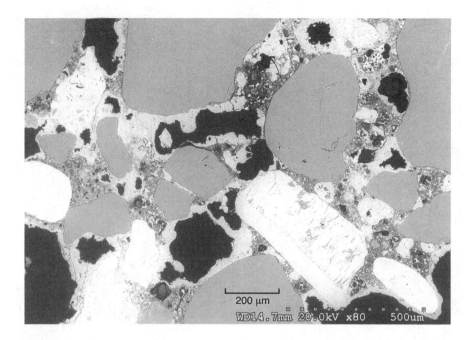

Figure 3.6 Backscattered SEM micrograph of a polished sample of poorly compacted render. The aggregate (sand) particles consist of quartz, [shelly limestone fragments of fragments of ironstone]. The "cement" now consists largely of calcium carbonate with minor calcium silicate hydrate due to the effects of atmospheric carbonation. The black areas are resin-filled voids.

the preparation of thin sections of rocks, minerals, and ceramics are described by Humphries (1992).

Valuable information about both the microfabric and macrofabric of rocks, sediments, and soils can be obtained nondestructively using techniques such as x-radiography (Hamblin, 1971) and x-ray micro-tomography (Mees et al., 2003).

3.1.5 Porosity and Permeability

The porosity of a soil, sediment, or rock is defined as the total volume of voids (pores) as a proportion of the total volume (solid plus voids), expressed as a percentage. The voids may be connected or unconnected, and a distinction can be made between primary porosity and secondary porosity. Primary porosity arises from voids that were present between the grains at the time of deposition, but in the case of a rock some of the original porosity may have been lost as a result of compaction and precipitation of cements within some of the voids. Secondary porosity results from dissolution of some of the original framework grains, or the early cements, during rock burial and diagenesis.

Porosity can be determined by gravimetric methods after water saturation, gas pycnometry, mercury injection, or nitrogen gas injection (Flint and Flint,

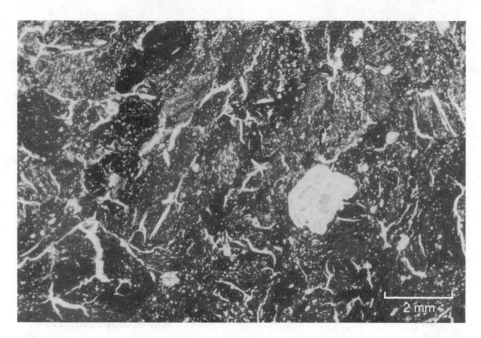

Figure 3.7 Optical transmitted light micrograph of a polished thin section of a household brick used to weigh down a bag containing a discarded firearm. The unusual textural features of the brick provide a useful criterion for comparison, alongside chemical and mineralogical data.

2002b). It can also be estimated by image analysis of serial thin sections or using techniques such as computer tomography of magnetic resonance imaging.

Permeability is the capacity of a sediment, soil, or rock to transmit fluids. It is partly dependent on the porosity and on the degree of connectivity between the pores (especially the size of pore throats). Permeability is normally determined experimentally using constant head and falling head permeameters.

Hydraulic conductivity is related to permeability and is a measure of the capacity of a rock or sediment to permit the passage of a fluid with specified physical properties.

In forensic work, the value of these properties lies in distinguishing between samples of the same rock type that have similar mineralogical and large-scale textural properties but different behavioral characteristics (e.g., perhaps due to fracturing or weathering). Examination of the distribution and nature of the porosity in thin section may also allow subtle textural differences to be identified.

3.1.6 Specific Surface

A property that is related to porosity, permeability, and particle size is *specific surface*, which is a measure of the exposed surface area of particles within a sample. It can be measured in several ways involving adsorption of liquids or gases (Pennell, 2002). Blott et al. (2004a) described a nitrogen gas adsorption technique that provides a proxy measure of three-dimensional particle shape and surface texture in bulk samples, or grain size fractions, of sediment and soil (Figure 3.8). In general, samples composed of near-spherical, smooth and nonporous particles (such as glass ballotini or well-rounded quartz grains) have low "surface area" values, whereas samples composed of highly irregular, rough or porous particles (such as volcanic tephra particles) have relatively high values. Some example data for different sand types from different parts of the world are shown in Figure 3.9.

3.1.7 Magnetic Characteristics

All rocks, sediments, and soils have magnetic properties that can be measured with great sensitivity and precision. Palaeomagentic "dating" of rocks has been carried out since the 1940s, but studies of environmental magnetism in relation to sediment provenance and depositional history began in the late 1960s, focusing initially on lake sediments (Thompson et al., 1975; Thompson and Oldfield, 1986). Since that time there have been numerous applications to fluvial, marine and aeolian sediments, as well as soils (Oldfield, 1998a; Dearing, 2000; Slattery et al., 2000). Early work concentrated on low field magnetic susceptibility measurements, but later studies have combined a variety of susceptibility measurements with investigation of other properties such as magnetic remanence (Dearing, 1999, 2000; Walden, 1999).

Figure 3.8 Coulter SA3100 instrument used to determine specific surface by nitrogen gas adsorption.

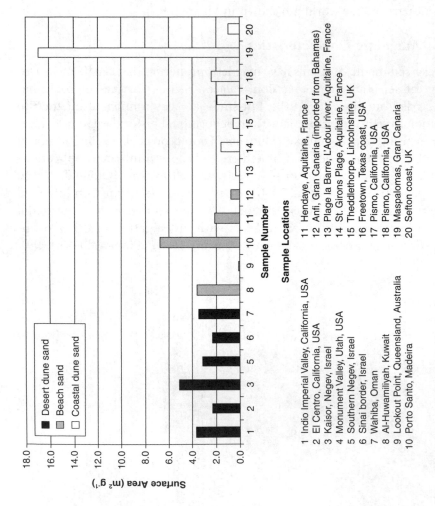

Figure 3.9 Comparison of surface area data, determined by nitrogen gas adsorption, for a range of sand samples collected from different parts of the world.

The magnetic signature is governed by a variety of factors, including the nature and concentration of detrital iron-bearing minerals, any metallic contaminants (e.g., from industrial sources), and the extent and nature of any postdepositional changes that have taken place (e.g., induced by fire or chemical diagenesis). Although magnetic susceptibility measurements have been used extensively in sediment provenance and tracing studies, to date there have been few reported applications in forensic casework.

3.1.8 Particle Size Distributions

All sediments and soils contain a range of particle sizes that can be described in terms of a *particle size frequency distribution* that summarizes the abundance of particles (in terms of particle numbers, weight, percentage weight, or volume) in each of a number of defined particle size classes.

Many different methods can be used to obtain particle size distribution data, including direct observation and measurement (e.g., under a microscope), computer image analysis, dry sieving, wet sieving, the sedimentation/pipette method, hydrometer method, settling tube, laser diffraction, Coulter-Counter, and x-ray sedigraph methods (Ingram, 1971; Textoris, 1971; Galehouse, 1971a; Head, 1984; Komar and Cui, 1984; McManus, 1988; Syvitski, 1991; Gee and Or, 2002; Chazottes et al., 2002; Sperazza et al., 2004; Pye and Blott, 2004a; Blott et al., 2004b; Blott and Pye, 2006). Each method has advantages and disadvantages, and choice of the one most appropriate to use is dependent on the nature of the materials of interest and the purpose of the analysis.

In general, dry or wet sieving are the most appropriate methods for analyzing the particle size distribution of materials that contain a high proportion of sand and gravel size particles, while laser diffraction provides a rapid and convenient method for the analysis of mixtures of sand and mud. In sieve analysis, the frequency distribution data are expressed in terms of weight percent or cumulative weight percent, whereas in laser diffraction analysis the size distribution is expressed in terms of volume percent or cumulative volume percent. Since the two techniques measure different aspects of size, and the algorithms used to calculate a size distribution from measured diffraction data make certain assumptions about the likely nature of the distribution, the results from the two techniques are not directly comparable (see Blott and Pye, 2006).

Particle size distributions have been determined in forensic investigations for many years, although early analyses and interpretations were relatively unsophisticated. Nickolls (1956, p. 64) noted that "sieving … is of considerable value provided that care is taken in selecting the control sample to be similar in type to the kind of material found in the crime samples." Examples were given by Nickolls of soil particle size distribution data obtained using a nest of seven mesh sieves. Nickolls also described the use of an absorption meter to characterize the turbidity of soil suspensions created by shaking a known mass of soil in water and allowing it to settle for varying time intervals. Dudley (1977) tested a Coulter Counter for the analysis of the silt fraction in soils and obtained reproducible results with samples as small as 200 mg. McCrone (1982) described a microscopic method for determining the size distribution of different minerals in even smaller soil and dust samples. This method involved measuring Martin's diameter during a linear traverse across slide mounts of the grains and reporting the data as numbers of particles in each 20 μm wide size class. Number frequency distribution curves were obtained from the data and average particle sizes calculated. A similar procedure for characterizing the size, shape, and composition of particles in small soil samples was described by Ganesh et al. (1989).

Junger (1996) used a nest of eight standard U.S. mesh sieves in a research investigation to assess variability within close-proximity soil samples; 3 g of sample material was used after cleaning to remove surface coatings. Lombardi (1999) used a set of small steel mesh sieves to analyze sand taken from the trouser cuffs and shoes of the murdered former Italian Prime Minister, Aldo Moro, and to compare it with samples taken from a beach close to Rome. Robertson et al. (1984) compared results from wet and dry sieving and concluded that wet sieving should be used in order to obtain reproducible results from soils. Wanogho (1985) and Wanogho et al. (1987a, 1987b) investigated a number of factors influencing dry and wet sieving results and concluded that all factors, including the mass of soil used and operating conditions of the shaker, should be standardized. Wanogho et al. (1987c, 1989) subsequently investigated the use of laser diffraction to analyze the <63 μm fraction and concluded that greater discrimination between samples could be obtained using a combination of wet sieving and laser diffraction analysis of the silt and clay fraction.

Sugita and Marumo (2001) reported a combination method which involved initial wet sieving through a 50 μm sieve to remove the fine fraction, dry-sieving of the >50 μm fraction using 0.2 and 2.0 mm mesh sieves, and quantification of the <50 μm fraction by measurement of the light transmittance during centrifugation of the soil suspension using a Horiba CAP A-300 particle size analyzer. This procedure was reported to enable discrimination of 87.9% in a suite of 73 samples collected from a 20 × 15 km area in the Nirasaki district of Japan.

Pye and Blott (2004a) described the use of a Coulter LS230 laser diffraction analyzer for the analysis of a range of soil and sediment types in the size range 0.04 μm to 2.0 mm using a total of 116 size classes. Highly reproducible results were reported for sample weights of only 50 mg. It was concluded that this method of analysis also offers advantages of speed, simplicity, and cost compared with other available methods when the primary purpose of the analysis is relative comparison. However, for the analysis of relatively clean sands, or sediments that contain a significant proportion (by volume) of low density particles, dry-sieving has advantages (Blott and Pye, 2006).

Figure 3.10 illustrates a comparison of samples of sand collected from the mouth and upper respiratory tract of a murder victim whose body was found on a beach at Skegness, Lincolnshire, with a sample taken close to the victim's head and another sample taken approximately 3 m away from the body. The victim had evidently been asphyxiated on the beach, but it was not certain whether this occurred at the position where the body was found. Comparative dry-sieving analysis and microscopic examination of samples taken from various locations in the vicinity of the body led to the conclusion that the sand in the victim's mouth and respiratory tract was most likely to have been derived from the location where the body was found.

Figure 3.11 compares particle size distribution curves, determined by laser diffraction, obtained from two locations on a van suspected of involvement in a hit-and-run incident and having collided with earth embankments on opposite sites of a road. One of the questioned samples from the van showed a close similarity (in terms of shape of the size frequency distribution curve and the summary parameters) with control soil taken from one side of the road, whereas the other questioned sample was very similar to soil taken from the bank on the opposite side of the road. The questioned samples also showed a very high degree of similarity in terms of color, chemical composition, and particle type assemblage. The mud distribution on the vehicle was consistent with it having struck two glancing blows, one on each embankment as it swerved across the road.

Figure 3.12 compares the particle size frequency distribution curves obtained by laser granulometry of a sample of dust, formed by cutting ceramic wall tiles with a tile saw on a concrete surface, with a sample collected from the sole of a shoe which had just been in contact with the dust source. The two size distribution curves have a broadly similar shape and the particle size summary parameters are similar, although not identical, the relative magnitudes of the two modes being reversed. Such differences can easily arise from small-scale heterogeneity in the source material, grain selectivity during the transfer process, or preferential retention of different sizes and shapes of particle.

In most environments, the surface sediments and soil show particle size and compositional heterogeneity to a varying degree. By way of example, Table 3.4 summarizes the variation observed in a group of 85 mixed sand and gravel samples collected from a 30 × 30 m grid on a beach at Sutton-on-Sea, Lincolnshire. A graphical representation showing the spatial variation in the median (D_{50} value) is shown in Figure 3.13. The variation is not random but organized across the grid, the coarser material being concentrated along the eastern (seaward) edge of the grid, corresponding to a prominent berm feature formed by wave action.

As a further example, Table 3.5 shows the variation in particle size parameters (determined by laser granulometry) found within 1 × 1 m grids at two experimental sites in Berkshire, England (Pye et al., 2006b). The mean,

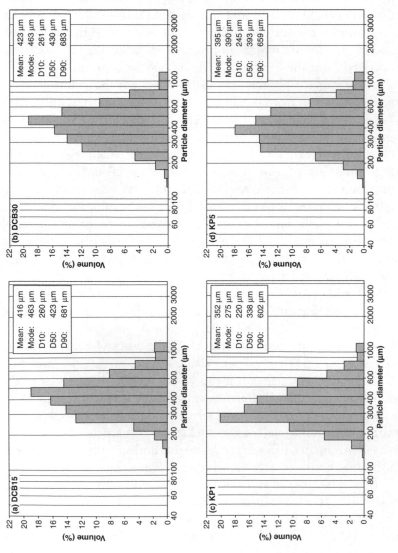

Figure 3.10 Comparison of particle size frequency distributions, obtained by dry sieving, from (a) the mouth and and (b) the upper respiratory tract of a murder victim found on a sandy beach in Lincolnshire. (c) A control sand sample taken from the beach adjacent to the position of the victim's head and (d) a control sample taken 3 m away from the body.

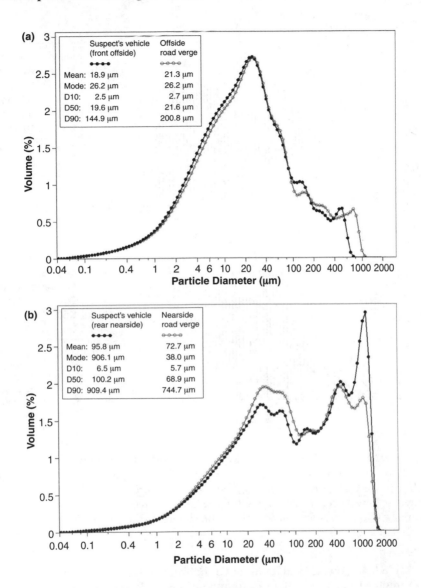

Figure 3.11 Comparison of particle size distribution curves and summary parameters, obtained by laser granulometry, for two samples of soil taken from the exterior of a vehicle suspected of being involved in a hit-and-run incident, and two control samples taken from embankments on either side of the road where the incident took place

median, and mode showed much less variability (expressed by range and the percentage coefficient of variation) at the Simon's Wood site, which is located within an area of woodland on thick sandy soils, than at the Arborfield Bridge site, which represents a concrete hard-standing area on which a relatively small thickness of "soil" had accumulated.

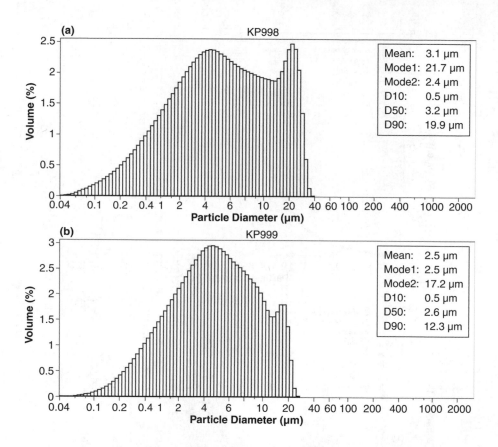

Figure 3.12 Particle size distributions, obtained by laser granulometry, of (a) a sample of dust from a pavement surface formed by cutting ceramic wall tiles with a powered tile cutter and (b) a sample recovered from the sole of a shoe known to have been in contact with the dust source. Note that, although the summary parameters are not identical, the shapes of the particle size distribution curves are generally similar.

Table 3.4 Average and Range of Grain Size Parameters for 85 Sediment Samples from a 30 × 30 m Diamond Grid on a Mixed Sand-Gravel Beach at Sutton-on-Sea, Lincolnshire, U.K.

	Mean	Max	Min	SD	CV (%)
Mean (μm)	797	2,675	350	378	47.4
Mode (μm)	877	9,600	215	1800	205.2
D10 (μm)	207	257	186	14	6.8
D50 (μm)	838	5,283	307	781	93.3
D90 (μm)	3688	13,249	928	1959	53.1
%Gravel	23.3	64.4	2.2	13.4	57.5
%Sand	76.7	97.8	35.6	13.4	17.5
%Mud	0.0	0.1	0.0	0.0	161.7

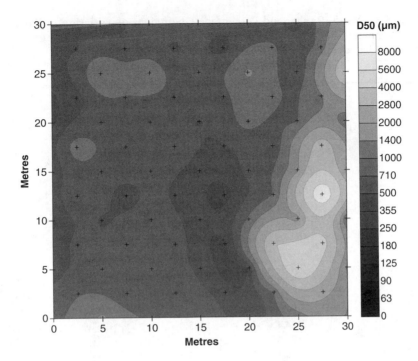

Figure 3.13 Spatial variation in the D_{50} (median size) within an experimental sampling grid on a mixed sand and gravel beach at Sutton-on-Sea, Lincolnshire. Sampling points are shown as black dots. Data were obtained by dry-sieving and processed using the GRADISTAT program.

Table 3.6 shows the degree of variation in particle size parameters found in a group of eight control soil samples collected from a track near Ilfracombe, Devon, compared with the data for two groups of soil samples taken from a vehicle which had been stuck in mud at this location, and four samples taken from a spade claimed by the car owner to have been used to dig the car out. With the exception of the mode, the particle size parameters for the mud samples from the suspect's spade showed a higher degree of variability than the scene samples, and a slightly higher degree of variability than the Group 1 car samples. However, the average values for mean, mode, median, and percentages of sand, silt, and clay obtained from the spade and from the scene control samples are similar to each other and also to the Group 2 samples taken from the car. The average values for the Group 1 samples taken from soil show several significant differences compared with the other three groups.

In all circumstances, an understanding of the variability within a scene of interest is essential, if the results of forensic particle size comparisons are to be interpreted correctly.

Table 3.5   Mean, Range, and Percentage Coefficient of Variation of Particle Size Parameters for Samples from Two Experimental 1m × 1m soil grids in Berkshire, U.K.

	Mean	Max	Min	CV (%)
Simon's Wood (n=9)				
Mean	111.8	127.1	97.5	8.5
Mode	185.3	185.3	185.3	0.0
D10	14.1	16.4	11.9	11.9
D50	160.4	177.2	149.8	5.3
D90	599.5	753.7	389.0	24.0
%Sand	72.8	76.4	70.6	2.8
%Silt	25.1	27.4	21.7	8.1
%Clay	2.1	2.3	1.9	7.1
Arborfield Bridge (n=9)				
Mean	206.3	293.7	160.7	19.4
Mode	897.2	1198.0	356.1	25.4
D10	22.7	33.0	16.8	23.0
D50	287.1	442.7	211.3	24.9
D90	1234.4	1483.0	951.4	12.8
%Sand	78.4	84.4	73.5	4.3
%Silt	20.1	24.6	14.5	15.5
%Clay	1.5	1.9	1.1	16.9

3.1.9   Shape Frequency Distributions

Particle shape is discussed more fully in Chapter 4 in relation to the characterization of individual particles, but it is appropriate to point out here that shape frequency distribution data can be obtained for bulk samples in a manner analogous to particle size distributions. The shapes of relatively large numbers of sediment grains can be quantified in two dimensions using some form of image analysis or in three dimensions using a dedicated combined size and shape particle analyzer, such as the Coulter Rapid Vue instrument. One of the most commonly used examples of the former type of method is Fourier analysis performed on projected (silhouette) images of sand grains imaged using an optical microscope and digital image capture system (e.g., Ehrlich et al., 1980; Pye and Mazzullo, 1994).

As discussed further in Chapter 4, Fourier analysis involves decomposition of the digitized projected outlines of a grain in terms of a series of harmonics of successively smaller wavelength. The lower order harmonics (1 to 4) provide information about the gross geometry (form) of a particle, whereas higher order harmonics (up to 20 or 24) provide information about other aspects of shape such as roundness and surface texture. An example graphical comparison of data for two groups of samples of sands from a location in Queensland, Australia, is shown in Figure 3.14. Limitations of

Table 3.6 Mean, Range, Standard Deviation, and Percentage Coefficient of Variation of Grain-Size Parameters for Soil Samples Taken from a Spade and Car Associated with a Murder Suspect, and Samples Taken from the Location where the Car was Abandoned

	Mean	Max	Min	SD	CV (%)
		Spade (n=4)			
Mean (μm)	35.7	43.4	28.7	7.6	21.4
Mode (μm)	67.2	80.2	55.2	11.4	16.9
D10 (μm)	5.2	6.1	4.5	0.7	14.0
D50 (μm)	40.4	53.3	30.8	11.0	27.1
D90 (μm)	199.9	271.2	156.2	52.9	26.5
%Sand	37.1	45.0	30.3	7.4	20.1
%Silt	59.4	65.6	52.0	7.0	11.8
%Clay	3.5	4.1	3.0	0.5	13.2
		Vauxhall Cavalier (Group 1) (n=6)			
Mean (μm)	28.7	35.2	23.5	5.0	17.4
Mode (μm)	52.6	60.6	31.5	12.1	23.0
D10 (μm)	4.4	5.0	3.7	0.5	10.9
D50 (μm)	29.6	35.6	24.5	5.0	16.9
D90 (μm)	207.5	310.7	139.2	69.7	33.6
%Sand	29.6	34.1	24.9	4.4	14.7
%Silt	66.4	70.9	62.4	3.9	5.9
%Clay	4.0	4.7	3.4	0.5	12.6
		Vauxhall Cavalier (Group 2) (n=5)			
Mean (μm)	47.6	81.1	23.7	24.2	50.9
Mode (μm)	40.4	80.2	23.8	26.7	66.0
D10 (μm)	4.8	7.6	3.1	1.9	39.8
D50 (μm)	47.8	84.2	24.0	25.7	53.8
D90 (μm)	395.8	623.2	160.3	200.4	50.6
% Sand	41.1	56.6	26.4	12.4	30.3
% Silt	54.1	67.0	40.1	11.1	20.5
% Clay	4.8	6.6	3.3	1.4	28.7
		Control Soils from Scene (n=8)			
Mean (μm)	35.5	40.9	29.1	4.1	11.6
Mode (μm)	65.3	116.4	26.2	26.5	40.6
D10 (μm)	5.0	5.6	4.5	0.4	8.1
D50 (μm)	37.8	43.8	30.4	5.6	14.7
D90 (μm)	258.0	371.6	177.6	61.8	23.9
%Sand	36.8	41.7	31.3	4.0	10.7
%Silt	59.8	65.0	54.7	3.8	6.3
%Clay	3.4	3.8	3.0	0.3	9.3

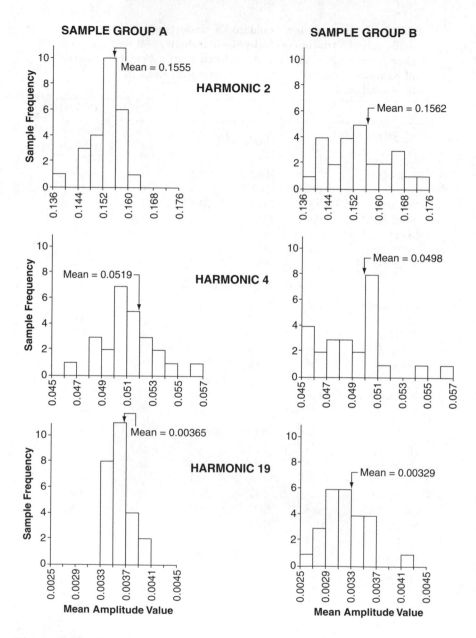

Figure 3.14 An example comparison of Fourier analysis shape data for two groups of Australian dune sand samples. Harmonics 2 and 4 provide information about the gross two dimensional form, whereas harmonic 19 provides information about particle irregularity, angularity, and surface texture. The differences between the mean values of each harmonic are statistically significant at the 99% level, reflecting differences in the degree of weathering experienced by the two groups of sands.

this data include the fact that the data obtained are only two-dimensional, and the grain outlines are to some extent "degraded" by the projection of a three-dimensional form onto a two-dimensional plane of observation.

Where possible, it is preferable to obtain three-dimensional information. Willetts and Rice (1983) described a method by which basic three-dimensional shape information can be obtained by photographing sand grains in two orthogonal orientations using a mirror inclined at 45°. More representative data can be obtained by imaging the grains many times in a series of random orientations while they are suspended in a turbulent flow. Automated analysis of this type has the advantage that large numbers of grains can be analyzed relatively quickly, and more representative sample statistics obtained. However, the resolution of many such systems is limited by the pixel imaging size employed, and they are more suited to sand-size particles than to clay and silt. An example of data obtained using the Coulter Rapid Vue system is shown in Figure 3.15. However, to date, there have been very few applications of quantitative shape frequency distribution analysis in forensic geology casework.

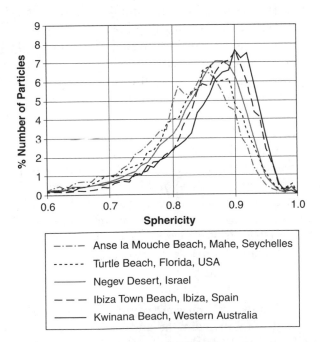

Figure 3.15 Comparison of sand grain sphericity frequency data obtained for a range of sands from different parts of the world obtained using a Coulter Rapid Vue image analysis instrument. Particles are imaged numerous times in random orientations while transported in a turbulent fluid flow, thereby giving an average measure of three-dimensional sphericity. Highly flattened or elongate particles may show a preferred orientation within the flow. A value of 1.0 equates to a perfect sphere.

3.2 Chemical Characteristics

3.2.1 pH

The pH is a measure of the hydrogen ion activity in a sample of water, soil, or sediment, and is expressed on a logarithmic scale that covers more than 14 orders of magnitude (representing hydrogen ion concentrations of 1 to 10^{-14} M). Pure deionized water has a neutral pH of 7.0. Most natural soils and surface water have pH values in the range 4.5 to 8.5, although lower (more acid) and higher (more alkaline) values are locally recorded.

Measurement of pH provides a rapid indication of the type of soil or sediment under consideration, for example, whether it is an acid, peaty type or an alkaline, more lime-rich type. The pH of soils is most commonly determined on a 1:1 or 2:1 slurry mixture of soil and deionized water (Black et al., 1965; Langmuir, 1971; Avery and Bascomb, 1974; Head, 1984; Thomas, 1996). Measurement can be undertaken using a variety of methods, but pH electrodes (electrometric measurements) are now most widely used and give the most sensitive results.

Dudley (1976a) investigated the discriminating power of pH for a range of soils from southern England using four colorimetric methods and found it to be "moderate." Dudley also investigated variation in pH results as a function of sample size. Duplicate samples of 13 soils were taken at each of six weights, namely 2.5, 1.0, 0.5, 0.25, 0.1, and 0.05 g. Of the 13 samples examined, 10 gave the same pH values for all sample weights, and no soil showed variation of more than 0.5 pH units compared with the result for the largest (2.5 g) sample. In a separate experiment, the pH values of six garden soils measured immediately after collection and drying were compared with measurements made on splits of the same soils which were left in a moist condition in sealed polythene bags for 14 days before being dried and then measured. The two sets of results were found to be identical, indicating that a delay of up to 14 days before measurement should not significantly affect pH results. However, if samples are left moist at room temperature over longer time periods, some change in pH values may occur.

3.2.2 Eh

In simple terms, Eh provides a measure of the *redox potential*, or oxidation–reduction status, of a soil or sediment, and is expressed in millivolts (mV). Redox potential is normally determined using a platinum electrode coupled with a suitable half-cell of known potential (Langmuir, 1971; Patrick et al., 1996). Eh varies with pH and temperature, but in general Eh falls logarithmically with the oxygen concentration. As Eh falls, a sequence of chemical reduction processes takes place, leading to the formation of secondary products. Low Eh is normally associated with waterlogging of soil pore spaces, low oxygen availability, and a tendency for reducing conditions to prevail, perhaps leading to the formation of secondary sulfide minerals. Therefore, Eh provides another means of comparing soils and providing information about likely environmental conditions in a soil sample from an unknown source. However, Eh values are prone to change rapidly after sample collection unless appropriate precautions are taken. Partly for reason, they have not been widely used in forensic work.

3.2.3 Electrical Conductivity

The electrical conductivity of a water sample or soil suspension (specified in micro-Siemens per centimeter, μS cm^{-1}) is a measure of the concentration of ions in solution and the ease with which an electrical current can be conducted through that solution, and is often used as a proxy measure for *total dissolved solids* (Rhoades, 1996). Deionized water has a very low electrical conductivity (close to zero), while most natural freshwaters have values in the range 30 to 500 μS cm^{-1}. The electrical conductivity of 1:1 soil–water slurries varies with soil type, ranging from <50 μS in the case of acidic podsolic soils to >5000 μS cm^{-1} in saline soils. This parameter is relatively straightforward to determine, using either an electrical conductivity meter or suitable solid-state electrodes, and provides a rapid method of estimating soil "salinity."

In forensic work, electrical conductivity measurements have most commonly been used as a screening tool to select soil or water samples which require more detailed analysis using other techniques. For example, Figure 3.16 illustrates the measured variation in the electrical conductivity of river water along a 40-km stretch of the River Avon in southwest England. In this instance, the measurements were used to identify areas where the river and riverbank sediments were likely to have experienced saline influence. Other examples of forensic usage include the rapid identification of sediment traces from suspected estuarine and marine sources, or which have been influenced by addition of treated sewage sludge.

3.2.4 Cation Exchange Capacity

Soils and sediments have a capacity to exchange a proportion of the cations (positively charged ions) present if placed in contact with a solution of differing composition. Most of the cation exchange sites are associated with clay minerals, sesquioxides, and organic matter, and the exchange processes in many soils are strongly dependent on pH (Sumner and Miller, 1996). Various methods have been developed for quantifying the cation exchange capacity (CEC) of soils, and the amount of exchangeable cations (e.g., Na$^+$, Mg^{2+}, K$^+$, and Ca^{2+}), but all are fairly time-consuming and in many cases show poor repeatability. For these reasons, this property has not been widely used in routine forensic soil comparisons.

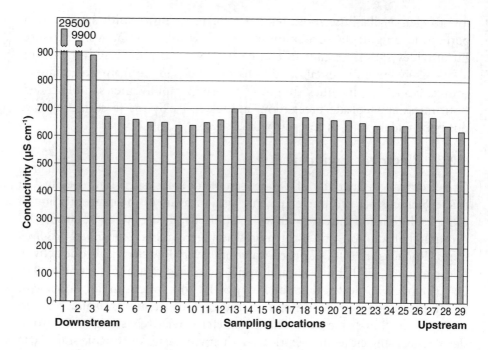

Figure 3.16 Variations in the electrical conductivity of river water along approximately 40 km of the Bristol Avon, sampled during an investigation into a murder at Bradford on Avon.

3.2.5 Anions

In addition to cations, soils and water contain anions (negatively charged ions) such as chloride, sulfate, nitrate, bicarbonate, phosphate, fluoride, and bromide. The concentrations of such ions can provide important indicators of the type of environment (e.g., chloride and sulfate-rich brackish or marine), or treatments that may have been applied to a soil (e.g., addition of nitrate- and phosphate-based artificial fertilizers or treated sewage sludge). The concentrations of anions, determined by a method such as ion chromatography or ion-selective electrodes (Willard et al., 1988), also provides an additional criterion for inter-sample comparisons in cases where there is little difference in bulk soil mineralogy or the concentrations of most major and trace elements.

In the case of soils and sediments, anion concentrations are usually based on 1:20 soil/deionized water ratio extractions. As little as 0.2 g of soil can be used, but results obtained from small samples can be subject to relatively large errors, and it is desirable to use larger quantities (1 or 2 g) where possible. Samples are shaken for 12 h and then centrifuged, prior to analysis of the supernatant liquid. Depending on the nature and concentration of the anions present, further dilution may be necessary.

3.2.6 Major and Trace Elements

By geochemical convention, *major elements* are those which are most abundant in Earth's crust and whose abundance is most frequently expressed in terms of weight percent oxide (% SiO_2, TiO_2, Al_2O_3, Fe_2O_3, including FeO, MnO, MgO, CaO, K_2O, Na_2O, and P_2O_5). Silica (SiO_2) comprises 57.3% of the bulk continental crust and alumina (Al_2O_3) constitutes 15.9% (Taylor and McLennan, 1985). The next abundant oxides are those of iron, calcium, magnesium, sodium, and potassium. However, some rocks, sediments, and soils may contain only low concentrations of these oxides (e.g., some sulfide ores, coals, and peats). *Trace elements* are those that typically have an abundance of <0.1% and their concentrations are most frequently expressed in terms of parts per million (ppm) or micrograms per gram (μg/g). However, in some rocks, sediments, and soils, the concentrations of certain trace elements may reach several percent (Alloway, 1995). High concentrations of trace elements are often associated with hydrothermally altered rocks and weathering residua (Rose et al., 1979). Volatile compounds, such as H_2O and CO_2, or anions, such as S^{2-}, Cl^-, and F^-, are also often present in such rocks and sediments and are sometimes reported in geochemical analyses.

Many different methods have been developed for the determination of major and trace element concentrations in rocks, sediments, and soils, including wet chemical methods, atomic absorption spectrometry (AAS), x-ray fluorescence spectrometry (XRF), neutron activation analysis (NAA), inductively coupled plasma spectrometry (ICP), and electron microprobe analysis (EMPA) (e.g., Hoffman et al., 1969; Zussman, 1977; Chaperlin, 1981; Johnson and Maxwell, 1981; Potts, 1987; Willard et al., 1988; Marques et al., 2000; Helmke 2002; Soltanpour et al., 2002; Wright, 2002; Jarvis et al., 2003, 2004; Parry, 2003; Thompson and Walsh, 2003; Nelms, 2005). For many applications in earth science, such as mineral exploration and geochemical mapping, XRF analysis has been preferred (Figure 3.17). For best results, major elements are normally determined by analysis of lithium metaborate fusion beads and trace elements by analysis of pressed powder pellets (Norrish and Hutton, 1969; Norrish and Chappell, 1977; Winspear and Pye, 1995a, 1995b; Karathanasis and Hajek, 1996). However, in routine exploration work, both major and trace elements have sometimes been determined by analysis of pressed powder pellets. This technique has also been used for the purposes of database building for environmental mapping and forensic applications (Hiraoka, 1994, 1997; Rawlins et al., 2002, 2003; Rawlins and Cave, 2004; Saye and Pye, 2006; Saye et al., 2006).

In recent years, the combined use of ICP-atomic emission spectrometry (ICP-AES) and ICP-mass spectrometry (ICP-MS) has also become popular (Figure 3.18). In many respects, these techniques are more suitable for use in forensic geoscience since the quantity of sample required for analysis is

Figure 3.17 Philips PW1480 XRF instrument with multisample changer used for the determination of major and trace element concentrations.

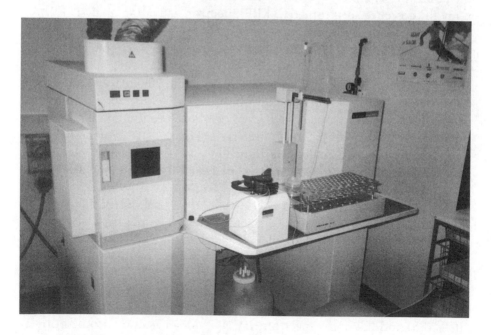

Figure 3.18 Perkin Elmer RL3300RL instrument used for elemental analysis by ICP-AES.

small (typically c. 0.25 g), and a large number of elements can be determined simultaneously with good sensitivity and precision (Jarvis et al., 2004; Saye and Pye, 2004; Pye and Blott, 2004b; Wray, 2005; Pye et al., 2006a, 2006b).

ICP analysis of rock, sediment, and soil materials is normally undertaken after dissolution of the powdered material using acids or alkalis; in some preparation procedures, a lithium metaborate fusion stage may also be employed (Jarvis, 1997; Walsh, 1997; Jarvis et al., 2003; Thompson and Walsh, 2003). Analysis is performed either by ICP-AES and ICP-MS separately, or in combination. However, it is also possible to analyze rocks, mineral grains, and fossils without crushing and dissolution, using laser ablation ICP-MS (Sylvester, 2001).

Table 3.7 shows major and trace element concentrations determined by ICP-AES and ICP-MS in five certified reference materials that are commonly analyzed alongside geological materials in forensic casework. Table 3.8 illustrates the average precision associated with repeated analyses of these reference materials during a single run (referred to by Pye et al., 2006a as Type 2 precision), the instrumental measurement precision associated with repeat measurements of single <150 μm fraction soil solutions analyzed sequentially within the same run (Type 1 precision), and the method precision associated with preparation and analysis of replicate soil samples analyzed in the same run (Type 3 precision). For the large majority of elements, the precision (expressed in terms of the % coefficient of variation) is better than 4% and in a majority of cases better than 2%. Owing to the fact that chemical composition often varies considerably with particle size, forensic comparisons should ideally be made on the basis of a standardized size fraction to reduce these variations (Pye and Blott, 2004b; Pye et al., 2006a). Ideally, the size fraction should be defined as narrowly as possible, but this requirement needs to be offset against the requirement to obtain sufficient material for analysis. Experience and experimentation have shown that the <150 μm fraction provides the best compromise in this respect, since in most cases it provides sufficient material for analysis from relatively small forensic samples, while providing a sub-sample which is fairly representative of the sediment/ soil as a whole (Pye et al., 2006c).

In some forensic casework, it is not possible to recover sufficient <150 μm soil material to allow ICP analysis to be carried out. There are also circumstances where it may be desirable not to remove soil material from the substrate on which it rests. In such circumstances, alternative means of obtaining bulk chemical information have to be considered. One way is to undertake multiple energy-disperse x-ray microanalyses of *in situ* soil material in the SEM. Qualitative EDXRA spectra and/or calculated elemental peak height ratios have long been used to make preliminary comparisons between bulk soil samples and suspended sediment aggregates (Daugherty, 1997;

Table 3.7 Certified (and Recommended) Values for Five Certified Reference Materials[a]

Element	Certified Reference Materials					Lower Limit of Quantitation
	ACE	BEN	GSN	MAG	SCo	
			ICP-AES Elements			
SiO_2	70.35	38.2	65.8	50.36	62.78	0.053
Al_2O_3	14.7	10.07	14.67	16.37	13.67	0.038
Fe_2O_3	2.53	12.84	3.75	6.8	5.14	0.004
MgO		13.15	2.3	3	2.72	0.003
CaO	0.34	13.87	2.5	1.37	2.62	0.01
Na_2O	6.54	3.18	3.77	3.83	0.9	0.009
K_2O	4.49	1.39	4.63	3.55	2.77	0.034
TiO_2	0.11	2.61	0.68	0.751	0.628	0.002
P_2O_5		1.05	0.28	0.163	0.206	0.01
MnO	0.058	0.2	0.056	0.098	0.053	0.0001
Be	12	(2)	5.4	3.2	1.84	0.55
Sc	0.11	22	7.3	17.2	10.8	2.34
Ni	1.5	267	34	53	27	8.19
Zn	224	120	48	130	103	3.43
Y	184	30	19	28	26	5.71
Ba	55	1025	1400	479	570	2.83
			ICP-MS Elements			
V	3	235	65	140	131	8.376
Cr	3.4	360	55	97	68	2.97
Co	0.2	61	65	20.4	10.5	0.4024
Cu	4	72	20	(30)	28.7	6.284
Ga	39	(17)	22	20.4	(15)	0.3386
Rb	152	47	185	149	112	0.5472
Sr	3	1370	570	146	174	0.982
Zr	780	265	235	126	160	0.4046
Nb	110	(100)	(21)	(12)	(11)	0.057
Mo	2.5	2.8	1.2	(1.6)	1.37	0.2312
Sn	13		(3)	(3.6)	(3.7)	0.928
Cs	3	0.8	5.7	8.6	7.8	0.102
La	59	82	75	43	29.5	0.2056
Ce	154	152	135	88	62	0.0702
Pr	22.2		(16)	(9.3)	(6.6)	0.031
Nd	92	70	50	38	26	0.13
Sm	24.2	12	19	7.5	5.3	0.0414
Eu	2	(3.6)	1.7	1.55	1.19	0.0096
Gd	26	(9.5)	5.2	5.8	(4.6)	0.0684
Tb	4.8	(1.3)	0.6	0.96	0.7	0.009
Dy	29	(6.4)	(3)	5.2	4.2	0.0328
Ho	6.5	(1.1)	(0.5)	1.02	0.97	0.0056
Er	17.7	(2.5)	(1.7)	(3)	2.5	0.0248
Tm	2.6	0.34	0.22	0.43		0.01
Yb	17.4	1.8	1.5	2.6	2.27	0.013

Table 3.7 Certified (and Recommended) Values for Five Certified Reference Materials[a] (Continued)

Element	Certified Reference Materials					Lower Limit of Quantitation
	ACE	BEN	GSN	MAG	SCo	
Lu	2.45	0.24	0.22	0.4	0.34	0.0032
Hf	27.9	(5.4)	(6.2)	3.7	4.6	0.0382
Ta	6.4	5.5	2.6	(1.1)	0.92	0.161
W	1.5	(29)	(490)	(1.4)	(1.4)	0.1366
Tl	0.9			(0.59)	(.72)	0.049
Pb	39	4	53	24	31	0.2674
Th	18.5	11	42	11.9	9.7	0.022
U	4.6	2.4	8	2.7	3	0.0258

Note: Values for the 10 major oxides are given in weight percent; all other values are in ppm ($\mu g\ g^{-1}$). Indicative lower limits of quantitation (10 sigma values) are also shown.

[a] Granite from Ailsa Craig, Scotland (ACE); basalt from Essey-la-côte, Nancy, France (BEN); granite from Senones, Vosges, France (GSN); marine sediment from Gulf of Maine, Boston, U.S. (MAG); and Cody Shale from Natrona County, Wyoming, U.S. (SCo).

De Boer and Crosby, 1997). An experimental investigation by Pye and Croft (2007) showed that, if standard operating and sample presentation conditions are maintained, fairly good levels of precision (5–10% CV) can be obtained for fine grained soil materials. The precision was found to vary for different elemental ratios, with the lowest average % CV being shown by the Si/Al and Si/K ratios, and the highest by the Si/O ratio. Coefficients of variation were also found to be significantly larger for "spot" EDXRA analyses than for analyses of small raster areas. In general, the larger the area analyzed, the better the precision.

3.2.7 Rare Earth Elements

The rare earth elements (REEs) are a group of 15 trace elements that belong to the *lanthanide* group of elements which have atomic numbers 57 to 71 and which occupy the sixth period in the Periodic Table (Henderson, 1996). The element yttrium (atomic number 39) is also normally considered as part of the rare earth group on account of its similar chemical behavior (Table 3.9). With the exception of promethium (atomic number 61), which has no long-lived nuclei and is therefore normally excluded from consideration, the REEs are not rare; for example, cerium is approximately five times as abundant as lead in the Earth's crust (Henderson, 1996). The REEs have proved to be of major value in provenance studies and in interpreting earth-forming processes. Rare earth minerals are also of considerable economic importance (Jones et al., 1996).

Although the measured concentrations of REEs can be compared directly, they are frequently normalized relative to the concentrations reported in chondritic meteorites (Table 3.9), and the normalized concentrations are

Table 3.8 Precision and Variability in the Chemical Composition of Reference Soil Samples and Certified Reference Materials, Determined by ICP-MS and ICP-AES, Expressed as Coefficients of Variation[a]

Element	Type 1 Precision		Type 2 Precision					Type 3 Precision	
	SW (*n*=9)	AB (*n*=9)	ACE (*n*=5)	BEN (*n*=5)	GSN (*n*=5)	MAG (*n*=5)	SCo (*n*=5)	SW (*n*=9)	AB (*n*=9)
ICP-AES Elements									
SiO_2	0.3	0.6	1.3	1.8	2.3	1.6	1.1	0.8	2.1
Al_2O_3	0.4	0.6	1.3	0.7	0.7	1.8	1.2	0.9	1.1
Fe_2O_3	0.8	1.0	2.2	2.8	3.7	2.6	1.7	1.4	2.2
MgO	0.3	0.5	1.4	2.7	3.8	2.2	1.7	1.5	0.3
CaO	1.4	1.0	3.1	2.7	3.8	1.9	1.5	0.6	1.6
Na_2O	1.4	0.7	2.1	2.7	3.3	2.2	1.0	2.0	2.8
K_2O	0.6	0.5	1.0	1.3	1.6	2.1	0.7	0.8	0.5
TiO_2	0.2	0.7	1.0	2.1	2.8	1.7	1.2	3.6	0.8
P_2O_5	1.6	0.9	4.7	1.1	0.9	0.5	0.6	3.0	2.0
MnO	0.3	0.6	1.4	1.9	2.5	1.7	1.1	0.5	1.2
Be	<LLQ	2.6	2.2	4.7	1.9	2.9	3.9	<LLQ	1.1
Sc	<LLQ	2.2	<LLQ	1.4	2.9	1.7	3.5	<LLQ	0.7
Ni	<LLQ	3.2	<LLQ	1.6	0.9	1.1	3.9	<LLQ	3.9
Zn	0.3	0.7	1.5	2.0	2.7	1.6	0.8	0.2	3.7
Y	2.6	0.8	2.5	3.4	2.5	2.4	3.3	23.7	1.9
Ba	0.4	0.6	1.8	2.3	2.6	1.6	1.1	3.3	4.2
ICP-MS Elements									
V	0.8	0.8	8.4	7.2	5.1	6.5	4.0	5.8	6.3
Cr	0.3	0.9	<LLQ	2.2	3.7	3.4	3.9	21.7	3.6
Co	3.8	1.4	<LLQ	2.1	2.5	3.1	2.7	13.9	1.6
Cu	6.1	3.1	<LLQ	2.3	4.4	2.5	4.5	5.7	4.8
Ga	1.9	1.0	1.7	2.3	2.7	3.1	3.6	2.8	1.2
Rb	1.0	0.8	2.1	2.7	3.2	3.1	3.2	1.0	1.8
Sr	1.0	0.5	2.4	6.2	2.2	2.4	2.8	1.6	2.6
Zr	0.4	0.6	1.7	2.0	2.6	3.3	3.1	4.0	10.0
Nb	1.1	1.1	2.0	2.4	2.9	3.1	3.3	6.1	3.2
Mo	8.0	2.7	2.4	2.2	2.3	1.9	4.5	3.7	3.3
Sn	13.9	2.3	2.1	3.6	3.9	2.9	5.3	7.5	5.5
Cs	1.5	1.3	2.1	2.8	2.8	2.8	3.3	1.9	1.8
La	0.8	0.8	2.7	2.3	2.7	3.4	3.6	11.8	7.9
Ce	0.8	0.7	2.7	2.4	3.2	3.6	3.7	13.6	4.2
Pr	0.6	0.8	2.8	2.4	3.0	3.2	3.7	12.6	6.0
Nd	0.5	1.0	2.6	2.2	2.8	2.7	4.0	12.0	4.8
Sm	2.0	1.4	2.7	2.0	3.1	4.0	3.3	10.0	2.8
Eu	2.4	1.8	2.1	2.2	2.7	3.7	8.1	4.1	0.9
Gd	2.2	1.5	2.4	2.4	3.1	4.0	2.7	2.7	2.7
Tb	2.1	1.0	2.5	3.1	3.2	3.0	3.3	1.9	1.6
Dy	1.3	1.7	2.8	2.6	3.1	3.1	3.0	7.2	1.1
Ho	1.2	1.4	2.5	3.0	3.7	3.7	4.0	16.0	0.6
Er	1.2	0.8	2.7	2.6	2.0	3.0	4.4	25.9	1.7

Table 3.8   Precision and Variability in the Chemical Composition of Reference Soil Samples and Certified Reference Materials, Determined by ICP-MS and ICP-AES, Expressed as Coefficients of Variation[a] (Continued)

Element	Type 1 Precision		Type 2 Precision					Type 3 Precision	
	SW (n=9)	AB (n=9)	ACE (n=5)	BEN (n=5)	GSN (n=5)	MAG (n=5)	SCo (n=5)	SW (n=9)	AB (n=9)
Tm	2.1	1.1	2.6	1.9	1.7	4.3	4.5	32.9	0.9
Yb	0.8	1.2	2.6	2.6	2.7	3.6	3.0	31.4	0.7
Lu	2.9	1.0	2.5	1.4	2.5	4.0	4.0	27.1	2.0
Hf	0.6	0.8	3.2	3.0	3.1	3.7	4.4	5.6	9.7
Ta	1.1	0.9	2.8	2.6	3.0	3.2	3.9	2.5	0.7
W	11.2	6.3	30.3	17.8	2.9	5.9	7.4	33.8	8.1
Tl	<LLQ	<LLQ	21.1	<LLQ	5.5	<LLQ	6.2	<LLQ	70.1
Pb	5.4	7.3	3.0	2.2	3.0	2.6	4.4	4.1	23.7
Th	1.9	0.7	2.7	3.0	3.4	3.4	4.4	10.8	6.3
U	0.9	1.1	2.8	2.7	3.1	4.0	3.7	6.6	3.8

Note: Types of precision identified are Type 1 (repeat consecutive measurements of the same sample solution), Type 2 (repeat measurements throughout an instrument run), and Type 3 (variability in replicate solution preparations).

[a] Soil samples are from Simon's Wood, Berkshire, U.K. (SW), and Arborfield Bridge, Berkshire, U.K. (AB). Certified reference materials are granite from Ailsa Craig, Scotland (ACE); basalt from Essey-la-côte, Nancy, France (BEN); granite from Senones, Vosges, France (GSN); marine sediment from Gulf of Maine, Boston, U.S. (MAG); and Cody Shale from Natrona County, Wyoming, U.S. (SCo).

Table 3.9   Rare Earth Elements, with Chondritic Normalizing Factors

Element	Symbol	Atomic Number	Chondritic Rare Earth Element Normalizing Factors
Yttrium	Y	39	2.1
Lanthanum	La	57	0.367
Cerium	Ce	58	0.957
Praseodymium	Pr	59	0.137
Neodymium	Nd	60	0.711
Promethium[a]	Pm	61	N/A
Samarium	Sm	62	0.231
Europium	Eu	63	0.087
Gadolinium	Gd	64	0.306
Terbium	Tb	65	0.058
Dysprosium	Dy	66	0.381
Holmium	Ho	67	0.0851
Erbium	Er	68	0.249
Thulium	Tm	69	0.0356
Ytterbium	Yb	70	0.248
Lutetium	Lu	71	0.0381

Note: N/A = not applicable.

[a] Promethium has no long-lived nuclei, and no natural abundance.

Source: From Taylor, S.T. and McLennan, S.M., in *The Continental Crust: Its Composition and Evolution. An Examination of the Geochemical Record Preserved in Sedimentary Rocks*, Blackwell Scientific Publications, Oxford, 1985, 312. With permission.

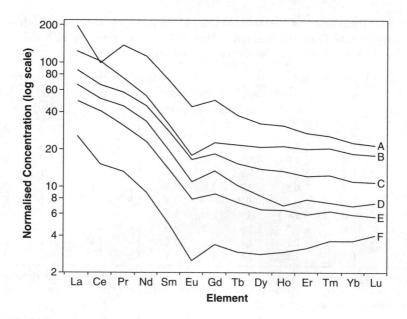

Figure 3.19 Examples of rare earth element profiles (chondrite normalized) in selected soil samples from southern England: (A) Warren Farm, Aldworth, Berkshire; (B) woods near Gatwick Worth Hotel, Crawley, Sussex; (C) roadside verge, Coaley, Gloucestershire; (D) river bank, Bradford on Avon, Wiltshire; (E) soil accumulation on hardstanding, Arborfield Bridge, Berkshire; and (F) woodland, Simon's Wood, Berkshire.

then plotted graphically (Figure 3.19). Many rocks and sediments show variable depletion of europium relative to the neighboring REEs, samarium and gadolinium. The degree of relative depletion (or enrichment) is expressed by the ratio Eu/Eu*, where Eu is the measured abundance of europium and Eu* is the theoretical concentration for no Eu anomaly; that is, a smooth REE pattern in the region Sm-Eu-Gd, calculated from $Eu_N/|(Sm_N)(Gd_N)|^{1/2}$. Values of Eu/Eu* of <0.95 indicate depletion, and values >1.05 indicate Eu enrichment (Taylor and McLennan, 1985). Some sediments and soils also show significant cerium depletion relative to the adjacent REEs, lanthanum and praseodymium (e.g., samples A and F in Figure 3.19). Such features can be useful forensic soil comparators.

3.2.8 Concentrations of Light Elements

Light elements are conventionally defined in earth sciences as those having an atomic number lower than 11. The concentrations of many of these elements, notably hydrogen (H), carbon (C), nitrogen (N), and oxygen (O), which are generally of greatest interest in soil studies, can be determined by wet chemical methods or by elemental analyzers (Bremner, 1996;

Nelson and Sommers, 1996; Tabatabai, 1996). Choice of procedure depends partly on whether determination of "total," or separate "organic" and "inorganic," elemental concentrations is required. The Carlo Erba and similar instruments are capable of simultaneous determination of total C, H, and N, plus O and S, depending on the model (Pella, 1990a, 1990b; Figure 3.20). For mineral soils, 5 to 10 mg of finely ground sample is typically required. Samples for CHN analysis are placed in a tin sample cup and introduced into a quartz reactor in a constant stream of helium. Oxygen is then introduced into the reactor, resulting in flash combustion at a temperature of 1700 to 1800°C. The gases resulting from combustion then flow through a heated Cu column to remove excess oxygen and then into a chromatographic column where they are separated and detected.

Determinations of CHNO and S can be useful in forensic studies in cases where samples contain a high proportion of organic matter. Such determinations are commonly made as an initial step when stable isotope ratios of the lighter elements are also to be determined.

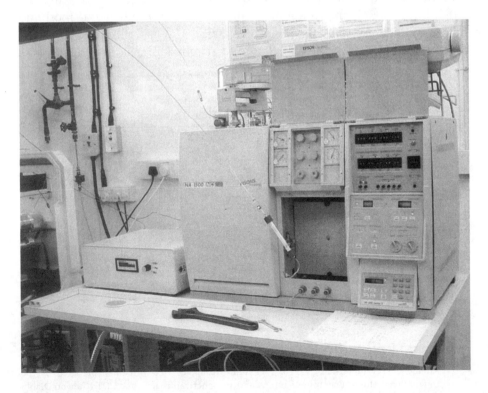

Figure 3.20 Fisons Carlo-Erba NA1500 elemental analyzer (photograph by D.J. Croft).

3.2.9 Light Element Stable Isotopes

Hydrogen, carbon, nitrogen, oxygen, and sulfur possess two or more stable isotopes, which do not undergo radioactive decay, and the stable isotopic ratios of these elements have been demonstrated to be useful indicators of many geological and biological processes (Faure, 1986; Faure and Mensing, 2004; Hoefs, 2004; Table 3.10). Light element stable isotope ratios are normally determined by mass spectrometry, and this type of analysis is widely referred to by the term isotope ratio mass spectrometry (IRMS). Several types of instrumentation and analysis procedure are available, including continuous helium flow isotope ratio mass spectrometry (CF-IRMS) and combustion-gas chromatography-isotope ratio mass spectrometry (C-GC-IRMS), which may or may not involve laser heating (Mattey, 1997; Figure 3.21).

    IRMS has been widely applied to studies of human and animal hair, teeth, and bone in a wide range of geological, archaeological, anthropological, and zoological contexts (Pye, 2004c). There have also been many applications of stable isotope ratios in relation to environmental forensic issues, such as pollution and deterioration of construction materials (Pye and Schiavon, 1989; Slater, 2003), and to questions of commercial product and food and drink authenticity (e.g., Bowen et al., 2005). IRMS techniques have also recently been widely applied in relation to criminal forensic investigations as

Table 3.10 Natural Abundances of Light Stable Isotopes

Element	Analyzed as	Isotope	Natural Abundance (%)	Terrestrial Range (%)	Reference Standard
Hydrogen	H_2	^1H	99.984	D = −450 to +50	V-SMOW ($^2H/^1H$ = 0.00015595)
		^2H	0.01557		
Carbon	CO_2	^{12}C	98.888	^{13}C = −120 to +15	V-PDB ($^{13}C/^{12}C$ = 0.00112372)
		^{13}C	1.1112		
Nitrogen	N_2	^{14}N	99.634	^{15}N = −20 to +30	Air ($^{15}N/^{14}N$ = 0.0036765)
		^{15}N	0.366		
Oxygen	CO_2 (O_2)	^{16}O	99.759	^{18}O = −50 to +40	V-SMOW ($^{18}O/^{16}O$ = 0.0020052)
		^{17}O	0.037		or V-PDB ($^{18}O/^{16}O$ = 0.0020672)
		^{18}O	0.204		
Sulfur	SO_2 (SF_6)	^{32}S	95.0	^{34}S = −65 to +90	CDT ($^{34}S/^{32}S$ = 0.0450451)
		^{33}S	0.76		
		^{34}S	4.22		
		^{36}S	0.014		

Note: Reference standards are V-SMOW (standard mean ocean water prepared by the International Atomic Energy in Vienna), V-PDB (belemnite from Pee Dee formation, prepared by the International Atomic Energy in Vienna), air (atmospheric nitrogen), and CDT (Canyon Diablo Troilite). Oxygen isotopes can be reported relative to V-SMOW (water and silicates) or V-PDB (carbonates).

Figure 3.21 Micro-mass VG Optima mass spectrometer used for CF-IRMS analysis (photograph by D.J. Croft).

a means of comparing samples of materials such explosives, drugs, packaging tape, and paint (Yinon, 1995, 2004; Reidy et al., 2005).

The potential of stable isotope ratio analysis for forensic soil comparisons has been investigated (Croft, 2003; Croft and Pye, 2003, 2004b; Pye et al., 2006b) and it has been demonstrated that it is often possible to discriminate between soils on the basis of ^{13}C and ^{15}N determined on bulk soil fractions. However, the degree of small-scale spatial variability varies between sites, depending on the nature of the "soil" and the types of organic matter involved. Studies at two experimental sites in Berkshire, England, indicated less small-scale (<1 m) variability in both ^{13}C and ^{15}N values determined on <150 μm soil fractions than on >150 μm fractions, and that the degree of variability differs significantly between sites (Pye et al., 2006b; Figure 3.22). Experiments at two other sites in the London area demonstrated that there can be statistically significant differences in both ^{13}C and ^{15}N values between samples collected at different times of the year, reflecting factors such as seasonal variations in vegetation growth, leaf litter fall, and microbial activity within the soil (Croft, 2003; Croft and Pye, 2004b). A detailed understanding of these processes, and the contributions to the "bulk" soil values, requires

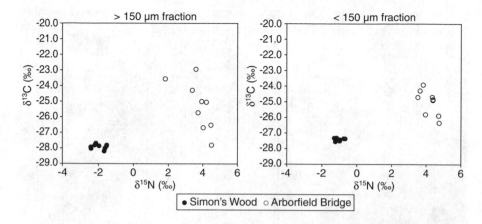

Figure 3.22 Bivariate plots of $\delta^{15}N$ vs. $\delta^{13}C$ for (a) >150 μm fractions and (b) <150 μm fractions of soils from 1 m experimental grids at Arborfield Bridge and Simons Wood, Berkshire.

analysis of individual organic and inorganic soil components sampled at different times of the year (Boutton, 1996; Cerling and Yang, 1996).

Table 3.11 presents a small illustrative data set from forensic casework relating to an armed robbery in West Yorkshire. Three replicate determinations were made on the <150 μm fractions of each of four samples taken from a suspect vehicle and two samples taken from areas with apparently matching tracks at the scene of the robbery. The data are compared as a bivariate plot in Figure 3.23. One of the tire track samples is clearly different to the samples from the vehicle in terms of both ^{13}C and ^{15}N values, whereas the other tire track control sample was similar in terms of these parameters.

3.2.10   Heavy Element Stable Isotopes

The stable isotopes of a number of heavier elements have been used in geological investigations as indicators of provenance and as natural tracers. The most commonly used are the ratios of $^{143}Nd/^{144}Nd$, $^{87}Sr/^{86}Sr$, and $^{206}Pb/^{204}Pb$, ^{207}Pb, ^{204}Pb, $^{208}Pb/^{204}Pb$ (Grousset et al., 1988, 1992; Borg and Banner, 1996; Ingram and Lin, 2002; Dia et al., 2006). As with the light stable isotopes, differing types of information can be obtained by determining the isotopic ratios of the bulk soil, sediment or rock, individual size fractions, various extracts/leachates, or individual solid constituents after separation (minerals, organic components, etc.).

In most soils, the dominant contribution to the Sr, Nd, and Pb isotope ratios comes from the local bedrock or drift geology that overlies it. The drift deposits may consist of sediment introduced by ice, water, or wind, mixed to a varying extent with local material formed by weathering of the

Table 3.11 Carbon and Nitrogen Stable Isotope Ratios of Four Questioned Samples from a Suspect Vehicle, and Two Control Samples from the Scene of an Armed Robbery

Sample	Location	^{13}C (‰)	^{15}N (‰)
KP2A	Control sample (crime scene)	−21.59	2.14
		−21.82	1.89
		−21.88	1.94
	Mean	−21.76	1.99
	SD	0.15	0.13
KP3A	Control sample (crime scene)	−27.04	4.85
		−27.17	4.56
		−27.15	4.77
	Mean	−27.12	4.73
	SD	0.07	0.15
KP4A	Mud sample from suspect vehicle	−26.43	4.20
		−26.65	4.15
		−26.57	3.97
	Mean	−26.55	4.11
	SD	0.11	0.12
KP5A	Mud sample from suspect vehicle	−26.74	4.64
		−26.67	4.59
		−26.67	4.66
	Mean	−26.69	4.63
	SD	0.04	0.04
KP6A	Mud sample from suspect vehicle	−25.82	5.23
		−25.78	4.91
		−25.92	5.06
	Mean	−25.84	5.07
	SD	0.07	0.16
KP7A	Mud sample from suspect vehicle	−26.09	5.21
		−26.36	5.45
		−26.13	5.09
	Mean	−26.19	5.25
	SD	0.15	0.18

underlying rocks. In such situations, the stable isotope ratios often show variation with depth, due partly to physical separation of different phases during soil formation and partly due to (usually airborne) additions to the surface (Dia et al., 2006).

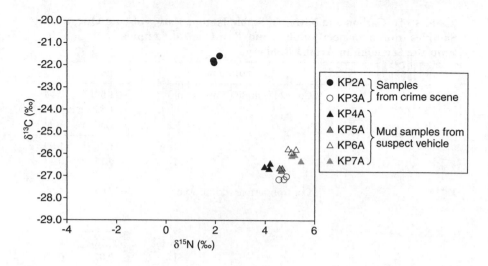

Figure 3.23 Bivariate plot of $\delta^{15}N$ vs. $\delta^{13}C$ for triplicate sub-samples of the <150 μm fractions of two reference samples taken from tire tracks at the scene of an armed robbery in West Yorkshire and four questioned samples taken from the wheel arches of a suspect vehicle.

Figure 3.24 illustrates the typical range of variation in Nd isotope ratio values found in rocks and sediments around the world, while Figure 3.25 provides a similar illustration for Sr isotope values. Detailed comparative studies normally require analysis of several leachate fractions as well as bulk samples (Pye, 2004c).

Owing to the fact that this type of isotopic analysis is relatively expensive to undertake, it has not been widely used in routine forensic soil comparisons. However, it has been used in certain serious crime cases where a key objective has been to identify the origin of unidentified bodies and where data relating to soil, vegetation, and various foodstuffs have been compared with data obtained from teeth, bone, and hair of the deceased.

3.2.11 Radioactive Isotopes

Many elements also have radioactive isotopes, or *radionuclides*, which decay over time to form daughter products (*radiogenic isotopes*), which may themselves be either stable or radioactive. During the process of radioactive decay, energy is released in the form of *alpha, beta,* and *gamma radiation*. The rate of decay is proportional to the number of radioactive atoms, and each radionuclide decays with a constant half-life. Radionuclides, which have a short half-life, decay more quickly and give off more intense radiation. Since the decay rate is independent of temperature, pressure, or chemical state, the age of a material can be estimated by measuring the proportions of different nuclides present. Measurement can be undertaken by several different techniques,

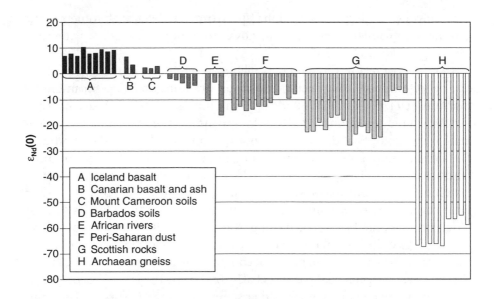

Figure 3.24 Values of $\varepsilon_{Nd}(0)$, a parameter (calculated from the $^{143}Nd/^{144}Nd$ ratio) for some rock and sediment samples from different parts of the world.

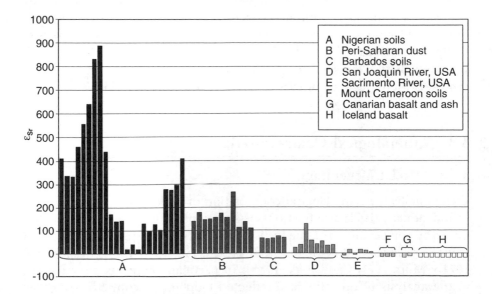

Figure 3.25 Values of the ε_{Sr} parameter (calculated from the $^{86}Sr/^{87}Sr$ ratio) for some soil and sediment samples from different parts of the world.

including mass spectrometry, alpha spectrometry, beta spectrometry, and gamma spectrometry (Faure, 1986; Lowenthal and Airey, 2001; Faure and Mensing, 2004; Dickin, 2005).

The ages of relatively old rocks (millions to thousands of millions of years) can be determined using methods such as uranium/lead (U/Pb), uranium/ thorium (U/Th), lead/lead (Pb/Pb), potassium/argon (K/Ar), argon/argon (Ar/Ar), rubidium/strontium (Rb/Sr), and samarium/neodymium (Sm/Nd) dating. The isotopes of uranium and thorium, and of other daughter products, can also be useful for dating samples ranging in age from a few tens of thousands to a few millions of years old. The ages of organic (carbon-bearing) materials up to about 50,000 years old can also be determined by radiocarbon (^{14}C) dating, whereas sediment ages ranging from a few tens of years to a few thousand years can be estimated using isotopes such as ^{210}Pb, ^{90}Sr, and ^{137}Cs.

The ages of individual mineral or organic constituents in rocks, sediments, and soils can provide a useful means of comparison and can be used to establish likely provenance. For example, Pell et al. (2000) used the U-Pb ages of detrital zircon grains, in conjunction with other sediment characteristics, to determine the provenance and sedimentary evolution of sands in the Simpson, Strzelecki, and Tirari Deserts of Australia. Radiocarbon dating has also been applied in several environmental forensics studies to determine the origin of organic pollutants in sediments and groundwaters (Kaplan et al., 1997; Kaplan, 2003).

Many sediments and soils in the vicinity of nuclear power stations, waste processing facilities, and weapons testing stations contain radionuclide contaminants, such as ^{137}Cs, ^{210}Pb, ^{238}Pu, ^{239,240}Pu, ^{241}Pu, and ^{241}Am (MacKenzie and Scott, 1982; Livens and Baxter, 1988a, 1988b). These contaminants can provide a useful regional or even location-specific "fingerprint."

3.3 Mineralogical Characteristics

3.3.1 Modal Mineralogy

Traditionally, the mineralogical composition of rocks and bulk samples, or of individual particle size fractions of sediments and soils, has been determined by point counting of thin sections using a polarizing petrological microscope (Krumbein and Pettijohn 1938; Tickell, 1965; Kelley, 1971; Galehouse, 1971b; Muir, 1977; Cady et al., 1986). In particular contexts, for example, in the analysis of ore minerals, reflected light microscopy has also been used to examine polished thin sections or polished blocks (Bowie and Simpson, 1977). Determination of mineral abundance has often been combined with quantification of particle/crystal size distribution, shape, and porosity, and with examination of the fabric and textural relationships within the rock, sediment, or soil (Harwood, 1988; Miller, 1988a).

Various methods of *point counting* have been proposed, but one of the most popular is the "line method" that involves counting the number of grains of each type encountered by the intersection of the cross hairs along equidistant linear traverses across the microscope slide (Galehouse, 1971b). Other authors have employed the "ribbon method" or "area method" that involves counting all grains intersected by, and contained between two parallel cross-hair lines (e.g., Lee et al., 2002), or the "Fleet method" that involves counting all grain within a field of view (Fleet, 1926). Another popular method involves automatic stage movement and counting of grains of each type that are encountered by intersection of the cross hairs at the center of the field of view (Chayes, 1949). When a master key is pressed, the stage advances by a preselected increment. Several traverses are made across each thin section until 250 to 300 points have been collected (Harwood, 1988). The counts for each mineral or particle type are then converted to a relative percentage and expressed as a *modal analysis*. The accuracy of the particle type "counts" obtained by this type of method is proportional to the relative abundance of each particle type in the sample and to the total number of points counted. Uncertainty errors rise steeply if less than 300 grains are counted (Figure 3.26).

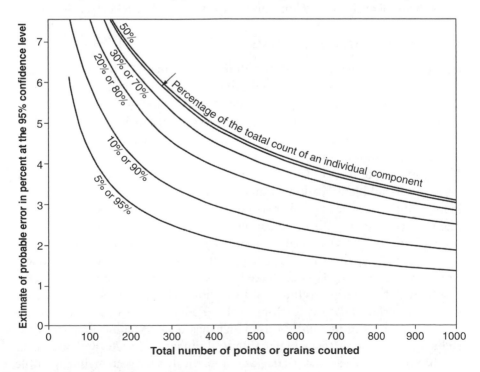

Figure 3.26 Estimated probable error as a function of number of points or grains counted for components present in various percentage abundances. (Plotted from data in Galehouse 1971b.)

In forensic work, 300 grains has been considered by several authors to be sufficient to allow differentiation between soils (e.g., Lee et al., 2002), although others have preferred 400 to 500 grains (McCrone, 1982; Junger, 1996). Experiments by Dryden (1931) indicated that increasing the count to 1200 grains increased the time required by a factor of four but resulted only in a twofold increase in accuracy compared with counting 300 grains, and Graves (1979) concluded that the extra time spent counting 1000 to 1500 grains was not necessary in routine casework, 300 grains being sufficient to differentiate between two soils with significant quantitative differences.

The results of modal analysis can be presented in tabular form but are frequently compared using triangular diagrams or statistical methods (Weltje, 2002; Weltje et al., 2004). Ternary diagrams can be used to show the relative percentages of quartz, feldspar, and rock fragments (QFR diagrams), or other components.

Microscopic examination has frequently been used to determine the mineralogical composition of small environmental and forensic dust samples (Hamilton and Jarvis, 1962; Palenik, 1979, 1988, 2000a; McCrone, 1982, 1992), where necessary supported by EMPA, SEM-EDXRA characterization, and other forms of microchemical analysis (McCrone and Bayard 1967; Merefield et al., 2000).

3.3.2 Heavy Mineral Analysis

Heavy mineral analysis is most commonly performed on the 31 to 500 μm size fraction of sediments obtained after heavy liquid separation (Hubert, 1971). In classical petrography, heavy minerals are often identified and counted along "ribbon" traverses across a grain mount in which the heavy minerals are set in matrix of Canada Balsam (Swift, 1971). Identification is based on a combination of color, translucency, shape, and surface texture. The most common types found in sediments are listed in Table 2.5. Where necessary to aid identification, grains can be impregnated in resin for examination with a polarizing microscope or polished for EMPA. At least 300 grains should be identified and counted.

Isphording (2004) described a case where heavy mineral assemblages provided evidence that pointed to the guilt of three Ku Klux Klan members who were accused, and later convicted, of the murder of a young black male. The body of the victim was found at a location in Baldwin County, Alabama, where the surface soils are developed on sediments of the Citronelle Formation, a Tertiary sedimentary body which has a heavy mineral assemblage characterized by a relatively high content of ilmenite, kyanite, and zircon, an absence of hornblende and garnet, and a relatively high tourmaline to rutile ratio. Analysis of soil samples taken from clothing and footwear associated with the defendants showed they had a clear Citronelle signature, and that they had lied about their movements on the night of the murder.

3.3.3 Automated "Mineralogical" Analysis by Computer-Controlled SEM

Since the development of computer-controlled electron microscopy and microprobe analysis in the 1970s, it has been possible to undertake automated compositional analysis of rocks and sediments involving thousands, or even millions, of analysis points. The potential of EMPA for the characterization and comparison of soil heavy minerals in a forensic context was illustrated by Smale (1973), but McVicar and Graves (1997) were the first to test the forensic potential of automated SEM-EDX analysis applied to sand grains. Their procedure to obtain a sample of separated grains for analysis was to scatter the sample on to a 150 μm mesh sieve and to press a piece of adhesive tape, attached to an SEM stub, onto the mesh, thereby transferring the grains. This resulted in a mounted sample containing approximately 5000 grains on a one-inch diameter SEM stub. Automated EDXRA analysis was then performed and the proportions of the detected elements in each particle recorded. At the end of the analysis, the particles were identified according to "classification rules" which had been established beforehand based on the elemental composition of reference minerals. In these experiments, different sand grain types could be identified with a precision (%CV) of <10% for components present in >5% abundance. Statistical comparisons indicated that four samples of the same color from different locations in Ontario could be successfully discriminated.

An example of a more modern automated system (QemSCAN) was described by Pirrie et al. (2004). This system is based on a Carl Zeiss EVO-50 SEM fitted with up to four energy-dispersive x-ray spectrometers. The availability of multiple spectrometers speeds up the process of obtaining x-ray spectra and allows a greater number of points to be analyzed in a given time period, compared with conventional systems which have only one x-ray spectrometer. The system described by Pirrie et al. (2004) was reported to be capable of characterizing approximately 1000 particles per hour. Data relating to particle size and shape could also be recorded, as well as backscattered electron images and color-coded "maps" showing variations in the internal composition of soil aggregates and sediment particles. Tests indicated that it was possible to differentiate seven soils from different locations in the Brisbane area, and to differentiate between distal and local contributions in atmospheric dust deposits.

A potential limitation of all such automated systems is that the identification of minerals or other, perhaps, noncrystalline, particle types is based on inference from the recorded elemental spectrum. In most instances, the x-ray count data are only qualitative, rather than quantitative, and sometimes it is not possible to differentiate between different species which have essentially the same elemental composition. For example, distinction between the iron oxide minerals haematite (Fe_2O_3), magnetite (Fe_3O_4), and wustite (FeO) can be difficult in the absence of high quality quantitative data, particularly when dealing with three-dimensional particles rather than polished sections. This problem is particularly severe in the case of minerals that form solid solution series whose members have a complex, potentially highly variable composition, such as amphiboles and pyroxenes. Another difficulty arises when particles are examined as uncovered mounted grains rather than in polished section; the apparent composition of the "grain" may reflect mixtures of surface coatings and the underlying grain. Moreover, grains that are touching can be wrongly counted as single grains, and particles that have highly irregular shapes can also be classified incorrectly. Such instances can be easily identified when manual EDX analysis of the particles is undertaken, but cannot be identified easily by automatic systems. On completion of each analytical run, which may take up to several hours, depending on the number of predetermined points to be analyzed, the acquired x-ray spectra are allocated to one of a number of categories which have been predefined by the analyst as being representative of the range of compositions exhibited by a particular mineral or other particle type. Errors can arise in this process.

Optical point counting methods can only be applied easily to sandstones, coarse to medium siltstones, and other rock types with equivalent particle/crystal sizes. The minerals in finer-grained rocks are difficult to resolve since they are smaller than the thickness of a typical thin section (30 μm). However, automated SEM-EDX analysis can be undertaken on particles as small as c. 10 μm. At conventional operating voltages and beam currents, the interaction volume between the incident electron beam and silicate or carbonate inorganic materials typically has dimensions of 3 to 4 μm. Since the incident electron beam may not be focused on the exact center of the particle being analyzed, the effective lower limit to the size of particle that can be characterized, without significant contamination from neighboring particles, is rather larger.

The mineral composition of fine-grained rocks and soils has usually been determined directly by bulk analysis methods such as x-ray powder diffraction (XRD, Figure 3.27; Wilson, 1987; Hardy and Tucker, 1988), supplemented where necessary by other methods such as infrared analysis (Farmer, 1974; Russell, 1987) and differential thermal analysis (Paterson and Swaffield, 1987). XRD has been used to quantify the proportions of different carbonate minerals present in limestones and as a supplementary technique to identify soil minerals and other crystalline components in numerous forensic investigations (e.g., Demmelmeyer and Adam, 1995; Lombardi, 1999; Brown, 2000; Brown et al., 2002; Ruffell and Wiltshire, 2004; Pye, 2004b; Stam, 2004).

Many forensic soil samples contain a significant proportion of x-ray amorphous matter which yields only diffuse peaks in the XRD traces, but for forensic comparison purposes it may be sufficient to identify similarities or differences

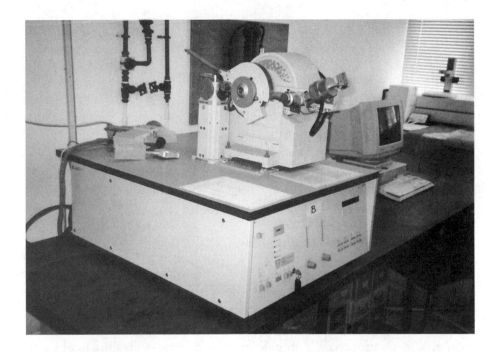

Figure 3.27 Philips bench-top x-ray powder diffractometer used for mineralogical identification.

in the shape of the x-ray traces, rather than attempting to identify precisely every component present (e.g., Figure 3.28). In other circumstances, it may be more appropriate to comparc peak height ratios, peak areas, or undertake a fully quantitative analysis using internal standards (Wilson, 1987).

3.3.4 Clay Mineral Assemblage

There are a considerable number of clay minerals found in nature, and numerous attempts have been made to classify them, but the main types encountered in soils and sediments fall into the following groups: (1) the serpentine-kaolinite group (including serpentine, chrysotile, berthierine, kaolinite, dickite, and halloysite); (2) the illite/mica group (including various polytypes of illite, mica, and glauconite); (3) the chlorite group (including chamosite and clinochlore); (4) the smectite group (including montmorillonite, nontronite, and saponite); (5) the vermiculite group; (6) interstratified minerals which contain layers of more than one of these types (e.g., illite-smectite, chlorite-vermiculite, and chlorite-smectite); (7) the sepiolite-palygorskite group; (8) the talc-pyrophyllite group; and (9) the imogolite-allophane group. In addition, the clay-size fractions of many soils also often contain associated minerals, such as iron oxides, quartz and calcium carbonate minerals, such as calcite.

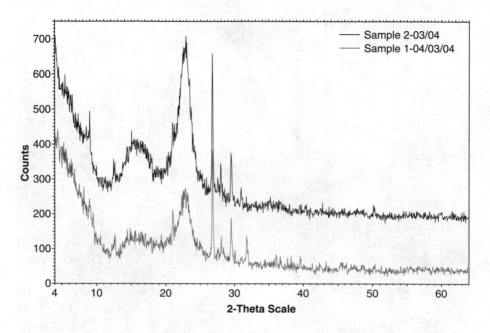

Figure 3.28 Comparison of x-ray powder diffraction traces obtained from a sample of dust formed by cutting ceramic tiles (upper) and a sample taken from the sole of a shoe that had made a single controlled contact with the dust (lower).

The nature of the clay mineral assemblage depends on a number of factors: (1) inherited clays derived from the parent rocks or sediments; (2) the nature of the clays formed *in situ* by weathering and/or diagenesis; and (3) additions introduced by air or water. Different clay mineral assemblages are stable under different physicochemical conditions, and the assemblages present in most soils bear a close relationship to the nature of the local climate and hydrological regime, as well as to the parent material from which they are derived. For example, many humid tropical soils are dominated by kaolinite, whereas illite, chlorite, and interstratified clays tend to dominate at higher latitudes. The minerals sepiolite and palygorskite are mostly restricted to arid zone soils, whereas allophane and imogolite, sometimes associated with gibbsite, are characteristic of soils formed from volcanic parent materials in humid climatic regimes. The assemblages of clay minerals often vary over relatively short distances, reflecting variations in parent material, microclimate and soil hydrology (e.g., Marumo et al., 1986, 1988).

XRD is a standard method for the characterization of clay mineral assemblages (Brindley and Brown, 1980; Moore and Reynolds, 1997), where necessary supplemented by electron microscopy, infrared analysis, thermal analysis, electron spin resonance, and other techniques (McHardy and Birnie, 1987; Righi and Elsass, 1996).

3.3.5 Bulk Luminescence Properties

Many minerals and other particles in soil or sediment will emit light if subjected to heating (*thermoluminescence*), electron excitation (*cathodoluminescence*), laser light excitation (*photoluminescence*), or vibration/friction (*triboluminescence*). Luminescence properties provide a potentially useful way of fingerprinting bulk samples, individual mineral grains, and a range of composite industrial products (McKeever et al., 1995; Nasdala et al., 2004a; Karakus, 2005). Several types of luminescence and related phenomena can also be used to date sediments, rocks, pottery, and related materials that may be of intrinsic interest or occur as fragments within soils and sediments.

The potential use of thermoluminescence in forensic work was identified in the 1970s (Ingham and Lawson, 1973), but there has been little significant use in forensic casework. The same is true of cathodoluminescence, which was initially developed by Sippel (1965, 1968) and evaluated from a forensic point of view by Dudley (1976c).

3.4 Bulk Organic Matter Characteristics

3.4.1 Organic Matter and Organic Carbon Content

An estimate of the total organic matter content of a soil can be obtained by determining the loss on ignition (LOI) at 550°C or through the use of H_2O_2 digestion (Black et al., 1965; Jackson, 1958; Gross, 1971; Wanogho et al., 1987a). The former method may lead to an overestimation due to the destruction of carbonates, sulfides, and some clay minerals, whereas the latter method may underestimate the true figure due to incomplete decomposition of refractory organic matter. Organic carbon content is most commonly determined by the wet-chemical potassium dichromate oxidation method of Walkley and Black (1934), or using a CHN elemental analyzer (Nelson and Sommers, 1996). More detailed characterization of different organic matter compounds, including humic and fulvic acids, is normally based on selective extraction a variety of methods of spectroscopic analysis (Swift, 1996; Thanasoulias et al., 2002).

3.4.2 Organic Compounds

Thornton (1974) and Thornton et al. (1975) sought to characterize soil compounds using enzymes. Dudley (1976b) investigated a colorimetric method to determine soil saccharide content and found good reproducibility and discriminatory potential. Andrasko (1979) used high pressure liquid chromatography (HPLC) to analyze polycyclic aromatic hydrocarbons (PAHs) in a suite of Swedish soils and obtained good reproducibility and a high degree of discrimination between soils taken from sites at least 0.5 km apart. Soil samples taken 5 to 10 m apart showed almost indistinguishable chromatograms, indicating that this method can provide a useful local scale "fingerprint." Nakayama et al. (1992) used pyrolysis gas chromatography (PyGC), first applied to soils by Wheals and Noble (1972), to study 113 Japanese soils and found good reproducibility for the same soil sample, and considerable variation between different soil sampling sites. Reuland and Trinler (1981) and Reuland et al. (1992) investigated organic extracts from soils in the U.S. using high performance liquid chromatography (HPLC) and found that the technique provides a reproducible means of correctly classifying and identifying soils. Siegel and Precord (1985) used a different HPLC method to study soils at 11 urban sites in Michigan and found that chromatograms of all the soil samples studied differed from each other quantitatively but not all could be differentiated qualitatively. They concluded that this method provides an excellent presumptive test but has not been proven to be "individualizing" (probative). Cox et al. (2000) reported a method for analysis of soil organic matter using Fourier transform infrared (FTIR) absorption spectrometry which proved capable of discrimination between four Oregon soil samples which could not previously be distinguished on the basis of color and organic matter determination by loss on ignition. A wide range of other natural and contaminant organic substances also occur in soils, especially those of urban areas, which can potentially be used for forensic comparison purposes; these include marker compounds such as methyl tertiary butyl ether (MTBE), polychlorinated biphenyls (PCBs), polychlorinated dibenzo dioxins and furans (PCDD/Fs), a variety of plant and petroleum hydrocarbons (Kaplan et al., 1997; Bruce and Schmidt 1994; Dawson et al., 2004) acid humas fractions (Thanasoulias et al., 2002), and pesticides (House and Ou, 1992). Reviews of organic substances in sediments and soils, and methods to characterize them, are provided by acid humas fractions [Thanasoulias et al., 2002; Schwarzenbach et al. (1993), and Swift (1996)].

3.4.3   Microbial Populations

Thornton and Crim (1986) noted that microbial populations in soil and sediments had essentially been ignored by the forensic community. Little has changed in the past decade, despite the fact that the general science in this area has moved on considerably and a variety of new techniques have been developed for charaterizing both living and fossil microbial populations. One such technique with considerable potential application in forensic studies is the signature lipid biomarker (SLB) technique which is based on the premise that all microorganisms have cell membranes composed of lipids, and in particular phospholipid fatty acids (PLFAs). PLFA recovered from individual isolates of bacteria have been used to characterize bacteria at the species level (Tunlid and White, 1992), and ratios of specific PLFA have been used to establish the nutritional or metabolic status of the bacteria (White, 1993). Sediments and soils in many different types of environment have now been characterized in terms of their PLFA profiles, which often preserve a record of bacterial activity over time (e.g., Pye et al., 1997). Lipid biomarkers have been used to determine the provenance of aeolian dust deposited in the oceans, and in a variety of other tracer contexts (Gagosian et al., 1981, 1982, 1987; Eglinton et al., 1993).

3.4.4 Pollen Assemblages

Most soil and sediment samples contain pollen grains and spores in greater or less amounts, although sometimes samples may be barren. Low concentrations and poor pollen preservation are often a particular problem in sandy sediments and soils. Any pollen present may be derived from locally growing living plants, from living plants growing some considerable distance away, or represent fossil pollen which has been recycled from older soils, sediments, or rocks. The state of preservation of individual grains can range from fresh, extremely well preserved to highly degraded and unidentifiable even at genus level (Bryant et al., 1990).

Assemblages of pollen and other botanical microfossils have been used in a considerable number of forensic investigations, usually in conjunction with other lines of evidence (Bryant et al., 1990; Mildenhall, 1990, 2004, 2006; Mildenhall et al., 2006; Graham, 1997; Lombardi, 1999; Brown, 2000, 2006; Brown et al., 2006; Horrocks and Walsh, 2001, Horrocks 2004). Procedures for sampling and preparing forensic samples for pollen analysis are described by Bryant et al. (1990) and Horrocks (2004). Essentially, they are based on more general pollen preparation procedures, but are modified to take account of the specific objectives of forensic investigations, the nature of forensic exhibits, and the fact that forensic samples are often much smaller, and therefore usually contain fewer pollen grains and other microfossils, than the sediment and soil samples normally taken for environmental research purposes. Most pollen identifications are undertaken using an optical microscope and X40 to X100 objectives, although additional examination at higher magnification using an SEM may be required to aid identification of problem grains. If quantitative information about pollen concentrations is required, a "spike" of an exotic spore or pollen type (e.g., *Lycopodium* or *Eucalyptus*) can be added in known concentration during the pollen preparation process.

In standard pollen analysis, it is standard practice to count at least identifiable 500 pollen grains and spores per sample (Bennett et al., 1990), and 600 is regarded by some authors as the minimum number of grains needed to obtain a sufficiently accurate estimate of the abundance of types present in relative percentages of more than 10%; for minor components, that is, below 5% representation, a count of at least 1000 grains is preferable (e.g., Birks and Birks, 1980; Moore et al., 1991). However, in questioned forensic samples, far fewer pollen grains are often present. Although a minimum of

500 grains has been counted in some forensic studies (e.g., Brown et al., 2002), in many cases only 150 to 250 grains (e.g., Horrocks et al., 1999; Mildenhall, 2004) have been identified and counted. In some instances, the number has been only 100 (e.g., Horrocks et al., 1998) or less. When the number of grains counted is small, percentage abundances are subject to large relative errors and in extreme cases have virtually no statistical meaning. Consequently, in such circumstances, it is more appropriate to make only qualitative comparisons between samples on the basis of presence/absence of particular pollen types, or an indication of apparent relative abundance (e.g., major component, minor component, or absent; Table 3.12).

Pollen and spore evidence can sometimes be highly specific in identifying contact between a suspect and a particular location, normally if one or more unusual (exotic) plant is present at the scene of interest. However, in many instances this is not the case and interpretations must be based on the number, nature, and apparent abundance of more mundane pollen types that may be present. Difficulties of interpretation can arise from the fact that some pollen types (e.g., pine and grass) can be transported very large distances from their source by wind or water, and secondary transfer of pollen between surfaces is a common occurrence. Pollen may accumulate on exposed surfaces (e.g., a car) over quite a long period of time and may be derived from many different locations; any pollen sample collected by surface washing or taping therefore stands a high chance of being composite, rather than representative of a single location at a single point in time. By comparison, there is much stronger likelihood that pollen extracted from a discrete lump of mud will be derived from a single restricted location.

A further difficulty can arise from the fact that pollen occurrences and concentrations often show a high degree of spatial and temporal variability, even within relatively confined sites, although this varies from one environmental setting to another. Bruce and Dettman (1996) reported a relatively high degree of agreement among the pollen and spore assemblages of replicate soil samples taken from a relatively small area of Australian coastal *Eucalyptus* woodland and an area of *Avicennia-Rhizophora* dominated mangrove woodland. Horrocks et al. (1999) also reported that the pollen assemblages of some Auckland, New Zealand, soil samples taken from, and between, consecutive shoeprints within a 115 × 6 m open grassy area were very similar to each other and also those from soil taken from the shoes which made the prints. Additionally, little difference was found between the pollen assemblages of 1 mm surface scrapings samples and samples taken at 20 mm depth. In an earlier paper, Horrocks et al. (1998) compared the pollen assemblages in the surface soil at this site with those in surface soil at 10 other locations with similar vegetation within the same geographical region. They concluded that the results showed that "localized areas of similar vegetation type, even

Table 3.12 Examples of Qualitative Inter-Sample Palynomorph Assemblage Comparison Matrices[a]

(a)	Washings from Trainers			Soil Samples from Scene of an Alleged Rape									
	Left	Right		1	2	3	4	5	6	7	8	9	10
					NR	NR		NR		NR			NR
Abies											■		
Betula											■	■	
Fagus		■											
Hedera											■		
Pinus							■		■		■	■	
Quercus													
Taxus				■									
Tilia											■		
Rosaceae											■		
Cirsium											■		
Filipendula		■											
Plantago lanceolata		■											
Poaceae	■	■							■		■		
Polygonum aviculare		■											
Rumex type		■									■		
Taraxacum type													
Pediastrum		■											
Charcoal		■									■		
Phytoliths		■											

Table 3.12 Examples of Qualitative Inter-Sample Palynomorph Assemblage Comparison Matrices[a] (Continued)

(b)	Mud Samples from Motorbike		Mud Samples from Pathway							
	KP9	KP10	KP1	KP2	KP3	KP4	KP5	KP6	KP7	KP8
		NR	NR			NR				
Alnus							■			
Betula							■			
Fraxinus							■		■	
Hedera					■		■			
Pinus				■			■		■	
Quercus							■			
Sambucus type							■			
Apiaceae							■			■
Cirsium type										
Cyperaceae										
Epilobium										
Poaceae	■									
Ranunculus type									■	
Sinapis type							■			
Taraxacum type									■	
Typha latifolia								■		
Filicales							■			
Pteridium							■			

Note: NR indicates no recovery of identifiable pollen grains.

[a] (a) Recorded types in 10 surface soil samples from the scene of an alleged rape and a pair of trainers from a suspect; (b) recorded types in eight samples of mud from a scene associated with a post office robbery and two mud samples taken from a motorbike allegedly used in the getaway.

within the same geographic region, have significantly different pollen assemblages." However, given that only 100 pollen grains were "counted" in the samples from each of the other 10 sites, the validity of this conclusion must be open to question.

The low abundances of pollen, and the extreme variability, which is sometimes found within restricted areas and closely associated items of forensic interest, are illustrated by two simple case examples. Table 3.12a shows a qualitative summary of pollen types identified in 10 samples of surface soil (0 to 25 mm depth) collected from the scene of a rape carried out in a churchyard (c. 25 × 20 m), compared with the pollen types identified in washings from the left and right trainers seized from an individual suspected of having committed the crime. In 9 of the 10 soil samples, identifiable pollen was either absent or present only in low abundance/diversity. A large qualitative difference in pollen abundance and recorded types was also found in the two washings samples from the left and right trainers. Only one pollen type was identified in common between the trainers and the soil samples from the churchyard, and that could only be identified as *Poaceae*, an extremely broad genus that includes many types of grass. Given the poor recovery of identifiable pollen, the results are inconclusive.

Table 3.12b shows another simple example relating to eight surface (0 to 10 mm) "mud" samples taken from a pathway where a motorbike, which had allegedly been involved in an armed robbery, was sighted, and two samples of mud splashes recovered from the underside of a motorbike seized later by the police. The samples from the scene again showed considerable variation in the abundance and diversity of identifiable pollen. The mud splashes from the motorbike yielded very little pollen which all belonged to the *Poaceae* genus. Once again, the results were inconclusive.

Standard pollen preparation procedures are destructive in that they usually require extraction and concentration of the palynomorphs, a process which involves dissolution of carbonate and silicate mineral grains by acid and alkali digestion. Consequently, if such procedures are employed, other potentially useful information can be lost. In some circumstances, it may be possible to circumvent this by using nonstandard pollen extraction and concentration procedures, such as centrifugation and flotation separation following thorough dispersion of the soil in water. Pollen slides can be prepared from some sediments and soil types without removal of all of the nonorganic matter, but in such cases identification and counting of pollen types is normally more difficult than in standard pollen slides.

3.4.5 Diatom Assemblages

Diatoms occur in many environments, including field soils and dust samples, but they are especially common in aquatic sediments. Since many diatom species are highly selective in terms of the pH, salinity, and nutrient status of the environments in which they live, analysis of diatom assemblages can sometimes provide useful environmental information. Diatom analysis can be simply qualitative, involving the listing of identifiable types in a suite of samples, or quantitative, involving counts on carefully prepared sub-samples that have been "spiked" using a marker in known concentration. Since considerable overlap in species composition is often found in samples from similar environments, analysis of the results is frequently aided by the use of multivariate techniques, which aim to identify assemblage "groups" (Cameron, 2004; Hartley et al., 2006). In most cases, diatom analysis alone is only capable of providing generic environmental type information, but combined with other lines of evidence it may help to define specific locations.

Properties of Individual Particles 4

4.1 Introduction

In many forensic situations, the amount of soil, sediment, or dust material recovered for analysis is too small to allow bulk methods of analysis to be used. There are also many situations where the objects of interest are single pieces of rock, brick, or concrete, or a relatively small number of gravel-sized particles. In such situations, the individual particles must be characterized individually and in as much detail as possible, in most cases nondestructively. Each particle should be characterized in terms of several different properties, since the larger the number of independent criteria available, the stronger the basis for comparison and interpretation, The nature of the properties which should be determined will depend on the type of particle involved, but in most circumstances will include the size, shape, surface texture, color, and any unusual surface markings. In specific circumstances, it may also be helpful to obtain information about other properties such as internal microstructure (fabric), luminescence properties, and chemical and mineralogical composition.

4.2 Particle Size

4.2.1 Aspects of Size

The "size" of an individual particle can be determined by a number of different methods of direct or indirect measurement (see Allen, 1997a, for a

119

review). The size of individual particles can be expressed in terms of several different properties, including:

1. weight
2. linear dimensions
3. cross-sectional area
4. perimeter
5. volume
6. surface area
7. equivalent spherical diameter

4.2.2   Weight

The weight of an individual particle can be accurately measured using an electronic balance. In the field, it is practical to weigh rock fragments and gravel particles up to 5 kg with an accuracy of 0.1 g or better, using a battery-powered digital balance, whereas in the laboratory, individual sand grains can be weighed with an accuracy of up to 0.001 g using an analytical balance. The weight to volume ratio provides a measure of the *particle density* (Flint and Flint, 2002a). The weight of a particle, or a ratio of the weight to total surface area or the maximum projected area, can also provide a useful indication of the relative inertia of a particle, and allow an assessment of the likelihood that it might experience movement when subjected to a natural applied stress (e.g., wind or water pressure).

4.2.3   Linear Dimensions

Many different linear dimensions can be determined for any given particle. The most appropriate dimension, or dimensions, to determine as an indicator of size will vary with the particle type and the purpose of the investigation. The size of a simple, regular geometrical object, such as a cube, is most commonly described in terms of the side length, although it may also be useful to determine the diagonal, or maximum corner to opposite corner, dimension (Figure 4.1a). If accurate calculations of area and volume are to be made, it is essential to measure the side lengths. In forensic work, it is often sufficient to measure the side lengths of relatively regular objects such as bricks, concrete blocks, and building stones. However, in the case of more irregularly shaped objects, such as many natural rock and sediment particles, no straight edges may be present and it may not be possible to determine side lengths. In such circumstances, it is conventional to measure three orthogonal dimensions that correspond to the longest (L = length), intermediate (I = breadth), and shortest (S = thickness) dimensions of the particle. If the requirement is simply to characterize the size and broad form of a particle larger than about 10 mm, the $L, I,$ and S dimensions can be determined using

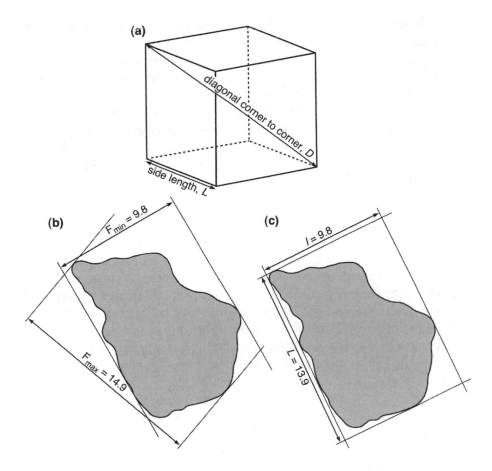

Figure 4.1 Alternative methods of quantifying the physical dimensions of an object: (a) a cube with side length L and diagonal corner-to-corner dimension D; (b) maximum and minimum *Feret* diameters, which need not be orthogonal; and (c) the maximum side length L and orthogonal side length I of the smallest rectangle (or rectangular box in the case of a three-dimensional object) within which the object can be enclosed.

a ruler, callipers, or measuring frame. For smaller particles, measurement usually requires a digital photographic image to be obtained and measurements made using suitable computer software.

In some cases, the measured L dimension may equate with the *maximum caliper diameter* or *maximum Feret diameter*, although this is not always the case. A Feret diameter can be defined simply as the perpendicular distance between parallel lines, tangential to the perimeter on opposite sides of a two-dimensional object when viewed in a particular orientation (Figure 4.1b). Several Feret diameters, which need not be orthogonal, can be measured for any projected image of a particle, and the maximum, minimum, and average values determined. In the case of a cube, the minimum Feret diameter is equivalent to the side length, L, which is also equal to I and S. However, with other shapes the minimum Feret diameter may not equate with any of the L, I, or S dimensions. The average Feret diameter determined by multiple measurements on a randomly oriented cube will be $1.5D$, where D is the side length. Although Feret diameters are extensively used in image analysis, in geological applications the L, I, and S dimensions have been more commonly used. For practical purposes, the L, I, and S dimensions can be defined as being equivalent to side lengths of the smallest rectangular box within which an object will fit (Figure 4.1c).

4.2.4 Cross-Sectional Area and Perimeter

In the case of a straight-sided object, the cross-sectional area and perimeter can be calculated based on measurements of the side lengths, but in the case of irregularly shaped objects, determination of these parameters usually requires image analysis performed on digitally captured images or calibrated photographs. In this case, the determinations are of *projected* area and perimeter, representing the "silhouette" of the three-dimensional particle transposed onto a two-dimensional plane. Normally, determinations will be made on the particle in three perpendicular orientations (displaying $L \times I$, $L \times S$, and $I \times S$, respectively, as shown in Figure 4.2).

In the case of particles smaller than about 4 mm, photographic images in three orientations can be obtained using a binocular microscope and digital image capture system, rather than a digital camera (Figure 4.3). Measurements are made in pixels and dimensionless ratios calculated without the need for calibration. However, if actual measurements of length, breadth, thickness, area, and perimeter are required, the system can be calibrated using a reference scale.

Size and shape measurements can also be made on digital images of sand and silt-size grains obtained in a scanning electron microscope (SEM). This is particularly useful for particles smaller than about 10 µm that are difficult to resolve by optical microscopy. The software for many modern SEM systems allows calliper dimensions and area to be determined directly on the SEM image in real time, or alternatively the image can be stored or exported for later analysis using more sophisticated software.

Most image processing systems calculate a derivative parameter that represents the diameter of a circle that has an equal area or perimeter to the particle under examination. This parameter is often referred to as the *equivalent spherical diameter*, although more correctly it is the *equivalent circular diameter*. If images are obtained in three orientations, an average value can be obtained to provide an estimation of equivalent spherical diameter.

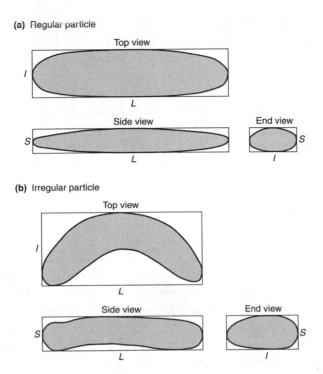

(a) Regular particle

Top view

_I_

_L_

Side view

_S_

_L_

End view

_S_

_I_

(b) Irregular particle

Top view

_I_

_L_

Side view

_S_

_L_

End view

_S_

_I_

Figure 4.2 The longest (L), intermediate (I), and shortest (S) orthogonal dimensions, corresponding to length, breadth, and thickness, of (a) a regular particle and (b) an irregular particle, seen in top view, side view, and end view.

Figure 4.3 Nikon SMZ800 stereomicroscope and Nikon DN100 Digital Net Camera, equipped with Eclipse Net software, used for image capture and image measurement.

Measurements made on projected grain images will inevitably show a degree of error owing to two factors. First, the projected outline of a particle may appear more regular than it actually is owing to the superimposition of three-dimensional features onto a two-dimensional plane. Second, there will be a degree of image distortion due to differences in the focal length between different parts of an object and the camera lens or scanning device; that is, those parts of an object that are further away from the camera appear smaller than those closer to the camera. This can be significant when images are obtained of an object in side view and especially in the case of elongated objects imaged in end view (e.g., Figure 4.4a).

4.2.5   Volume

Volume can be considered to provide the best descriptor of particle "size" (Wadell, 1932), but in practice it is a difficult property to determine accurately, especially for small particles. In the case of particles >4 mm, volume can be measured directly by determining the volume displaced when the particle is immersed in water or another fluid (Krumbein and Pettijohn, 1938; Flint and Flint, 2002a). However, if the object of interest is porous, the surface must first be sealed or covered (e.g., with cling-film), prior to immersion. For smaller particles, this method is impractical. An alternative method that can be applied to particles with a wide range of sizes utilizes x-ray tomography or laser scanning (Garboczi, 2002). However, analyses are relatively time-consuming and expensive to undertake. For routine forensic description/comparison purposes, it is usually satisfactory to estimate the volume of larger particles using the immersion method, and to report the results alongside values for the L, I, and S dimensions, weight, and particle density (e.g., Table 4.1).

4.2.6   Surface Area

Surface area is relatively easy to calculate for regular geometrical shapes, such as a sphere, cube, or rhomboid, but is difficult to quantify for irregularly shaped particles. In the case of large particles (>40 mm), an estimate may be obtained by enclosing the particle in aluminum foil and subsequently measuring the area of the foil after removal. Gas adsorption techniques may also be used to estimate surface area of single large particles, but the results are strongly influenced by factors such as surface texture and porosity (Allen 1997b; Pennell, 2002). For smaller particles, it is only feasible to make average estimates of surface area on bulk samples.

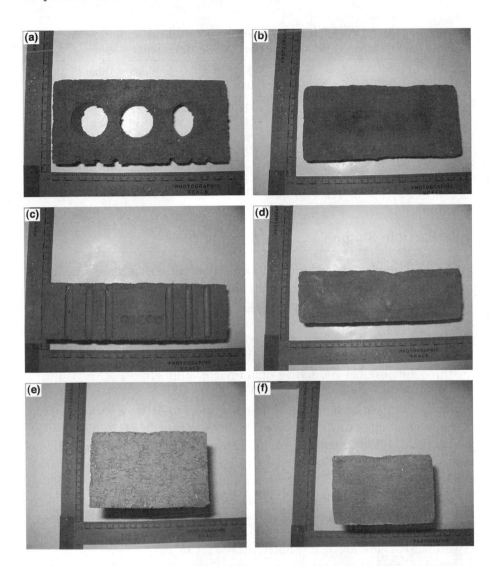

Figure 4.4 Six faces of a brick photographed individually to record textural features and surface markings. Note the distortion in photograph (a) which arises partly because of varying differences in distances between camera lens and different parts of the object; the top and bottom of the central hole are essentially coincident with each other, whereas the sides of the left-hand and right-hand holes can be clearly seen.

Table 4.1 Physical Characteristics (Dimensions, Dimension Ratios, Weight, Volume, and Density) of Six Contrasting "Rock" Fragments

Particle	Material Type	L (mm)	I (mm)	S (mm)	I/L	S/I	Weight (g)	Volume (ml)	Density (g cm^{-3})
1	Quartzite	46	43	22	0.93	0.51	62.69	22.0	2.85
2	Brown chert	64	50	48	0.78	0.96	134.36	52.0	2.58
3	Black chert	58	50	19	0.86	0.38	70.95	27.0	2.63
4	Concrete	64	55	52	0.86	0.95	304.48	136.0	2.24
5	Brick	54	27	26	0.50	0.96	39.90	18.0	2.22
6	Coal	57	39	25	0.68	0.64	33.89	23.0	1.47

4.3 Particle Shape

Shape is an important fundamental property of an object but is one of the most difficult to specify for any but the simplest of shapes. In a general sense, *shape* can be taken to be the external expression of an object defined by a surface envelope that encloses the volume occupied by the object. Two-dimensional objects can be considered to be infinitely thin and to have an infinitely small volume. However, most geological objects of forensic interest are three-dimensional and have a quantifiable volume.

In the geological literature, the terms particle *morphology, shape, form*, and *surface texture* have been used by different authors to describe various aspects of shape, resulting in considerable confusion (Wentworth, 1922b, 1922c, 1933; Wadell, 1932, 1933; Krumbein, 1941; Flemming, 1965; Pryor 1971; Pettijohn et al., 1972; Whalley, 1972; Teller, 1976; Barrett, 1980; Orford and Whalley, 1983, 1987, 1991; Orford, 1990; Benn and Ballantyne, 1993; Benn, 2004). However, following Blott and Pye (2007), in the following discussion, *morphology* is used as an over-arching term to include all aspects of the three-dimensional expression of an object. It has two main components: *shape* and *surface texture* (Figure 4.5). There are many different aspects of shape, but four may be regarded as being of particular importance (Blott and Pye, 2007): *form*, which is an expression of the relative magnitude of the principal axial dimensions of an object, reflected by its degree of *flatness* and/or *elongation*; *sphericity*, which describes the degree to which the external envelope of an object approximates that of a true sphere; *roundness*, which relates to the degree of sharpness of individual edges and corners on an object; and *irregularity*, which relates to the number, size, and distribution of projections and indentations. Surface texture relates to the overall smoothness/roughness of the particle surface and also to the nature, abundance, and distribution of individual small-scale features which may be present on the surface.

The shapes of some particles can be described qualitatively by reference to a defined geometrical shape (e.g., sphere, cube, and prism), an organic

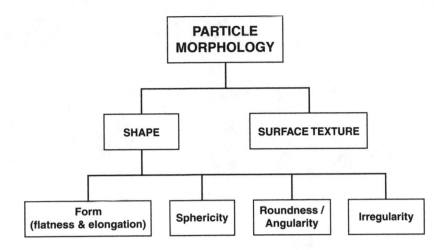

Figure 4.5 The principal aspects of particle morphology.

analogue (e.g., kidney-shaped or *reniform* and worm-like or *vermiform*), or some other readily recognized object (e.g., the letter *T* or the letter *Y*, as shown in Figure 4.6). However, many such terms are subjective and are of little assistance when detailed comparisons are required between particles of similar general shape, or when the particles under consideration have no readily recognizable shape. In order to describe, compare, and distinguish such particles, it is necessary to obtain quantitative descriptors of the shape in at least two, and preferably three, orientations.

4.3.1 Form

Particle form relates to the relative relationship of the overall dimensions of a particle that determine whether it can be described as *equant*, *flattened*, *elongated*, or *both flattened and elongated* (Figure 4.7). The basic aspects of particle form can be quantified on the basis of the L, I, and S dimensions of a particle and their ratios. However, such parameters provide no information about degree of sphericity or roundness, and it is not possible to differentiate, for example, between a cube and a sphere of equal L.

The essential aspects of the form of an object can be described by a combination of two ratios: S/I, which provides a measure of relative flattening, and I/L, which provides a measure of elongation (Blott and Pye, 2007). These two ratios can be plotted either on a bivariate graph of the type initially proposed by Zingg (1935) or a trivariate diagram of the type proposed by Sneed and Folk (1958). Zingg divided his form diagram into four fields and classified particles (pebbles) as *spherical*, *discoid*, *rod-like*, and *bladed* (Figure 4.8) on the basis of the S/I and I/L ratios. Sneed and Folk (1958) used a combination of the S/L ratio and the parameter $(L-I)/(L-S)$, which was later referred to as a

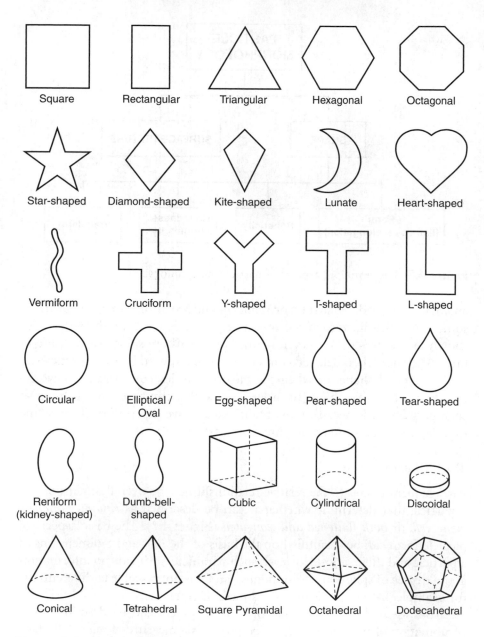

Figure 4.6 Some examples of readily recognizable two- and three-dimensional shapes.

disc-rod index (Illenberger, 1991), to divide their form triangular diagram into 10 fields (Figure 4.9). Blott and Pye (2007) proposed a simpler classification system based on the *S/I* and *I/L* ratios, which can be plotted either on a Zingg-type diagram or on a triangular diagram. Using this system, individual particles can be described, for example, as *non-elongate, non-flattened* or *highly*

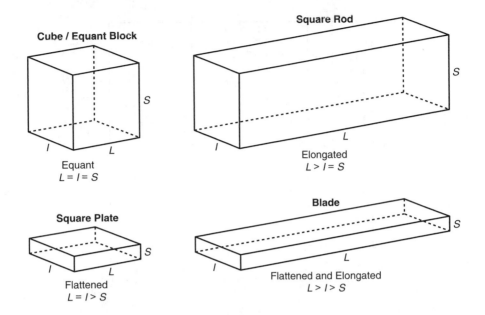

Figure 4.7 The four end-member cases that can be identified in relation to particle form.

elongate, moderately flattened (Figures 4.10a and 4.10b). Objects can also be described by single terms such as *blades, plates, rods,* etc. (Figures 4.10c and 4.10d). If sufficient particles are available in a sample, their *S/I* and *I/L* ratios can be plotted as frequency histograms (Figure 4.11).

In the engineering literature, the ratio of *I/L* to *S/I* has widely been used as a general form factor:

$$F = \frac{p}{q} \tag{4.1}$$

where
    q = the elongation ratio, *I/L*
    p = the flatness ratio, *S/I*.

Values of this expression range from 0 to infinity for extreme forms. The formula was modified by Williams (1965) so that values range from −1 to +1:

$$W = 1 - \frac{LS}{I^2} \quad \text{when } I^2 > LS$$

$$\tag{4.2}$$

$$W = \frac{I^2}{LS} - 1 \quad \text{when } I^2 \leq LS$$

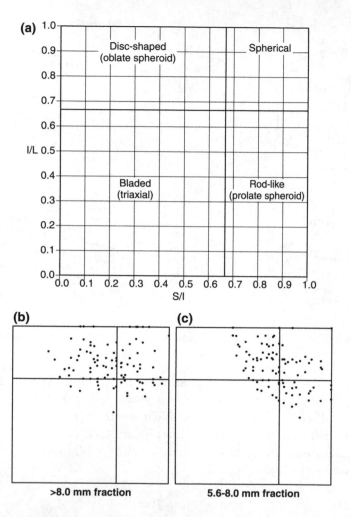

Figure 4.8 (a) Diagram proposed by Zingg (1935) used to classify particles into four groups based on the I/L and S/I ratios; (b) and (c) show plotted data for two size fractions of gravel particles (>8.00 mm and 5.6 to 8.00 mm) from a sample of commercially sold aggregate of forensic interest.

Values of <1 indicate a particle that is relatively elongated, whereas values of >1 indicate particles that are relatively flattened. However, the measure does not adequately differentiate between equant and bladed forms.

    Several authors have developed more complex summary form factors or shape factors (e.g., Wentworth, 1922b; Krumbein, 1941, Pye and Pye, 1943; Corey, 1949; Ashenbrenner, 1956; Sneed and Folk, 1958; Dobkins and Folk, 1970), but for forensic purposes these offer little advantage over the simple form ratios described earlier. The main use of form factors lies in their use as an aid to predicting the behavior of particles during sediment transport.

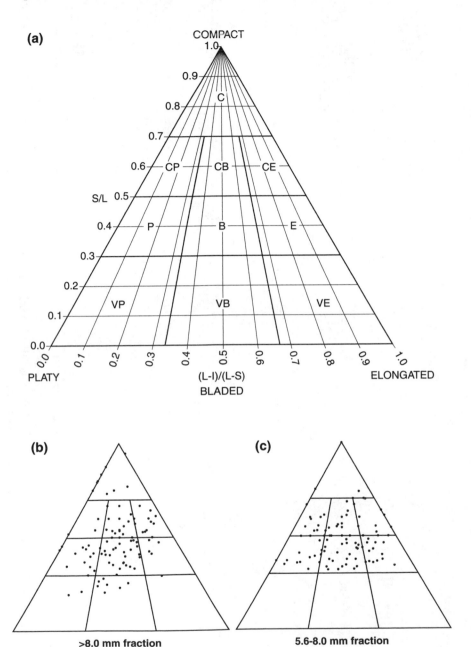

Figure 4.9 (a) Triangular diagram devised by Sneed and Folk (1958) to classify particles in relation to three form end-members. The ten form classes are C = compact, CP = compact platy, CB = compact bladed, CE = compact elongated, P = platy, B = bladed, E = elongated, VP = very platy, VB = very bladed, and VE = very elongated; (b) and (c) show plotted data for the two size fractions of gravel shown in Figure 4.8.

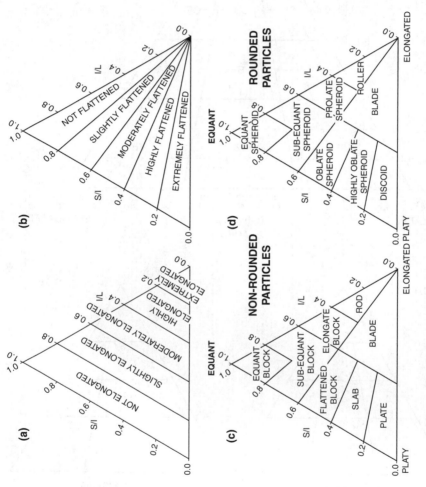

Figure 4.10 An alternative triangular form classification scheme and associated terminology, as proposed by Blott and Pye (2007).

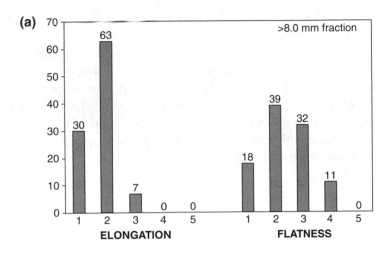

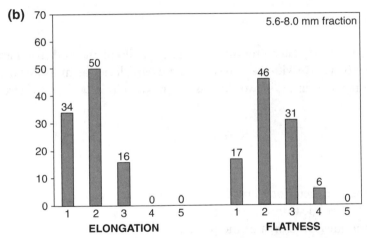

Figure 4.11 Frequency histograms (number) of the elongation and flatness parameters determined from measurements of L, I and S on 100 particles from each of two size fractions of gravel (>8.00 and 5.66 to 8.00 mm).

4.3.2 Sphericity

Wadell (1932) proposed a method of determining true grain sphericity, which involved measuring the grain surface area and volume:

$$\psi = \frac{s}{S} \qquad (4.3)$$

where
  s = the surface area of a sphere of the same volume as the particle
  S = the actual surface area of the particle.

A sphere will have a ratio of 1 and all other solid bodies will have a smaller value.

However, as noted earlier, surface area is impractical to measure, and volume is difficult to determine accurately for small particles. Consequently, several alternative methods of estimating sphericity have been proposed.

Cox (1927) developed one of the first methods of estimating sphericity using projected grain images viewed in two dimensions. He compared the area of the particle with that of a circle with the same perimeter. He called this a *roundness* parameter, K, defined as:

$$K = \frac{4\pi A}{P^2}$$

(4.4)

where
  A = the particle area
  P = the particle perimeter.

Wadell (1933) later proposed a slightly different method, comparing the perimeter of a circle with the same area as the particle, to the measured perimeter. Because it was calculated in two-dimensions, he called this the *degree of circularity*, ø:

$$\phi = \frac{c}{C} = \sqrt{\frac{4\pi A}{P^2}}$$

(4.5)

where
  c = the perimeter of a circle with the same area as the particle
  $C = P$ = the measured perimeter of the particle
  A = the measured area of the particle.

To avoid the measurement of the particle perimeter, which can be time-consuming, Tickell (1931) ratioed the area of the grain to the diameter of smallest circumscribed circle:

$$\phi = \frac{4A}{\pi (D_c)^2}$$

(4.6)

Wadell (1933, 1935) later produced a similar formula, based on the ratio of the diameter of a circle with the same area as the particle, to the diameter of the smallest circumscribed circle, a measure he defined as the *projection sphericity*:

$$\phi = \frac{d_c}{D_c} = \sqrt{\frac{4A}{\pi (D_c)^2}}$$

(4.7)

where
  d_c = the diameter of a circle equal in area to the projected area of the particle when it rests on one of its larger faces, more or less parallel to the plane of the longest and intermediate diameters
  D_c = the diameter of the smallest circumscribed circle around the particle projection
  A = the measured area of the particle.

Riley (1941) proposed a simplified parameter, which he referred to as the *inscribed circle sphericity*, ϕ_0, to describe the sphericity of particles viewed as two-dimensional projected images (Figure 4.12):

$$\phi_0 = \sqrt{\frac{D_i}{D_c}} \qquad (4.8)$$

where
  D_i = the diameter of the largest inscribed circle
  D_c = the diameter of the smallest circumscribed circle.

Rittenhouse (1943) published a two-dimensional visual comparator containing standard outline images for grains with differing values of calculated Wadell projection sphericity, ϕ. However, the comparator contains a large

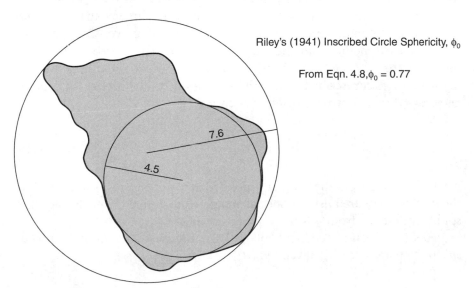

Riley's (1941) Inscribed Circle Sphericity, ϕ_0

From Eqn. 4.8, $\phi_0 = 0.77$

7.6

4.5

Figure 4.12 Visual representation of Riley's (1941) method of determining inscribed circle sphericity, ϕ_0.

number of grain outlines and the method suffers from a high degree of operator variation.

A measure of two-dimensional grain "roundness", which is frequently used in image analysis, is the *roundness parameter, R,* given by:

$$R = \frac{P^2}{A} \tag{4.9}$$

where
  P = the perimeter of the grain
  A = the area of the grain in the plane of projection.

Like Riley's parameter, this parameter is more correctly considered as an *outline circularity parameter.*

These methods were developed primarily for application to sand-size grains viewed as projected images. However, they can be applied in three dimensions if the particles are photographed in two or three different orientations. In the case of gravel-size particles, this can easily be achieved if the particles are held in position using Blu-Tack® and three photographs taken for image analysis (Figure 4.13). In the case of sand-size particles, the grains can be mounted in a diagonal array on an adhesive carbon disc attached to a 1 cm SEM stub (Figure 4.14). The grains can then be imaged in three orientations using a binocular microscope and image capture system, prior to examination of the surface textures in the SEM. An average of the three measurements can then be used as a measure of three-dimensional sphericity.

Several other authors, principally concerned with gravel-sized particles, have proposed methods for the estimation of sphericity based on measurements of the L, I, and S dimensions, rather than projected area and the diameter of circumscribed or inscribed circles. Krumbein (1941) proposed an *intercept sphericity* parameter, Ψ_i, defined by:

$$\Psi = \sqrt[3]{\frac{IS}{L^2}} \tag{4.10}$$

where L, I, and S are the longest, intermediate, and shortest dimensions of the particle measured using a special apparatus to define the enclosing ellipsoid. Values range from 0 to 1 for a perfect sphere.

Folk (1955) and Sneed and Folk (1958) proposed a *maximum projection sphericity* parameter, Ψ_p, given by:

$$\Psi_p = \sqrt[3]{\frac{S^2}{LI}} \tag{4.11}$$

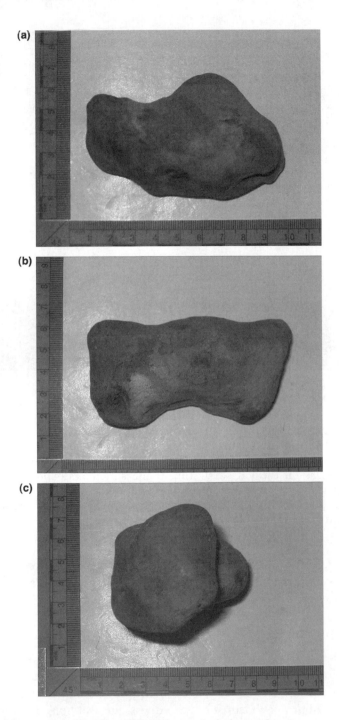

Figure 4.13 An irregular (knobbly) chert clast from the Quaternary drift of east Berkshire, U.K., photographed in three orthogonal orientations which reveal the *L* & *I*, *L* & *S*, and *I* & *S* dimensions, respectively, to the camera.

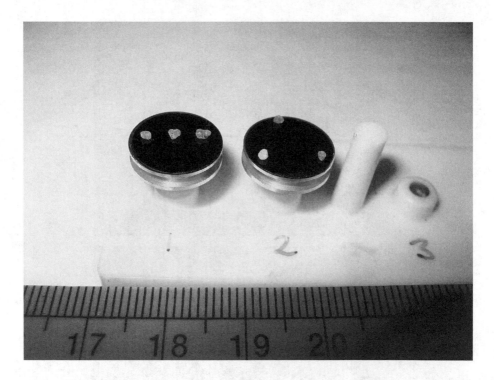

Figure 4.14 Quartz sand grains mounted on 10 mm diameter aluminum SEM stubs using doubled-sided adhesive carbon discs for the purposes of particle shape analysis using a stereomicroscope and Eclipse Net image processing software. If it is considered sufficient to photograph and/or measure the grains in two orthogonal orientations, the grains can be mounted in a linear array, as on the left-hand stub. If measurements/photographs in three orientations are preferred in order to provide average values for elongation, flatness, and sphericity, the grains must be mounted in a triangular array, as on the right-hand stub.

where L, I, and S are the longest, intermediate, and shortest dimensions measured on the maximum projected image of the particle. Values also range from 0 to 1.

4.3.3   Roundness

Wadell (1932, 1935) regarded sphericity and roundness as independent attributes, with roundness related only to the relative sharpness of edges and corners. According to this view, a sediment particle may have a high degree of sphericity but a low degree of roundness. For example, a complex polyhedron such as a tetra-kaikahedron has a high degree of sphericity (Wadell sphericity value of 0.97), but also has zero roundness because the faces meet at an angle (Ashenbrenner, 1956).

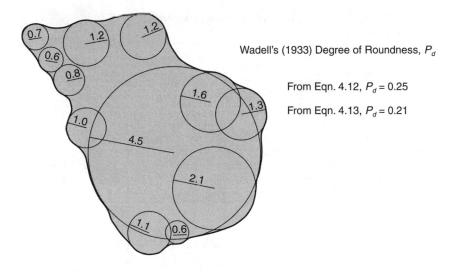

Wadell's (1933) Degree of Roundness, P_d

From Eqn. 4.12, $P_d = 0.25$

From Eqn. 4.13, $P_d = 0.21$

Figure 4.15 Visual representation of the method proposed by Wadell (1933) to quantify degree of roundness, P_d.

Wadell (1933) proposed a method of calculating the *degree of roundness*, P_d, based on measurements of the radii of the largest inscribed circles that would just fit into the grain corners seen in the projected image of a grain (Figure 4.15). Two slightly different formulae were proposed:

$$P_d = \frac{\sum r/R}{N} \tag{4.12}$$

$$P_d = \frac{N}{\sum R/r} \tag{4.13}$$

where
  R = the radius of the maximum inscribed circle
  r = the radius of curvature of individual corners
  N = the number of corners including those whose radii are zero.

The Wadell method of corner roundness determination is very time-consuming and is frequently beset by ambiguities relating to the number, location, and size of circles that should be fitted to the corners of the projected grain profile. Consequently, most workers have preferred to compare observed grain profiles with a visual comparator of the type published by Russell and Taylor (1937), Krumbein (1941), Powers (1953), and Lees (1964, 1965).

Table 4.2   Comparison of Alternative Particle Roundness Classification Schemes

	Russell and Taylor (1937)		Pettijohn (1949)		Powers (1953)		Blott and Pye (2007)	
	Class Limits	Arithmetic Midpoint	Class Limits	Geometric Midpoint	Class Limits	Geometric Mean	Class Limits	Geometric Mean
Very angular	—	—	—	—	0.12–0.17	0.14	—	—
Angular	0.00–0.15	0.075	0.00–0.15	0.125	0.17–0.25	0.21	0–0.13	0.09
Sub-angular	0.15–0.30	0.225	0.15–0.25	0.200	0.25–0.35	0.30	0.13–0.25	0.18
Sub-rounded	0.30–0.50	0.400	0.25–0.40	0.315	0.35–0.49	0.41	0.25–0.50	0.35
Rounded	0.50–0.70	0.600	0.40–0.60	0.500	0.49–0.70	0.59	0.50–1.00	0.71
Well rounded	0.70–1.00	0.850	0.60–1.00	0.800	0.70–1.00	0.84	—	—

Russell and Taylor (1937) recognized five grain roundness classes whose class limits and arithmetic midpoints, defined by Wadell roundness values, are shown in Table 4.2. Their class limits were not systematically selected, but their unequal size reflects the fact that the eye can more easily distinguish degrees of roundness when the roundness values are low (i.e., where relatively small, sharp corners are involved). Pettijohn (1949) modified the scheme by changing the class limits and defining the midpoints of each interval using a geometric scale (Table 4.2). However, even this scheme was considered by Powers (1953) not to provide small enough divisions at the lower roundness end of the scale. Powers provided a further modified scheme that divided the interval from 1.00 to 0.12 into six classes based on a 0.7 ratio scale. This provided an additional *very angular* category with a geometric mean value of 0.14 (Table 4.2). The lower scale limit of 0.12 was justified on the grounds that, "with the exception of crystals, particle roundness less than 0.12 is not differentiated" (Powers, 1953, pp. 118–119). Powers' visual comparator displayed three-dimensional photographic images of particles that were formed from modeling clay to represent six categories of roundness and two classes of sphericity (i.e., equancy; Figure 4.16). This system, in its original or modified

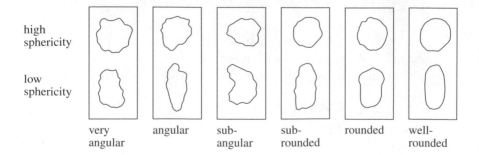

Figure 4.16 A simplified two-dimensional grain outline version of the visual Roundness Comparator developed by Powers (1953).

form, has been widely applied, especially to sand-grains, for more than half a century (e.g., Folk, 1955; Pettijohn et al., 1972, p. 586). However, workers concerned mainly with gravel-size fragments, or with highly angular particles such as those typically found in mechanical weathering debris and crushed aggregates, have preferred to use the visual comparator developed by Krumbein (1941), which has nine roundness classes corresponding to Wadell roundness values of 0.1 to 0.9, intermediate values being assigned to broken pebbles. However, in practice, a distinction between the nine classes, especially the more rounded ones (>0.6), can be difficult to make. Moreover, all of the images provided in the comparator are for near-equant particles (i.e., a similar category of sphericity), and differences in assessment of degree of roundness can arise due to differences in overall particle form (relative elongation or flatness).

Blott and Pye (2007) proposed a simplified classification of particle roundness that defines four classes based on a logarithmic scale to base 0.5 (Table 4.2, Figure 4.17, and Figure 4.18) ranging between end member forms of perfect angularity (= 0) and perfect roundness (= 1.00). This system is applicable to both sand and gravel particles, and has the advantage of simplicity, speed of application, and a relatively high degree of reproducibility. The logarithmic scale to base 0.5 was preferred to the logarithm to base 0.7 scale used by Powers, because in practice it is difficult to distinguish six categories with sufficiently high precision, and because the assigned interval boundaries of the 0.5 logarithmic scale equate better with visual perceptions of *rounded*, *sub-rounded*, *sub-angular*, and *angular* particles.

4.3.4   Angularity

Lees (1964, 1965) expressed the view that particle *angularity* is more than simply the inverse of corner roundness, as measured by Wadell (1932, 1935) and represented by the Krumbein (1941) and Powers (1953) visual comparators:

```
Clearly angularity is not just the absence of roundness
but is a distinct concept. Roundness refers only to
the character of the corner after it has been modified
(e.g. by attrition), but is not concerned by the
angular relationship between the planes bounding that
corner. The roundness concept is thus unable to deal
satisfactorily with crushed aggregate or with any
naturally brecciated material such as scree deposits
and volcanic pyroclasts. For example, when assessment
of crushed aggregates by reference to Krumbein's
roundness chart (Krumbein, 1941) is attempted, one
finds that the majority of particles would have to
be given a value of 0.1 and yet be clearly different
in angularity. (Lees, 1965, p. 6)
```

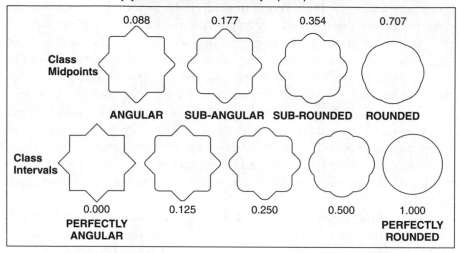

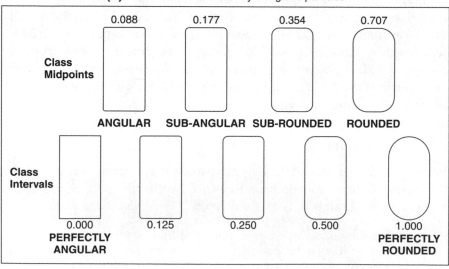

Figure 4.17 A visual grain outline comparator for the classification of particles in terms of roundness/angularity: (a) for relatively equant particles and (b) for relatively elongated particles. Class intervals and midpoints correspond to values calculated using the method of Wadell (1933).

(a)

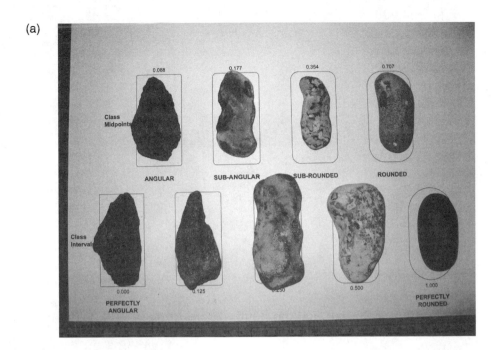

(b)

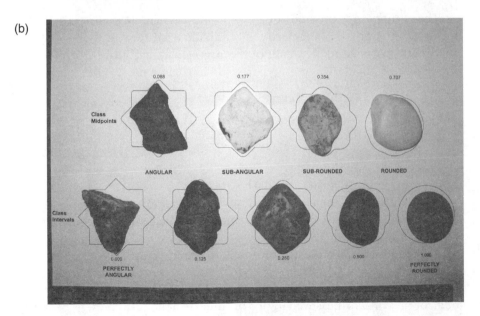

Figure 4.18 Actual gravel particles corresponding with each of the roundness/angularity class end-members and midpoints shown in Figure 4.17.

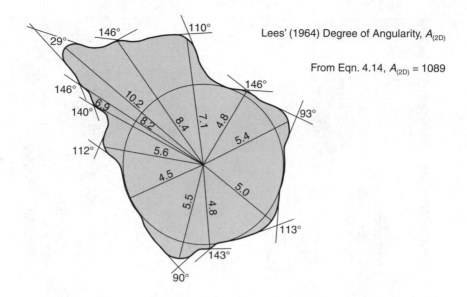

Figure 4.19 Visual representation of the method proposed by Lees (1964, 1965) to quantify degree of angularity, A_{2D}.

Lees proposed a method for quantifying particle angularity that takes into account both the angularity of individual corners and the distance between the corner and the center of the largest inscribed circle (Figure 4.19). The degree of angularity of each corner is calculated by:

$$A_i = (180° - a)\frac{x}{r} \qquad (4.14)$$

where
  A_i = the degree of corner angularity
  a = the measured angle between lines tangential to each corner
  x = the distance to the tip of the corner from the center of the maximum inscribed circle
  r = the radius of the maximum inscribed circle.

The total degree of grain angularity (A_D) is indicated by the sum of values for all of the corners measured in two or three orthogonal planes, respectively. Values range from 0 to >1500. The method of calculation is rather time-consuming, and it is more convenient to make reference to the visual angularity comparator that Lees presented (Lees, 1965).

In practice, a visual distinction between many of Lees' angularity classes is difficult to make (especially between the lowest eight, least angular categories), and the concept of particle angularity is heavily influenced by the overall particle form (i.e., the number and relative magnitude of grain projections).

Sukumaran and Ashmawy (2001) proposed two new parameters, which they termed a *shape factor, SF*, and an *angularity factor*, AF, based on analysis of grain outlines using a QUANTIMET image analyzer. The projected "true" outline of each particle was first approximated by drawing chords around the perimeter of the particle. Each chord linked two points on the perimeter at each of 40 equally spaced angles radiating from the centroid of the grain. The angle of each chord was then compared with that of a chord between equivalent points on an ideal circle, the difference between the two being referred to as a *distortion angle*. The shape factor was defined as the deviation of the global particle outline from a circle. Since the distortion angles may be either positive or negative, the sum of the values is an indication of relative deviation from a circular shape. The angularity factor was defined in terms of the number and sharpness of the corners (on the discretized inscribed polygon). Angularity of each corner was defined as the difference between 180° and the measured internal angle of the corner. The sum of the squares of the differences was then used in order to amplify the influence of sharper corners. After normalization, values of the normalized angularity factor range from 0 for a circle to >107 for very irregular particles. The method is objective and highly repeatable on the same grain outline, but the measure incorporates other aspects of shape in addition to "corner roundness" in the sense of Wadell (1932). For example, 2:1, 3:1, and 4:1 ellipses, which have no corners, have AF values of 4, 11, and 29, respectively. The number of radii selected to define the chords and angularity need not be fixed at 40, but varied according to the degree of irregularity of the particles under consideration.

4.3.5   Irregularity

Blott and Pye (2007) defined a further aspect of particle shape, irregularity which they distinguished from form, sphericity, roundness and the Lees concept of angularity. Irregularity is applicable both to particles which have sharp, angular corners and those whose corners and edges are rounded to varying degree. It is defined as the degree to which the external shell of a particle departs from a regular plane, which may be curved or composed of a series of flat faces. Many natural gravel- and sand-size particles can loosely be described as "knobbly," with the surface characterized by a number of projections (positive relief features, or "bumps") and indentations (negative relief features, concavities, or "hollows"). The projections and indentations may be rounded, angular, or a mixture of the two. For example, many chert pebbles, derived from the Chalk, and which form a significant component of Quaternary drift deposits found in southern Britain, have this type of shape (Figures 4.20a and 4.20b). Many coralline and other biogenic carbonate gravel fragments also have highly irregular shapes (Figure 4.20e). Other gravel- and sand-size particles, which frequently have irregular, "knobbly" shapes, include silicified organic matter (woody material, faecal pellets, teeth and bone fragments), glauconite, and other clay pellets of authigenic origin.

A measure of particle irregularity can be provided by quantifying the number and magnitude of surface projections and indentations visible in a two-dimensional image of a grain relative to the diameter of the largest inscribed circle (Figure 4.21). The irregularity index (I) is then calculated as follows:

$$I_{(2D)} = \sum \frac{y - x}{y} \tag{4.15}$$

where

   x = the distance from the center of the largest inscribed circle to the nearest point of any concavity
   y = the distance from the center of the largest inscribed circle to the convex hull, measured in the same direction as x.

The total degree of irregularity is indicated by the sum of values for all of the concavities measured in the plane of projection. Where the distance to the convex hull is difficult to measure or estimate, y can be calculated by measuring the distance to the projections adjoining any concavity:

$$y = \frac{a \cos A + b \cos B}{2} \tag{4.16}$$

where

   a and b = the distances from the center of the largest inscribed circle to the tip of the projections either side of the concavity
   A and B = the angles between a and x and b and x, respectively.

In some circumstances, it may be sufficient to make measurements in one plane of projection, but if measurements are made on the particle viewed in three orthogonal orientations (c.f. Kuo et al., 1996), a better measure of the three-dimensional irregularity, $I_{(3D)}$, can be obtained by summing the values obtained from the three individual projections (Blott and Pye, 2007). Values range from 0.0 to >20.0 for extremely irregular forms. Most natural sediment particles have values in the range 0.0 to 1.0.

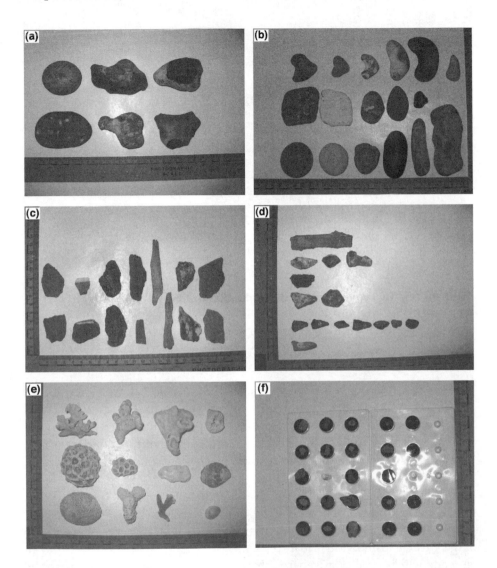

Figure 4.20 Examples of gravel particle assemblages with contrasting shapes: (a) a well-rounded, sub-rounded, and fractured quartzite/chert assemblage from Quaternary fluvio-glacial drift deposits, Berkshire, south-central England; (b) predominantly sub-rounded chert, chalk, and quartzite assemblage from Aldeburgh on the Suffolk coast; (c) an angular assemblage of predominantly crushed dolerite used as an ornamental garden dressing; (d) a mixed assemblage of angular fragments of crushed road stone, sub-angular and sub-rounded chert clasts, concrete, brick, and mortar fragments from a roadside location in Berkshire; (e) irregular, branching clasts of coralline gravel from a tropical beach in North Queensland; and (f) 21 sub-angular and sub-rounded particles of fluvial gravel, composed mainly of dolerite, sandstone, and burnt shale, recovered from the trachea and lungs of a murder victim (mounted on SEM stubs).

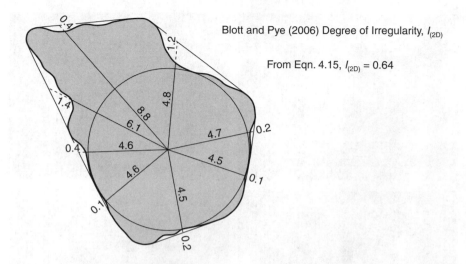

Figure 4.21 Visual representation of the method of quantifying degree of particle irregularity, I_{2D}, as proposed by Blott and Pye (2007).

4.4 Surface Texture

The term *surface texture* refers to the nature, density, and distribution of small-scale topographic features present on the surface of a particle and which are independent of size and overall shape. Krumbein and Pettijohn (1938, p. 304) recognized three categories of surface features on larger particles (>2 mm): degree of *smoothness*, degree of *polish* or *gloss*, and *surface markings*.

Degree of smoothness, or its inverse, roughness, may be regarded as a micro-topographic property that indicates the absence or low frequency of relatively large projections or indentations. In the case of pebbles, "relatively large" relates to individual features which are larger than about 0.1 mm in size and which "feel" relatively rough to the touch, and may be seen with the naked eye or using a hand lens (Figures 4.22a, 4.22b, and 4.22c). In the case of sand particles, the features that give rise to relative roughness or smoothness range in size up to 0.1 mm and can only be seen with the aid of an optical microscope or SEM.

Polish or gloss is an optical property that has to do with the regularity of light reflection, and, to some extent, adsorption and forward scattering (diffraction). A high degree of irregular light scattering gives rise to a visually dull surface, whereas a high degree of reflectance (backscattering) gives rise to a visually bright, glossy (or glassy) surface. A polished or glossy surface need not be smooth, although more commonly than not it is. The visual appearance of a particle is governed not only by reflectance, but also by

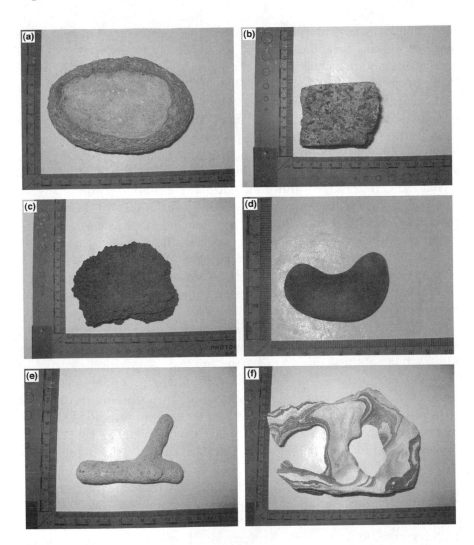

Figure 4.22 Examples of single gravel-sized clasts with distinctive morphological properties: (a) a piece of orange brick with adhering gray mortar on one side which has been rounded by abrasion in a marine beach environment; (b) a parallel-sided piece of a broken concrete paving slab; (c) a lump of volcanic tephra with a very rough surface texture; (d) a lunate, well-rounded chert clast from a high energy marine beach environment; (e) a piece of well-rounded branching coral from a moderate-energy sub-tropical beach; and (f) a piece of honeycomb-weathered banded sandstone from an arid desert environment.

adsorption and forward scattering (e.g., transparent, colorless particles may appear visually different to translucent or opaque particles with the same degree of surface smoothness or roughness). Partly for this reason, detailed examination of surface textural features is often better carried out using a combination of reflected light microscopy and SEM, since the two techniques yield complementary information. The SEM has greater depth of field, larger range of working magnification, and the capability to undertake chemical microanalysis of individual areas of the grain surface (Krinsley and Margolis, 1969; Krinsley and Doornkamp, 1973; Trewin, 1988; Mahaney, 2002; Goldstein et al., 2003; Figure 4.23 and Figure 4.24).

Surface features may be large or small relative to the size of the particle. Large features that may occupy a high proportion of the grain surface include *conchoidal fractures, multiple stepped surfaces, dish-shaped depressions, grooves, striations, large pits, ridges,* and *crystalline overgrowths* with or without planar crystal faces. Small-scale features include *small pits* and *grooves* of various types. Grain surfaces that are characterized by large numbers of closely spaced pits and intervening small projections are often described as having a *frosted* appearance, whereas surfaces that have a smaller number of larger irregularities are often referred to as *pitted* (Krumbein and Pettijohn, 1938).

Surface textural features may be intrinsic to a grain or represent extrinsic phenomena such as adhering particles and coatings. The surfaces of many particles display a combination of the two types, reflecting the nature of the particle source, transport history, and subsequent weathering history. Under certain conditions, secondary silica or aluminosilicate material may precipitate on parts or all of the grain surface, giving rise to a "waxy" appearance under the optical microscope (Folk, 1976). In many situations, useful information can be obtained by examining the particles before and after removal of surface particles and coatings. Procedures for grain pretreatment prior to SEM examination are discussed, for example, by Smart and Tovey (1981, 1982).

During the 1960s and 1970s, considerable research was undertaken on small-scale surface textural features, especially those on quartz sand grains. The initial work was undertaken on grain replicas viewed in a transmission electron microscope (TEM) (Krinsley and Takahashi, 1962; Porter, 1962), but the advantages of direct observation in the SEM were quickly recognized (Krinsley and Donahue, 1968; Krinsley and Margolis, 1969; Krinsley and Doornkamp, 1973). This work was founded on the hypothesis that surface textural features and assemblages record information about the origin, transport history, and subsequent weathering or diagenetic history of the grains. However, subsequent experience showed that many of the quartz surface textural features are not unique to a specific process or set of environmental processes or conditions, and the procedures used to identify and classify the features are subject to a relatively large degree of operator variation (Culver et al., 1983). However, if studies of intrinsic grain surface textural features are combined with investigation of other characteristics such as mineralogical inclusions, coatings, adhering particles, and surface contaminants, they can be of value in general geological (Elzenga et al., 1987; Mahaney et al., 2001; Mahaney, 2002) and forensic investigations (Pye, 2004b).

(a)

(b)

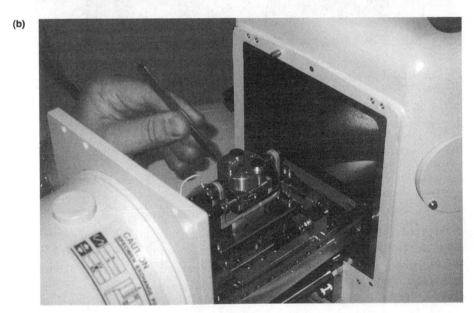

Figure 4.23 (a) Hitachi S-2600N variable pressure scanning electron microscope and PGT energy-dispersive x-ray microanalysis system; (b) sample
chamber and moveable stage on a Hitachi S-3500 variable pressure SEM.

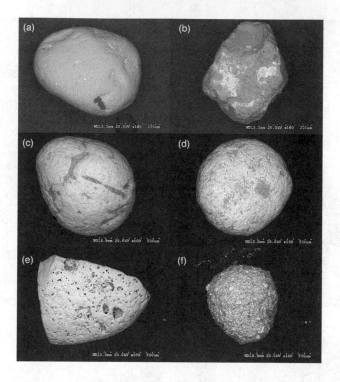

Figure 4.24 Examples of sand-size grains imaged using the backscattered electron mode in a VP-SEM. This mode of imaging allows both topographic and compositional information to be obtained together in the same image. (a) Well-rounded quartz grain with small dark patch of organic coating; (b) sub-angular quartz grain with conchoidal fracture surface and patchy coating of iron oxyhydroxide (pale gray/white); (c) well-rounded quartz grain with inclusions of albite feldspar; (d) well-rounded calcium carbonate ooid with small patches of adhering clay; (e) broken fragment of a larger, better-rounded fragment of industrial slag; and (f) sub-rounded clay pellet aggregate.

Experimental studies have shown that surface textural characteristics have a relatively small influence on the hydraulic behavior of sediment grains, but they are known to have a significant impact on engineering behavioral properties. Barksdale and Irani (1994) used a surface roughness scale to quantify the surface texture (roughness) of aggregate particles that employed visual examination and an index scale (SR) ranging from 0 for glassy particles to 1000 for very rough particles. An alternative image analysis method described by Janoo (1998) involves measuring the perimeter of a particle (in 2D) and comparing it to the convex perimeter to give *a roughness ratio, R*:

$$R = \frac{P}{P_c} \tag{4.17}$$

where
  P = the measured perimeter
  P_c = the convex perimeter.

For general descriptive purposes, sand grains can be classified on the basis of their surface texture viewed using a binocular microscope. A fourfold classification is generally sufficient: *very smooth* = *glassy* or *polished*, *smooth* = *waxy*, *rough* = *frosted*, and *very rough* = *pitted*. If a quantitative measure is required, the number of projections and indentations along a unit length of the projected particle outline can be determined by examination of digital projected grain images or photographs. Depending on size of particle and the scale of the "roughness" features of interest, the particles can be viewed either using a binocular microscope or a SEM. In the case of gravel particles, a similar fourfold classification of *very rough*, *rough*, *smooth*, and *very smooth* can be made on the basis of a simple "touch" test.

The degree of surface roughness will reflect both the nature of the material and its degree of transport or weathering exposure. In certain circumstances, high levels of surface roughness may also result from the growth of authigenic crystals on the surface of a particle.

More detailed examination of individual surface textural features normally requires SEM examination. Many different features have been identified on quartz sand grains by different authors (Table 4.3), but some are ambiguous and difficult to distinguish. The presence or absence of particular features can be recorded, either in terms of presence/absence or relative abundance (e.g., *absent*, *rare* (<5% of grains), *sparse* (5 to 25%), *common* (25 to 75%), or *abundant* (>75%). Results can then be compared in terms of frequency histograms or comparison matrices (Table 4.4). Statistical analysis of the results is only worthwhile if sufficient grains and samples have been examined and enough counts made of the features of interest (Pye, 2004b). Unfortunately, in many previous SEM surface textural studies, the number of grains examined has been too small to provide statistically meaningful results. Many studies have examined less than 50 grains, and some as few as 10 grains, e.g., Bull and Morgan 2006. In the author's view, at least 300 sand grains, and preferably many more, should be examined to provide a meaningful basis for statistical comparison. If this criterion is met, a useful first step is to calculate the mean, standard deviation, and range of frequencies for each surface characteristic in each sample and/or group of samples under consideration. The data can then be compared using a combination of univariate and multivariate statistical techniques, such as principal components analysis (Elzenga et al., 1987) or canonical analysis (Culver et al., 1983).

Semiquantitative grain surface textural analysis is time-consuming and expensive to undertake, and many investigators have concluded that it provides a poor effort to return ratio. An alternative is to adopt a simplified approach

Table 4.3 Quartz Grain Surface Textural Features Recognized by Four Different Authors

Krinsley and Doornkamp (1973)	Higgs (1979)	Culver et al. (1983)	Mahaney (2002)
Precipitated upturned silica plates	Small irregular pits	Large conchoidal fractures	Low relief
Silica plastering	Medium irregular pits	Small conchoidal fractures	Medium relief
Smooth precipitation surface	Large irregular pits	Large breakage blocks	High relief
Capping layer	Small conchoidal fracture	Small breakage blocks	Chattermarks
Quartz crystal growth	Medium conchoidal fracture	Arc-shaped steps	High frequency fractures
Chemically etched V-forms	Large conchoidal fracture	Random scratches and grooves	Low frequency fractures
Conchoidal fracture	Straight steps	Orientated scratches and grooves	Crescentic gouges
Mechanical V-forms	Arcuate steps	Parallel steps	Arc-shaped steps
Straight or slightly curved grooves	Fracture plates/planes	Non-orientated V-shaped pits	Linear steps
Dish-shaped concavities	Parallel striations	Meandering ridges	Deep troughs
Silica coating reproducing underlying structure	Imbricated grinding features	Dish-shaped concavities	Mechanically upturned plates
Irregular solution–precipitation surface	Adhering particles	Upturned plates	Lattice shattering
Dulled solution surface	Meandering ridges	Micro-blocks (chemical or mechanical)	Abrasion features
Deep surface solution	Straight scratches	Roundness–rounded	Dissolution etching
Disintegration by solution or by salt crystal growth	Curved scratches	Roundness–sub-rounded	Preweathering surfaces
Large-scale chemical decomposition	V's	Roundness–sub-angular	Weathered surfaces
Mechanically formed upturned plates	Angular outline	Roundness–angular	Precipitation features
Flat cleavage face	Rounded outline	Facet	Adhering particles
Cleavage-planes (semiparallel lines)	Low relief	Cleavage flake	V-shaped percussion cracks
Rounded grains	Medium relief	Precipitation platelet	Edge rounding

Adhering particles

High relief
Orientated etch pits

Anatomizing etch pattern
Solution pits

Solution crevasses
Scaling
Silica globules
Silica flowers
Silica pellicle
Crystalline overgrowth

Carapace
Chemically formed V-shaped, orientated pits
Cleavage plane
Silica precipitation–amorphous/cryptocrystalline
Silica precipitation–euhedral
Solution pits and hollows
Dulled surface from solution of silica
Chattermarks (chemical or mechanical)
Star cracking
Low relief
Medium relief
High relief

Overprinted grains
Clay coatings

Fresh surfaces
Craters

Sawtooth fractures
Rolled grains
Slickenside surfaces
Cracked grains
Abrasion fatigue
Parallel ridges
Elongate depressions
Smoothed-over depressions
Bulbous edges
Shock lamellae
Shock-melted grains
Fracture faces
Radial fractures
Subparallel linear fractures
Conchoidal fractures
Curved grooves
Straight grooves
Sharp angular features

Table 4.4 Example of a Simple Matrix Comparing the Relative Abundance of Quartz Grain Types in Three Questioned Samples and Five Control Samples from a Scene of Interest[a]

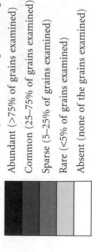

	Questioned Samples			Control Samples from Scene				
Quartz Grain Morphological Type	1	2	3	4	5	6	7	8
Sub-angular with chemically modified conchoidal fractures								
Well-rounded with randomly orientated abrasion pits								
Highly angular with conchoidal fractures								
Sub-rounded with clay and iron oxide coatings								
Euhedral crystals with chipped/sub-rounded edges								
Polycrystalline with blocky texture								
Sub-angular and sub-rounded with partial calcite coating								

[a] Only questioned sample 3 shows a significant degree of similarity to the control samples (most notably sample 8).

- Abundant (>75% of grains examined)
- Common (25–75% of grains examined)
- Sparse (5–25% of grains examined)
- Rare (<5% of grains examined)
- Absent (none of the grains examined)

where the presence or absence of a limited number of grain attributes is recorded, and the different combinations of features present are used to define a number of grain "types" which can be designated simply as Type I, Type II, Type III, etc, e.g., Bull and Morgan (2006). The features of interest and grain types are likely to vary from situation to situation, and often only be specified after reconnaissance examination of questioned and/or control samples. Such examination need not be restricted to quartz grains, but can include a wide range of particle types, including those of anthropogenic origin. The grain types can be specified on the basis both of surface textural features and other criteria such as inclusions or coatings. The latter are more easily identified if the grains are examined in the SEM using the backscattered electron (BSE) mode (Figure 4.24). Results can be compared visually or using nonparametric statistics. However, if poor precision is to be avoided, great care must be taken to ensure that the features used to identify the different particle types are unambiguous, and that sufficient grains are examined and classified (Pye, 2004b, 2007).

4.5 Characterization of Particle Morphology Using Fourier Analysis, Fractal Analysis, and Fourier Descriptors

Since the early 1970s, several methods have been developed that allow objective description of particle morphology at a variety of scales, including overall form, roundness, and surface texture. One of the earliest was Fourier series decomposition, often referred to simply as *Fourier shape analysis* (Schwarz and Shane, 1969; Ehrlich and Weinberg, 1970; Ehrlich et al., 1980). The procedure involves digitizing the projected outlines of grains in two dimensions and decomposing the digitized grain perimeter into a number of fundamental harmonic wavelengths (Figure 4.25). The radial distance between the particle centroid and a series of defined points on the grain boundary (defined by a polar coordinate system) is determined and expressed as a function of the polar angle (Full and Ehrlich, 1982). The lower order harmonics reflect the broad-scale form of a grain, whereas the higher order harmonics provide information about roundness and surface texture (Ehrlich et al., 1980). Data are obtained from a sample of several hundred grains taken from a standard size fraction and the percentage frequencies of occurrence plotted for each harmonic amplitude. A maximum entropy method is then used to define interval boundaries in the resulting harmonic frequency histograms, and to identify the most informative harmonics (Full et al., 1984).

The technique suffers from being only two-dimensional, and results may be influenced by the preferred orientation of nonequant grains on the imaging plate (Tillman, 1973). There are also methodological difficulties with the technique, notably the identification of the particle centroid and difficulties

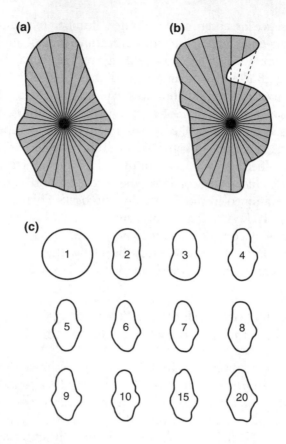

Figure 4.25 Visual representation of the principle of grain shape character-ization by Fourier analysis: (a) a regular particle with selected radii between the particle centroid and perimeter; (b) an irregular particle with large re-entrant which poses difficulties for conventional Fourier analysis; (c) selected harmonics which combined describe the shape of the grain shown in (a).

in specifying a unique set of coordinates for the grain perimeter in the case of particle shapes with re-entrants (Figure 4.25b). However, it has been used with reported success in a number of sediment provenance and trans-port studies (Dowdeswell, 1982; Mazzullo et al., 1983, 1988; Dowdeswell et al., 1985; Mazzullo and Magenheimer, 1987; Haines and Mazzullo, 1988; Mazzullo and Ritter, 1991).

Several authors have considered that fractal analysis is a better technique for irregular particles with re-entrants and features such as pore spaces (Orford and Whalley, 1983, 1987, 1991; Moore and Donaldson, 1995; Vallejo, 1995). However, fractal analysis can fail to distinguish between particles with greatly different shapes; for example, the fractal dimensions of a smooth circle and a square rhombus are almost identical (Sukumaran and Ashmawy, 2001).

Moreover, user-friendly software programs are not readily available. Consequently, the technique has not been extensively used in sedimentological research or in forensic soil investigations.

Thomas et al. (1995) and Bowman et al. (2001) recommended using *Fourier descriptors* as another alternative way of characterizing grain shape. The Fourier descriptor method (Wallace and Wintz, 1980) defines various points on the grain boundary in terms of their positions relative to each other, rather than in terms of distance and polar angle from the grain centroid. Definition of the centroid is unnecessary, and problems with grain re-entrants do not arise. The data are processed using Fast Fourier Transform (FFT) methods and the resulting average logarithm of modulus is plotted against harmonic number. Samples can be classified into groups using discriminant analysis (Thomas et al., 1995).

Bowman et al. (2001) used three lower order Fourier descriptors, denoted "Signature Descriptors," to provide measures of *elongation, triangularity,* and *squareness,* an additional descriptor, denoted *asymmetry,* to provide a measure of particle irregularity, and a series of higher order descriptors to provide information about surface texture. However, despite the fact that the potential of this method has been recognized for 25 years (Clark, 1981), there has been only limited uptake in sedimentological studies, and none reported specifically in the context of forensic geology. In large part this is due to the unavailability of commercial, easy to use software packages.

4.6 Three-Dimensional Particle Shape Analysis Using X-Ray Tomography and Laser Profilometry

X-ray computed tomography (X-ray CT), sometimes alternatively known as X-ray micro-tomography, provides a nondestructive method of visualizing the distribution, size, and shape of solid objects within a three-dimensional matrix, for example, a lump of rock or concrete, a wax block, or part of the human body (Garboczi, 2002; Mees et al., 2003). Digital information can be obtained about the broad aspects of the geometry and surface topology, and digital volume and surface area calculated (Figure 4.26). Once a three-dimensional digital profile of a particle has been obtained, spherical harmonic analysis (a 3D equivalent of 2D Fourier analysis) can be undertaken to characterize various aspects of the particle shape. The smallest particle that can be routinely analyzed is c. 250 μm (Garboczi, 2002), although smaller particles may be identified and located using high resolution (HR-) and ultra-high resolution (UHRXCT) systems. The technique has been applied to a variety of problems in the earth sciences, for example, studies of mineral distribution and form variation within rocks (including diamond source rocks and meteorites), shape analysis of calcareous microfossils, mammalian teeth and bones,

Figure 4.26 A 3D x-ray computer tomography image showing three solid particles within a light element organic matrix (block of wax-impregnated lung-tissue from a drowning victim).

and pore size distribution in relation to the hydraulic properties of soils and rocks (Carlson et al., 2003; Rogasik et al., 2003). Variants of the technique, such as synchrotron computed micro-tomography (SCMT), also allow elemental analysis to be undertaken on particles of interest which are contained within a solid body (Jones et al., 2003). A well-publicized forensic palaeontology application of HRXCT was the demonstration that *Archaeoraptor*, which had been suggested by *National Geographic* magazine to be the "missing link" between birds and extinct dinosaurs, was a fake which had been assembled from up to five different specimens and species (Rowe et al., 2001).

An alternative method of characterizing 3D particle shapes that has been reported to give satisfactory results is optical tomography. An algebraic reconstruction technique (ART) is used to synthesize 3D particle morphologies from 2D shape descriptors obtained from optical tomography projections. Comparisons with data obtained by XRCT indicated good agreement for sand-size particles (Giodarno et al., 2006).

Scanning optical (laser) instruments have also been widely used to provide information about surface morphology, including roughness and the distribution of various marks on bones of archaeological or palaeontological interest, or machined surfaces in engineering applications. These techniques have been collectively referred to as 3D microprofilometry (Kaiser and Katterwe, 2001). Line scans or area scans can be performed using a laser stylus profilometer to provide a 3D surface model. However, their potential for describing the surface texture of rocks and sediment particles has not yet been fully explored.

4.7 Color

The color of individual particles can be described in qualitative terms (e.g., white, black, brown, and red) or more precisely in numerical or graphical terms using a computer-controlled spectrophotometer system of the type described in Chapter 3 to quantify the color of powdered samples of soil and sediments (Croft and Pye, 2004a). It is possible to analyze individual sediment and rock particles as small as 5 mm in diameter using this type of system, and smaller particles (0.1 mm or less) can be analyzed using specialized micro-spetrophotometry systems.

Figure 4.27 shows a number of differently colored beach gravel particles for which individual reflectance curves and L*, a*, and b* parameter plots are compared in Figure 4.28. The data were obtained from a circular area with a diameter of 3 mm. Particles that have a relatively coarse particle/crystal size, that is, have a "speckled" appearance in hand specimen, show considerable spatial variability if analyzed in this way, but measurements on finer grained, relatively homogeneous materials show good reproducibility. In the case of larger particles that show color banding or speckling, color determinations can be made separately on each band or zone.

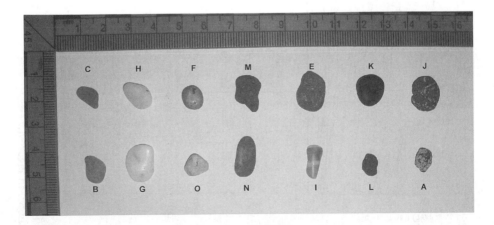

Figure 4.27 (See color insert following page 46.) An assemblage of rounded fine gravel particles from a marine beach in eastern Fuerteventura, Canary Islands.

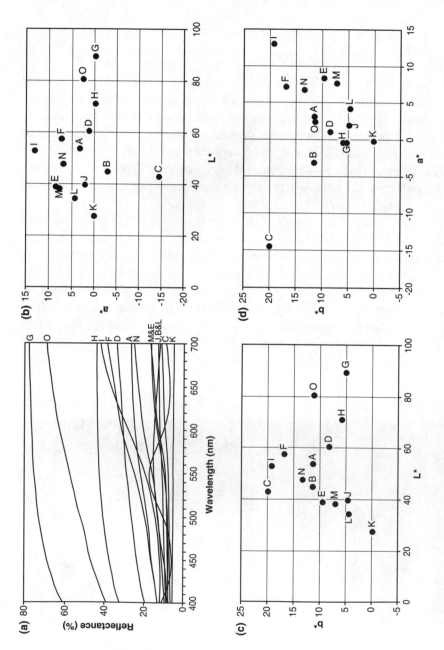

Figure 4.28 Comparison of reflectance curves and L*, a*, and b* color parameters for each of the gravel particles shown in Figure 4.27.

4.8 Luminescence Properties

Cathodoluminescence (CL) has been used as an imaging tool in sedimentary geology for almost half a century (Long and Agrell, 1965; Sippel and Glover, 1965), particularly in relation to studies of carbonate rocks and sediment grains, but also applied to quartz, feldspars, heavy minerals (Sippel, 1968; Tovey and Krinsley, 1980; Miller, 1988b; Barker and Kopp, 1991; Milliken, 1994), and refractory ceramics (Karakus, 2005). CL emissions can be displayed as color images in a dedicated CL microscope or as gray level images in an SEM equipped with a suitable detector (Figure 4.29). "Color" CL images can also be obtained in the SEM using red, green, and blue filters (Marokowitz and Milliken, 2003; Reed, 2005), or by pseudo-coloration coding of a grayscale image.

CL emissions are strongly controlled by trace elements and structural defects present within the crystal lattice and can provide useful information about patterns of crystal growth, diagenesis, fracturing, and rehealing

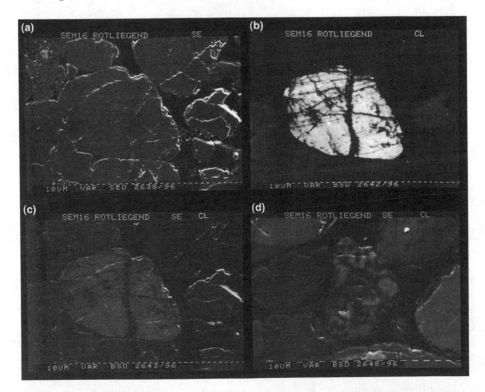

Figure 4.29 SEM images of part of a polished section of quartz sandstone obtained using (a) the SE detector, (b) the CL detector, and (c) and (d) combined signals from the SE and CL detectors. (a), (b), and (c) show the same quartz grain which contains healed fractures. (Philips 501 BSEM with KE Developments CL detector.)

(Marshall, 1988; Boggs et al., 2001; Bernet and Basset, 2005; Boggs and Krinsley, 2006). Dudley (1976c) carried out a pilot study to assess the potential of CL to assist mineral identification in forensic studies and concluded that it could usefully complement other microscopic techniques, but to date the technique has received little application.

Although thermoluminescence (TL) and optically stimulated luminescence (OSL) have been used mainly as a means of age dating (Aitken, 1985; Wintle, 1996, 1997; Wintle and Murray, 1997; Stokes, 1999; Fattahi and Stokes, 2003), at a simpler level, photon emission spectra can be used to compare individual mineral grains or samples of rock, pottery, or glass. Most work has been undertaken on quartz or feldspar grains, or rocks and pottery containing these minerals, volcanic glass and ceramic materials (Rendell et al., 1985; Huntley et al., 1988; Duller, 1997). However, despite the potential, TL and OSL have not been applied to any significant extent in forensic geological investigations.

4.9 Composite Characterization of Particles and Objects

4.9.1 Gravel-Size Particles, Rocks, and Other Large Objects

Detailed characterization of any particle usually requires a joint approach that combines quantitative measures and verbal description based on visual observation. As an initial step in any forensic geology investigation, each particle or object of interest should be examined with the aid of a magnifying lens, high intensity light source, and low power binocular microscope, in order to note any potentially distinguishing features, prior to photographing the object and making any dimensional measurements. Many features of interest are not amenable to numerical characterization. For example, notable features of the particle shown in Figure 4.22a are that it is well rounded, generally regular, has an overall orange color. The shape, color, and texture suggest that it is a brick that has been rounded by abrasion, probably in an aqueous environment such as a marine beach or river. Identification as a brick is supported by the presence of an elongate depression, representing the remains of a "frog," on the side opposite the camera, and the presence of gray mortar on one surface suggests it once formed part of a wall.

Each of the other particles shown in Figure 4.22 also has distinctive features of shape, surface texture, internal texture, and color which are more readily represented by verbal description than quantitative measurement, and which play an important part in the interpretation of the nature and probable origin of each particle.

In order to provide a complete record of the external features present, any object or particle of forensic interest should be photographed in six orientations which equate to the four sides and two ends (this applies both to straight-sided objects such as an unworn brick or a rounded natural cobble; e.g., Figure 4.4). Detailed notes and high magnification photographs should be taken of any features considered to be of potential diagnostic significance, such as inscriptions, surface coatings, and paint spots. If size permits, detailed examination of the surface texture and adhering particles, including those of organic origin such as diatoms and pollen, should then be undertaken in an SEM equipped with a large chamber and EDXRA facilities. Since it is undesirable to coat the object with carbon, an environmental SEM (ESEM) or variable pressure SEM (VP-SEM) should be used for this purpose (Figure 4.23).

Figure 4.20 shows a number of contrasting gravel particle assemblages recovered from different locations of forensic interest. Differences in shape, texture, and overall particle character are obvious. Figure 4.20f consists of 21 gravel and coarse sand-size particles mounted on SEM stubs, which were recovered from the trachea and lungs of a murder victim who had been shot, apparently after having his face immersed in a shallow stream or other water body containing sand and gravel. The body was then transported, together with that of another victim, to a different location where both were set on fire. However, the gravel particles survived. Detailed analysis of the particle shapes, surface textures, compositions and the assemblages of adhering diatoms suggested that the source may have been the margins of a shallow, relatively fast-flowing stream. Comparison with control samples taken from streams, paths, car parking areas and other locations identified by Police intelligence led to the identification of a possible source location.

4.9.2 Sand-Size Particles

Small granular samples encountered in forensic investigations frequently contain a mixture of sand and silt-size particles, some of which represent common minerals (quartz, feldspar, mica, and heavy minerals) and others that are composite (lithic fragments), amorphous, and organic or of anthropogenic origin. The precise origin of many of the latter types of particle can be difficult to identify unless suitable reference samples are available. Consequently, for the purposes of initial inter-sample comparison, it is often convenient simply to identify a number of particle "types" which have certain distinguishing features (in terms of size, shape, texture, color, and patterning), whether or not their exact identity is known at that stage of investigation. Samples can then be compared using a simple presence/absence matrix to determine whether two samples are broadly similar in terms of particle "types" in common (e.g., Table 4.5). This form of analysis is referred to by the author as *particle typology analysis*. If samples or particular particle types appear to display potentially significant features in common, more detailed analysis can then be performed.

Table 4.5　Example of a Simple Particle Type Presence/Absence Matrix

	Soil Samples from Crime Scene				Samples from Suspect Vehicle					
	KP2	KP3	KP4	KP5	MT510	MT511A	KP50	KP51	ELW2	ELW6
Clear quartz										
Translucent quartz										
Brown quartz										
Rose quartz										
Iron-stained quartz										
Chert										
Opaque heavy minerals										
Angular limestone										
Ooids/well-rounded limestone										
Cement/concrete										
Brick										
Black spherules										
Ironstone										
Rootlets/stem debris										
Other organic material										
Fibres										
Plastic										
Slag (black, gray, or brown)										

Legend: ■ abundant　▨ present　□ absent

There are many examples of the use of particle typology analysis in the forensic geological literature. For example, Lombardi (1999) identified the presence of 34 mineral and rock types in beach sand found in the trouser cuffs of the murdered Italian Prime Minister, Aldo Moro, and in the car in which his body was found in the center of Rome. In addition, the samples contained several types of planktonic and benthic microorganisms, identifiable littoral plant remains, and particles of bitumen that originated from a marine oil spill. Comparison with control samples from different parts of the Tyrrhenian Sea coast allowed a relatively small part of the shore to be identified as the probable source.

4.10 Elemental Composition

4.10.1 Alternative Analytical Techniques

The elemental composition of individual particles can be determined by a variety of methods, choice of which is controlled by particle size, type, and the need or otherwise to undertake non-destructive testing. In the case of homogeneous samples that are relatively large, it may be acceptable to remove a sub-sample for destructive chemical testing using a technique such as x-ray fluorescence (XRF) or solution ICP-AES/ICP-MS. An alternative, partially destructive method is to cut the rock fragment, soil aggregate, or sediment grains of interest and to prepare a polished section which can be used to provide information about internal texture/fabric and elemental composition using a technique such as wavelength-dispersive (WD) or energy-dispersive (ED) electron microprobe analysis (EMPA), energy-dispersive x-ray microanalysis in the SEM (SEM-EDXRA), or laser ablation inductively coupled plasma mass spectrometry (LA-ICP-MS). If specific information is required about trace elements or the composition of thin surface coatings, other techniques may be useful, including ion microprobe mass analysis (IMMA) (Hinton, 1995), secondary ion mass spectrometry (SIMS) (Bisdom, 1981; Bisdom et al., 1983), and proton-induced x-ray emission (PIXE) (Fraser, 1995; Kuisma-Kursula, 2000).

4.10.2 Electron Microprobe Analysis

Quantitative electron microprobe analysis is probably the most useful technique for routine application because it offers the capability to analyze small areas (<5 μm) within individual minerals with high accuracy and precision. Areas of interest can be identified relatively easily using a combination of optical microscopy and BSE imagery, and for most material types the technique is minimally destructive (i.e., is easily repeatable).

Figure 4.30 CAMECA S100 microprobe equipped with wave-length dispersive x-ray spectrometers.

The electron microprobe was first developed in the 1950s and became a routine technique in the early 1960s, since when there have been many technological refinements (Long, 1977; Reed, 2005). Applications have included elemental "fingerprinting" and correlation of volcanic glass shards (Westgate and Gorton, 1981; Hunt and Hill, 2001), detrital garnets in sediments and sedimentary rocks (Morton, 1985), artificial glass fragments, and paint flakes and soil mineral grains for forensic purposes (Smale, 1973). Modern microprobe systems, such as the CAMECA S100 system (Figure 4.30), can often undertake both WDS and EDS x-ray analysis, and have both secondary electron (SE) and backscattered electron (BSE) imaging capability. Figure 4.31, Figure 4.32, and Figure 4.33 provide an example of data obtained in a forensic investigation that sought to compare calcium carbonate ooids recovered from a suspect vehicle and an "alibi" location suggested by the defense as a potential source.

4.10.3   SEM-EDXRA Analysis

Computer-controlled SEM-EDXRA systems have been used for approximately 25 years to obtain quantitative and semiquantitative elemental composition data and to undertake automated "modal" mineralogical assessments of rocks and sediments (Minnis, 1984; Nicholls and Stout, 1986; McVicar and Graves, 1997; Kennedy et al., 2002; Goldstein et al., 2003; Pirrie et al., 2004; Pownceby, 2005). The same systems can be used to analyze and map the spatial distribution of elements within individual particles. In many situations, however, more useful information can be obtained if manual observation and analysis is undertaken. The human eye is able to identify subtle variations in texture and gray level that may correspond to significant features such as surface coatings, internal zonations, and inclusions that may be of diagnostic or discriminatory importance. Frequently, qualitative or semiquantitative SEM-EDXRA is sufficient for this purpose. Investigations can be carried out either on sectioned grains or on three-dimensional grains or aggregates (de Boer and Crosby, 1995; Evans and Tokar, 2000; Pye and Croft, 2007). Reference library examples of EDX spectra obtained from the more commonly encountered minerals are provided in Welton (1984) and Reed (2005). Representative elemental spectra can be obtained from single particles of 5 μm size or less, allowing characterization in circumstances where LA-ICP-MS may be difficult or impossible (Figure 4.34).

Since SEM-EDXRA is essentially non-destructive, it can be combined with other forms of analysis, such as Raman spectroscopy (Boughriet et al., 2004), and the evidence preserved for further examination.

4.10.4 Laser Ablation ICP-MS Analysis

An overview of various LA-ICP instrumentation, analysis techniques, and their applications is provided by Vogt and Latkoczy (2005; Figure 4.35). Applications have included comparison of volcanic glasses (Westgate et al., 1995; Becker et al., 2000), studies of trace metal variability within ooids (Freile et al., 2001), corals, other carbonates (Perkins et al., 1991; Pearce et al., 1992; Sinclair et al., 1998), and detrital garnets (Fedorowich et al., 1995), investigations of CL banding within crystals (Cox, 2004), and comparisons of crime scene evidence (Watling et al., 1997).

The sampling volume in LA-ICP-MS is generally one to two orders of magnitude larger than that associated with EMPA, which may be either an advantage or a disadvantage depending on the heterogeneity of the material and the degree of spatial averaging desired. LA-ICP-MS does not require polished samples, although markedly irregular surfaces should be avoided. LA-ICP-MS is partially destructive, resulting in the formation of relatively large ablation pits, lines or "raster" areas on the surface of the specimen, depending on the analysis mode chosen (Figure 4.36). However, the method is less destructive than solution ICP-AES and/or ICP-MS, which has been used extensively in geological analysis (Wray, 2005) and other investigations, including forensic examination of synthetic glass fragments (Montero et al., 2003). Comparisons of glass analysis data have concluded that the laser ablation method is simpler, faster and produces comparable discrimination to external calibration and isotope dilution solution ICP-MS (Trejos et al., 2003).

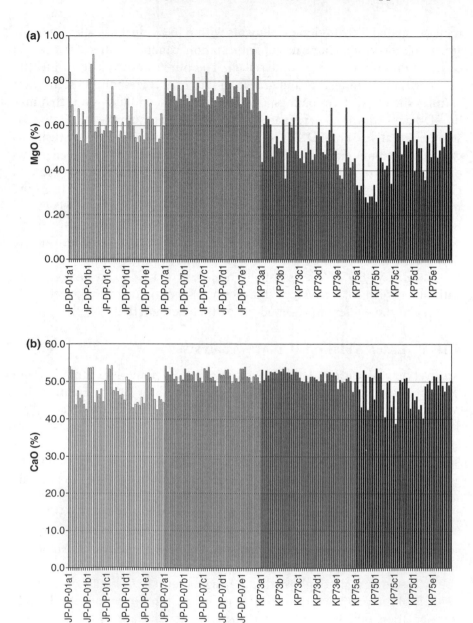

Figure 4.31 Example of microprobe data obtained from carbonate ooids in four samples of forensic interest, two (KP73 and KP75) from a vehicle allegedly involved in a murder plot, and two (JP-DP01 and JP-DP07) from an innocent location suggested by the defense to be the source of ooids on the vehicle. Data are shown for nine spot analyses on each of five ooids in each sample. The variation of CaO content in each group of ooids is broadly similar, but significant differences between samples, and between some individual ooids, are evident in terms of the MgO, FeO, and MnO contents.

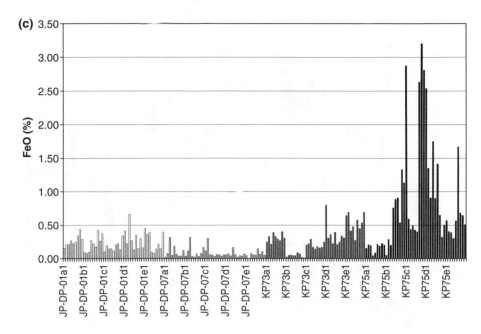

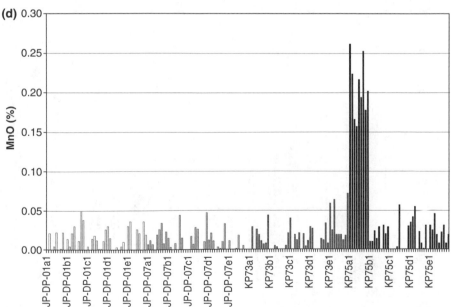

Figure 4.31 (Continued)

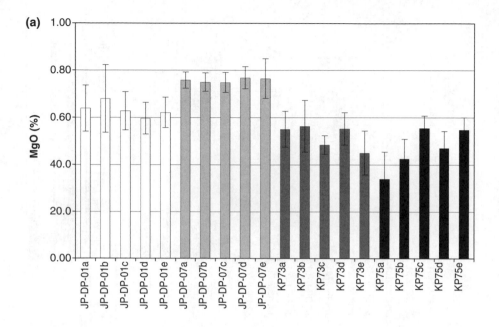

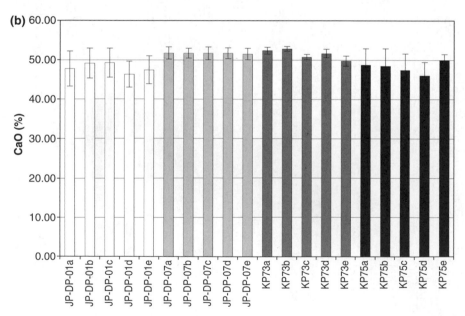

Figure 4.32 Mean concentrations and one standard deviation error bars for each of the ooids shown in Figure 4.31. Differences between the vehicle samples and the comparison samples in terms of MgO and FeO content are more clearly seen.

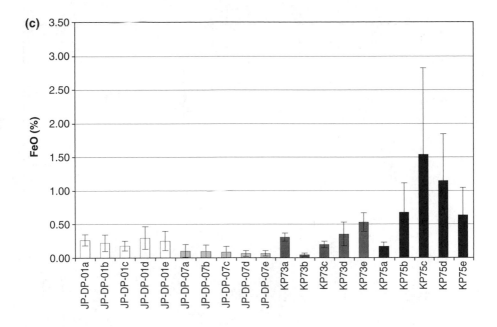

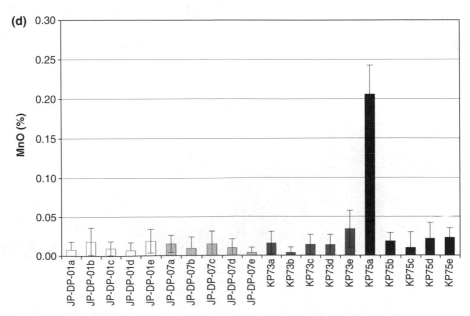

Figure 4.32 (Continued)

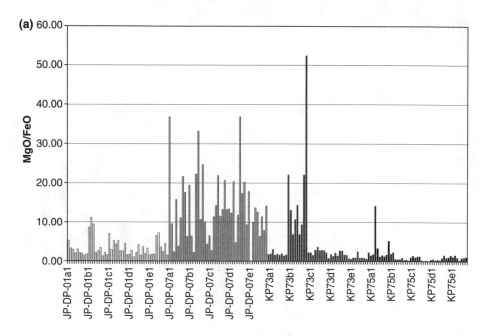

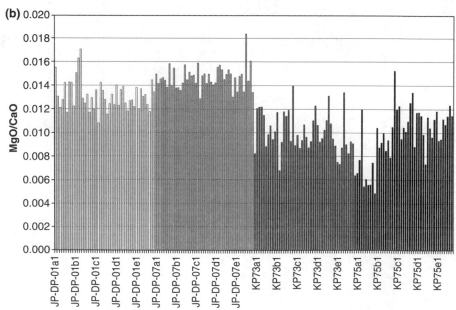

Figure 4.33 Comparison plots of the MgO/FeO and MgO/CaO ratios for the individual ooid analyses shown in Figure 4.31. The two comparison samples can be distinguished from each other on the basis of the MgO/FeO ratio and both comparison samples can be distinguished from the vehicle samples on the basis of the MgO/CaO ratio.

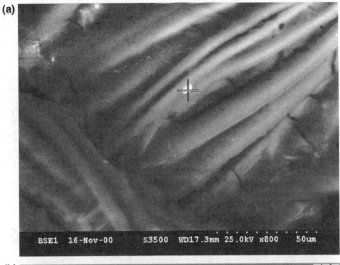

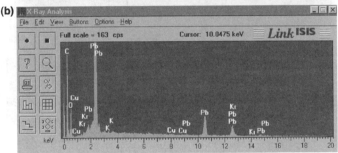

Figure 4.34 (a) Backscattered VP-SEM image showing a small lead particle (white) on the surface of a blood-stained shirt collar that had been buried in the ground for several months prior to discovery; (b) qualitative EDXRA spectrum obtained from the particle shown in (a). Other similar lead particles present on the shirt showed traces of copper. Examination of control samples from the pit in which the shirt was buried failed to identify any similar Pb and Pb-Cu particles. The composition and form of the Pb/Pb/Cu particles, combined with the identified presence of propellant particles, provided support for the hypothesis that the wearer of the shirt had been hit in the face/head by a copper-jacketed lead bullet. No evidence was obtained to indicate that the local soil/sediment contained such particles which had contaminated the shirt.

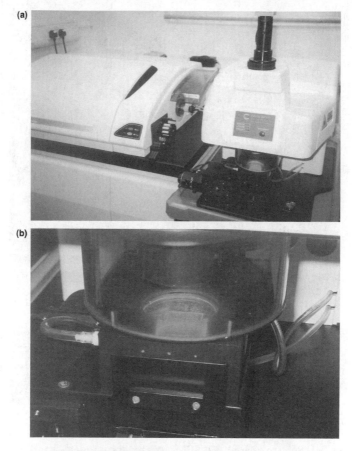

Figure 4.35 (See color insert following page 46.) (a) New Wave laser ablation ICP interface used to obtain *in situ* elemental data from solid samples; (b) closer view of the ablation chamber shown in (a).

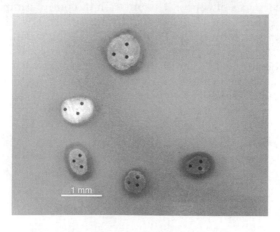

Figure 4.36 BSE SEM image of group of a calcium carbonate ooids taken after laser ablation analysis.

4.11   Isotopic Composition

4.11.1   Stable Isotope Ratios

It may be possible to obtain both light and heavy isotope ratio data from quite small single particles or rock fragments if the elements of interest are present in sufficiently high concentrations. Particles can be analyzed either by conventional solution or thermal analysis techniques, which are inevitably destructive, or by laser-ablation methods, which are partially destructive. Multi-collector magnetic sector field LA-ICP-MS instruments have been commonly used where sample size is small and there is a need to avoid total destruction of the sample. The main isotopes that have been analyzed in this way are those of oxygen, carbon, strontium, and lead (Vennemann et al., 1992; Christensen et al., 1995; Aleon et al., 2002; Ghazi and Millette, 2004).

4.11.2   Isotopic Dating

Laser and probe-type methods can also be used to obtain age estimates for individual sediment grains and minerals extracted from rock samples. Several methods and minerals have been used, including ^{40}Ar/^{39}Ar laser-probe dating of detrital white micas, U-Th-Pb dating of zircons (Froude et al., 1983; Goldstein et al., 1997; Pell et al., 1997, 2000; Machado and Simonetti, 2001), ion-probe Hf-isotope dating of zircons (Knudsen et al., 2001), and Rb/Sr dating of single quartz grains and aggregates of quartz grains (Powers et al., 1979). The main use of such data has been to identify the likely sources of sediments based on a consideration of the distribution and exposure history of rocks of known geological age and composition. However, there are no published examples of forensic casework applications.

4.12   Mineralogical Identification and Characterization of Noncrystalline Materials

The mineralogical composition of an individual crystal or sedimentary particle can be determined by a variety of methods that relate to the crystallographic structure, including classic optical petrography (Milner, 1962a, 1962b; Tickell, 1965; Gribble and Hall, 1985), single crystal x-ray diffraction (Putnis, 1992), electron diffraction (McLaren, 1991; Putnis, 1992), and laser Raman spectroscopy (Nasdala et al., 2004b). In specific situations, such as in studies of gems and glasses, other forms of spectroscopy, including electron spin resonance (ESR), nuclear magnetic resonance (NMR), infrared (IR) spectroscopy, ultraviolet (UV) spectroscopy, Mossbauer spectroscopy, and x-ray spectroscopy, may also be of assistance (Farmer, 1974; Amthauer et al., 2004; Beran et al., 2004; Geiger, 2004; Kohn, 2004; Mottana, 2004). Such techniques have been widely applied in forensic investigations relating to the theft, smuggling, and imitation of gemstones and industrial diamonds. However, they can also be useful in characterizing more mundane minerals, such as quartz, feldspar, or calcite, if the particles/crystals in question are of special evidential interest.

Raman spectroscopy provides a means of nondestructive characterization of a wide range of materials, including minerals, glasses, and biomaterials of various types (Edwards, 2001, 2004; Edwards and Hassan, 2004; Harvey et al., 2002). Raman scattering is insensitive to the presence of water and hydroxyl groups, therefore avoiding the need to dehydrate the specimen prior to analysis. Both rough and smooth surfaces can be examined without the need for any surface coating. Objects are examined in a dedicated Raman microscope, most frequently equipped with an Nd^{3+}/YAG laser (Long, 2002). Operating conditions are varied according to material type, but a wave number of 1064 nm in the near infrared range frequently excites the sample with minimal fluorescence (Edwards, 2004). The sampled areas typically range in diameter from c. 8 to 100 μm. The technique has been used to test the authenticity of archaeological artefacts and art objects, to establish the provenance of ivory, and to investigate the mode of preservation of mummified human remains, including those of the Alpine Iceman (Williams et al., 1995; Gniadecka et al., 1999; Edwards, 2001; Edwards and Hassan, 2004).

4.13 Micro-Fabric of Rocks and Soils

As described in Chapter 3, micro-fabric may be defined as the internal arrangement of mineral particles and other constituents within a soil, sediment, rock, or man-made material such as concrete. Even quite small soil or sediment aggregates and rock particles (<2 mm) can be impregnated in epoxy resin blocks and thin sectioned in one or more orientations. For routine work on geological materials, standard thin sections measuring 20 × 30 mm and 30 μm in thickness are prepared (Ireland, 1971; Harwood, 1988; Miller, 1988a; Humphries, 1992; Figure 4.37 and Figure 4.38). However, for some purposes, and especially in soils work, much larger sections (up to 20 × 30 cm) may be required (Jongerius and Heintzberger, 1963).

In order to preserve the structure of moist, unconsolidated soils and sediments, the moisture is gradually replaced with acetone, prior to resin impregnation (Murphy, 1982, 1985, 1986). The sections may be covered with a glass cover slip if no analysis is to be undertaken, or left uncovered and finely polished if SEM imaging and elemental analysis is required. In some circumstances,

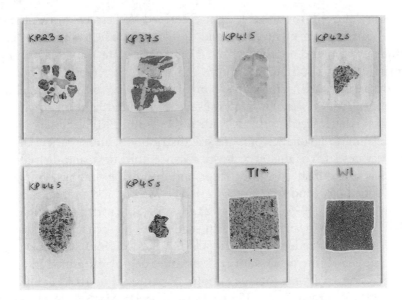

Figure 4.37 Standard geological thin sections prepared from a series of gravel particles recovered from a murder victim (top left) and various comparison samples of igneous rock.

etching or staining of the section may enhance features of particular interest (Norton et al., 1983), and resolution of fine micro-structural detail may be aided by the preparation of ultra-thin (c. 5 µm) sections (Bowles, 1968).

If only one, or a small handful of sand-size particles are of interest, the grains can be placed in a small rubber mould and embedded in epoxy resin,

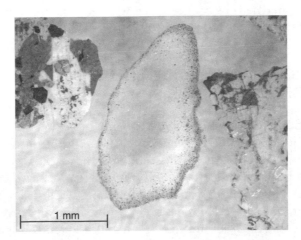

1 mm

Figure 4.38 Transmitted light micrograph of three of the smaller gravel particles recovered from adhesive tape wrapped around the head of a murder victim.

following a procedure similar to that described by Petraco and Gale (1984) for the examination of multi-layered paint chips. The resin block is then carefully ground and polished by hand on a glass plate using diamond paste (Taggart, 1977).

Comparisons are normally made on the questioned soil/sediment and control samples from a scene of interest using transmitted light microscopy, supplemented where necessary by BSE microscopy and EMPA. Observations are made relating to texture, color, pore size distribution, and a variety of micro-fabric features such as the presence or absence, thickness, and distribution of clay, calcium carbonate, organic matter, and iron oxide coatings on individual grains, root holes, and fractures (Brewer, 1976; Bullock and Murphy, 1980; Fitzpatrick, 1984, 1993; Bullock et al., 1985). Soil micro-morphology has proved useful as a means of comparing soil traces on archaeological artefacts, such as gold coins and jewellery, and alleged find locations (Canti, 2005).

The internal micro-fabric of rock samples can be examined in a similar way and compared with control samples of standard images in suitable reference texts (Bennett et al., 1980; O'Brien, 1981; MacKenzie et al., 1982; Adams et al., 1984; Adams and Mackenzie, 1998; Krinsley et al., 1998; Vernon, 2004). A great variety of rock textures and micro-fabrics exists, reflecting the diversity of igneous, metamorphic, and sedimentary processes. The internal microstructures of man-made materials, such as concrete and bricks, also show a wide diversity and can provide a useful criterion for comparison.

In situations where it is undesirable to section a rock or sediment particle of forensic interest, certain information about the internal structure can be obtained using non-invasive imaging techniques. Traditionally, this has been achieved by x-radiography (Hamblin, 1971), and more recently using techniques such as magnetic resonance imaging (MRI) and x-ray computer aided tomography (XCAT) (Borgia et al., 1996; Chen et al., 2002).

4.14 Identification and Characterization of Organic Particles

Organic particles in soils, sediments, rocks, and dusts can be of many different origins and be either modern or ancient (i.e., derived from sources that are hundreds to millions of years old). Pollen grains, seeds, and other plant remains may be either modern or old, whereas fragments of coal are inevitably old. The identification of individual biological particles, or small fragments of peat, coal, etc., often relies on a combination of morphological and chemical criteria. Peat, coal, and coke fragments frequently display distinctive internal structures and microscopic/spectroscopic characteristics which help classification and source identification. Reflectance microscopy, fluorescence microscopy, micro-photometry, SEM examination, and TEM analysis can each provide useful information to assist characterization and comparison of a questioned material (Davis and Vastola, 1977; Juckes and Pitt, 1977; Teichmuller and Wolf, 1977; Herman, 1998).

Modern pollen grains, seeds, and similar plant material can be characterized by Fourier Transform infrared analysis in an FTIR analytical microscope, and potentially by DNA analysis (see papers in Coyle, 2004). Natural fibers and hairs of various types are also frequently found in soils, sediments, and dust samples. Their identification is a specialist operation that normally requires a combination of microscopic and spectroscopic techniques (Petraco and Kubic, 2004).

Sampling and Sample Handling

5

5.1 The Nature of Samples and Their Limitations

In routine forensic casework the soil examiner is often required to make a comparison between a sample of *questioned* soil present on an item of investigative interest and one or more soil samples collected from a known source. Various terms have been used to describe the latter type of sample, including *control sample*, *comparison sample*, and *reference sample*. However, these terms should not be regarded as entirely synonomous. In the author's opinion, the term *control sample*, in the context of forensic geological work, should be restricted to samples of soil, sediment, or rock that are collected from known geographical locations at a known time and date. These locations may include a crime scene and one or more "alibi" locations suggested by a defendant.

The questioned sample may, for example, be a lump of mud obtained from a shoe suggested to have been worn by a suspect at the time and scene of a crime. In such circumstances, it may be useful to know if similar mud is also present on other shoes owned by the suspect, or perhaps by members of his family and work colleagues. Samples taken from such other items of footwear can be described as *comparison samples*. In some circumstances, particularly, where minerals, rocks, or fossils are concerned, it may also be helpful to compare a questioned sample with "type examples" held in a museum or other archive collection. Such samples, which are normally well documented in terms of their physicochemical and/or biological characteristics, are referred to as *reference samples*.

In many situations, the questioned sample of interest is quite small, and the forensic examiner has little or no control over its size or character; it may be difficult to determine the degree to which it is representative of the source from which it is derived. This is one of the major respects that makes soil and rock examination, analysis and interpretation for criminal forensic purposes different to that undertaken for environmental forensic or more general survey purposes. In the latter types of work, the investigator usually has much more control over the size and nature of the samples that are taken for analysis and comparison. A further difficulty in criminal trace evidence work is that the soil examiner does not always have the opportunity to collect the control or other comparison samples personally. In many cases, such samples are taken from scenes by crimes officers and sometimes by police officers or others. Consequently, there may be uncertainties about the "representativeness" of these samples. The trace evidence soil examiner must always bear in mind these issues when assessing the results of any laboratory comparisons that are undertaken.

5.2   General Sampling Guidance

There is an abundance of published information and official guidance relating to soil sampling for purposes such as soil survey, environmental assessment, and geochemical prospecting (e.g., International Standards Organization, 1993–2002; Cline, 1944; U.S. Department of Agriculture, 1972; Mason, 1992; Salminen et al., 1998; Orton, 2000; Webster and Oliver, 2001; U.S. Environmental Protection Agency, 2002), but relatively little written specifically for forensic geology and soil investigations (e.g., Skinner, 1988; Murray and Tedrow, 1992, pp. 81–94; Forensic Science Service, 1994; Wade, 2003; Murray, 2004, pp. 113–122).

In the majority of environmental investigations and geochemical mapping studies, the prime objective is to collect a suite of samples that is representative and unbiased. Samples may be collected as a spatial array to produce maps of soil properties, statistical frequency tables, and as time series to monitor changes in soil or sediment properties over time. The question of what "representativeness" actually means has been the subject of considerable discussion in the environmental forensics literature (Gilbert and Pulpisher, 2005; Nocerino et al., 2005; Petrisor, 2005; Ramsey and Hewitt, 2005; Warren, 2005). The answer, perhaps rather unsatisfyingly, is that "representativeness" is context-dependent, reflecting the aspect of interest and the nature of the question being asked.

In general environmental survey work, emphasis is often placed on the need to obtain an unbiased set of samples which are sufficiently numerous to give an accurate impression of the nature of the background population which is being sampled. Random sampling is often advocated as the best way of avoiding systematic bias, although the number and distribution of the points may be "weighted" to take account of spatial variations in topography, geology, or land-use. Purposeful sampling may also be employed to test specific hypotheses, but relatively rarely. In criminal forensic investigations, however, purposeful sampling assumes much greater importance.

In order to address the initial question of whether there is a high degree of similarity, or "match," between a questioned sample and a crime scene, samples are normally taken from the scene in a purposeful manner. For example, samples should be taken from any obvious footwear impressions or other areas of ground disturbance, or from specific locations where assessment of the scene suggests an offender is likely to have stood or walked, such as a flower bed below a window used for forced entry or the ground next to a deposited body. For comparison, samples should also be taken at a number of other, effectively randomly chosen points within the scene or its immediate surrounding area.

If the results of the initial comparisons demonstrate a high degree of similarity between the questioned sample and the controls from the scene, the next question to be addressed is whether the crime scene samples can be distinguished from others in the surrounding area, or any locations that may have been visited by the suspect. This question must be addressed by a combination of random sampling and further purposeful sampling, for example, from possible "alibi" locations suggested by the suspect.

Truly random sampling requires no prior knowledge and may be undertaken simply by obtaining geographical coordinates using a random number generator. However, in forensic work, a combination of purposeful and stratified random sampling generally provides the most productive and time/cost efficient option where there is a need to test specific hypotheses, for example, to establish whether "alibi" locations suggested by a defendant can be excluded as a source of questioned soil. Purposeful and stratified random sampling requires an initial assessment of background information relating to the geology and soil patterns within the areas of interest, and to the known movements of suspects and their associates (e.g., family members who may have driven the suspect's car).

In many major crime investigations, a considerable period of time may elapse between the committing of a crime and a suspect being arrested. Since some soil properties may change with time, notably those involving organic components such as pollen, seeds, and microbes, soil sampling should be undertaken as soon as possible after a crime has been committed, even if questioned samples are not available at that time. However, analysis of the samples need not be undertaken until questioned material becomes available. Clearly, no purposeful sampling of other potentially relevant sites, such as alibi locations, can be undertaken until a suspect is arrested. Whenever there is a time delay between control sampling and seizure/analysis of questioned samples, the potential effect on the results of the comparisons needs to be borne in mind.

In the case of a murder scene or body deposition site, control soil sampling should normally be carried out after removal of the body and before serious disturbance of the site which may involve cutting down of the vegetation, finger-tip searches, and possibly digging by police search and forensic teams. To obtain maximum information and an optimum sampling regime, the forensic soil examiner should preferably be asked to visit the site and take samples in person. However, if this is not possible, a suitably trained scenes of crimes officer should take the samples using standard protocols and ensure that an adequate photographic and documentary record of each individual sampling location is obtained.

5.3   Sampling Strategies for Control Samples

5.3.1   Crime Scenes

Crime scenes differ greatly in their nature and consequently the sampling strategy must be adapted to suit each individual case. Some of the most commonly encountered types of environment used as body deposition sites, and for other serious crimes, include woodland (Figure 5.1 and Figure 5.2), agricultural fields and pasturelands close to tracks and hedgerows (Figure 5.3 and Figure 5.4), moorland (Figure 5.5), quiet country lanes (Figure 5.6), beaches (Figure 5.7), areas of wasteland (Figure 5.8), alleyways (Figure 5.9), waste tips (Figure 5.10), and urban streets (Figure 5.11). The nature, number, and location of control samples that should be taken will vary according to the specific nature of the environment at each crime scene. In some situations, it may be appropriate to use only purposeful sampling, but in others a combination of purposeful and random sampling may be required.

If footwear marks are evident in the soil or on some other surface (Figure 5.12), samples should either be taken directly from these areas, after photographing, or recovered later from plaster casts of the marks (e.g., Bull et al., 2006). The same principle applies to vehicle tire marks and impressions. If only small areas of bare soil are exposed within the area around the crime scene, such as flower bed below a window through which forced entry has been gained, samples should be taken from these areas at the points where an offender is most likely to have stood (Figure 5.13).

The number of samples that should be taken will depend on the size, number, and heterogeneity of potential soil source areas at and around the crime scene. The number of samples taken should never be fewer than three, and preferably at least five, even at a very small crime scene; the representativeness of a single sample is always open to doubt. Larger or more variable crime scenes will require a larger number of samples, perhaps more than fifty at a very large or complex scene.

Figure 5.1 Woodland environment near Crawley, Sussex, where the battered body of a young woman was found. Note the piles of builder's debris and polythene bags filled with waste. Such materials can provide forensically useful sources of "unusual" particles in woodland and agricultural areas.

Figure 5.2 (a) Tree-lined ditch alongside an unmade country track near Laken-heath, Suffolk, where the bodies of two young girls were discovered; (b) former position of the bodies within the ditch.

Figure 5.3 A muddy track surrounded by agricultural fields south of Blackpool, Lancashire, where the body of a young male was discovered.

Figure 5.4 A shallow grave in the corner of a field, shielded from a main road near Pulborough, Sussex, where the body of a young girl was discovered.

Figure 5.5 An area of moorland on Exmoor, Somerset, where a body was found wrapped in black polythene.

Figure 5.6 A field entrance area adjacent to a quiet country road near Ruan High Lanes, Cornwall, where a female dog walker was murdered.

Figure 5.7 The upper beach at Skegness, Lincolnshire, where the body of a teenage boy was discovered.

Figure 5.8 An area of wasteland near Heathrow Airport Terminal 4 from which homemade mortar bombs were launched by the IRA.

Figure 5.9 An alleyway adjacent to houses near Gravesend, Kent, where the body of a teenage girl was discovered.

Figure 5.10 A waste-tip near Reading, Berkshire, where a night watchman was beaten to death.

Figure 5.11 Pavement in a London suburb where a young woman was hit on the head with a blunt instrument, leading to her subsequent death.

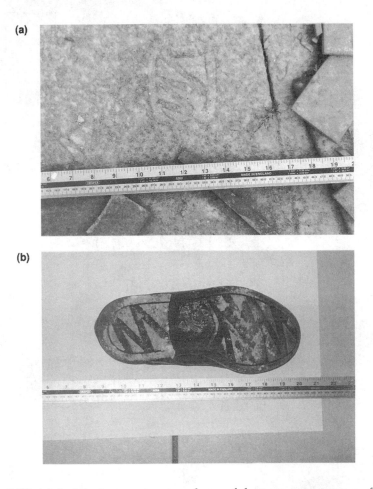

Figure 5.12 (a) A shoe impression in a layer of dust on a concrete surface; (b) the sole of a shoe with a tread pattern matching that of the impression.

In situations where an area of investigative interest contains no specifically identified features of interest, for example, a field within which a suspect is believed to have walked while burying and then recovering, an object such as a firearm, soil sampling of the whole area can be undertaken using a variety of systematic (grid-based) or random strategies (Figure 5.14 and Figure 5.15). If the objective is to produce predictive maps based on geostatistical techniques such as kriging (Webster and Oliver, 2001; Yakes and Warrick, 2002), there are advantages in adopting a regular grid-based strategy with a large number of sampling points. However, such an approach involves significant analytical cost, and for many purposes it may be sufficient to take samples at a much smaller number of regularly spaced or randomly chosen locations. Figure 5.14 shows a number of alternative sampling strategies which involves

(a)

(b)

Figure 5.13 (a) The scene of an armed break-in near Lincoln; (b) shoe marks in soil on the windowsill, the soil being derived from the flower bed below.

progressively larger numbers of samples. Clearly, the larger the number of samples, and the wider their distribution, the better the picture which can be obtained of variability within the area of interest.

If there are clear linear features at, or close to, a crime scene, such as a footpath or track, samples should be taken at intervals along that feature. If footmarks or tire tracks are evident, these should be sampled directly (Figure 5.15f). Muddy potholes, large lumps of mud fallen from agricultural

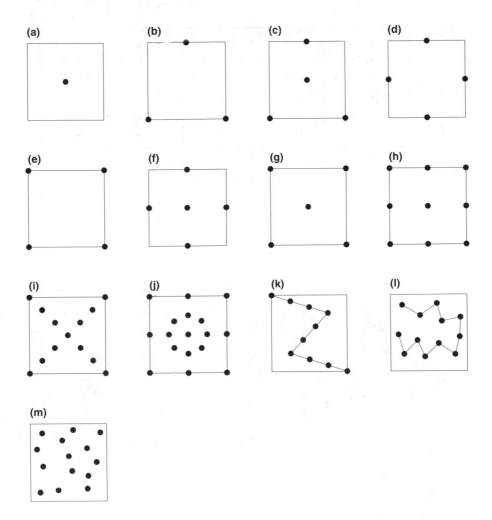

Figure 5.14 Some alternative sampling strategies for a discrete area such as a small field.

vehicles, or other potential opportunities for soil transfer should be purposefully sampled in a similar manner.

If no footwear or other marks are evident around the object of interest (e.g., a body) at the center of the scene, and the area appears relatively homogeneous, samples can be taken in a systematic manner, for example, radiating out on all sides in a cross-pattern (Figure 5.15h). However, if there are distinct environmental zones within the area, it may be more appropriate to take samples from each zone along a series of transects. Figure 5.15l shows an example of a beach and sand dune system with three distinct environmental zones from which samples can be taken, either at regular intervals or randomly selected points along one or more transects.

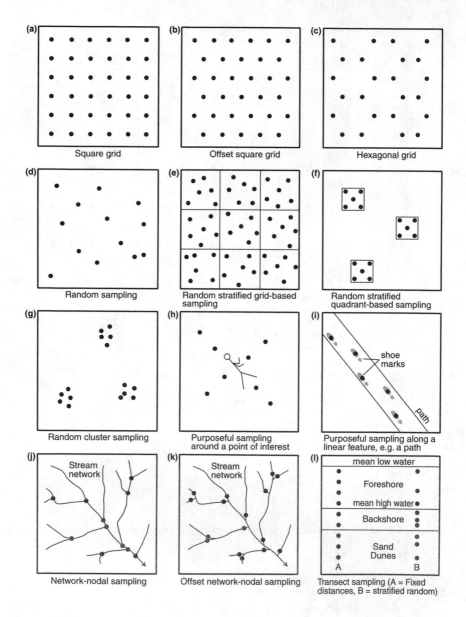

Figure 5.15 Additional alternative sampling strategies.

If the scene of interest is a typical suburban house and garden, the sampling strategy can again be entirely purposeful, a combination of purposeful and systematic, or purposeful and random. Purposeful sampling should be undertaken close to any features of identifiable interest, such as a site of recent digging activity, or areas with clearly unusual character. In the absence of such features, systematic sampling of the lawn and flower borders

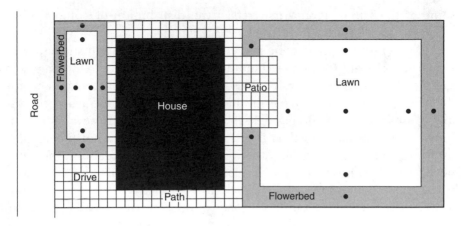

Figure 5.16 A suggested sampling strategy for the garden of a small suburban house.

in a modified cross-pattern is usually adequate to characterize the soil pattern in the garden (Figure 5.16).

Particularly at larger scenes, and those where the surface clearly shows significant local scale variability (as on many unmade tracks and lay-by surfaces), it is important to quantify the local scale variability by taking samples in groups of three in a triangular array (e.g., Figure 5.17). The locations of

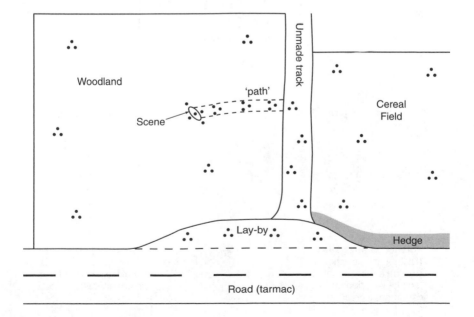

Figure 5.17 An illustrative sampling strategy for a crime scene in a rural environment.

some of the sample groups may reflect the distribution of potholes, tire-tracks, or areas of visually different surface character.

In geochemical prospecting and environmental pollution studies, individual samples are sometimes combined to produce one mixed composite sample for analysis (Figure 5.18). This method provides an average value for the whole area sampled and has the advantage of keeping the analytical cost low compared with analyzing the samples individually. However, for most forensic purposes, this method is unsuitable, since it does not provide any information about local scale variations and, in most situations, the composite sample is unsuitable for comparison with questioned samples. Even where a questioned soil sample is likely to be a mixture from several point locations, for example, acquired on a pair of boots by walking around a field,

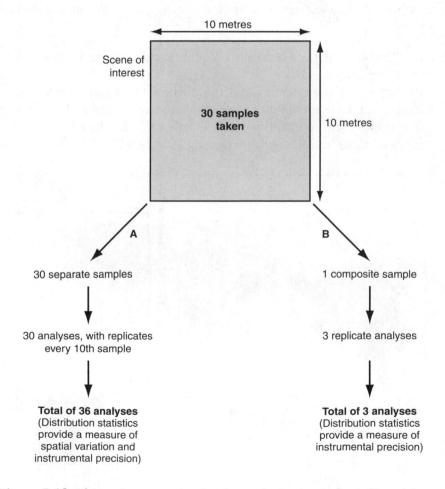

Figure 5.18 Alternative strategies for the analysis of samples collected from a scene of interest; in most situations, Strategy A is preferable.

the best strategy is to analyze the control samples individually and to calculate the average values, ranges, standard deviations, or coefficients of variation for each attribute of interest.

5.3.2 Sampling Across a Wider Area

When an absence of suitable database information makes wider sampling necessary in order to evaluate the likely significance of an apparent similarity between a questioned sample and crime scene samples, the first step is to review the available information relating to the drift geology and soil types in the region of interest. Areas of broadly similar soil type, developed on the same general parent material, can then be identified as priority areas for further sampling and comparative analysis. A number of other areas with different geological parent material or soil type should also be selected for sampling in order to test the hypothesis that they will have different physical and chemical properties to those in the crime scene area. However, cost and time limitations generally do not permit the undertaking of a systematic or truly random sampling exercise across an entire region, only to demonstrate that the vast majority of samples are different to those of interest at the crime scene.

Within the areas identified as being most likely to have soil properties similar to those of the crime scene, samples can be taken from locations identified on a random or randomly stratified basis. At each identified location, three samples should ideally be taken in a triangular array, at 1 m spacing, as described earlier. If budget limitations do not permit this, a single sample can be taken, but this is less satisfactory since no information is obtained about local scale variation. In choosing the additional control sampling sites, consideration should also be given to any available information about locations known to have been visited by the suspect, or, if no such information is available, about the relative accessibility of various locations to pedestrian or vehicular traffic. For example, if the issue of interest is the transfer of gravelly chalk mud onto the wheel arches and suspension struts of a car, it makes sense to focus additional control sampling on trackways, lay-bys, hard-standings, and other locations where such material has been used as a surfacing material. Reference materials from known suppliers and stratigraphic levels in working and abandoned quarries should also be obtained for comparative analysis.

5.4 Size and Type of Sample

The type and size of samples that should be taken will also be dependent on the nature of the environment in question, the type of soil or sediment involved, and the nature of the activity that has taken place at the scene of interest. For example, if the objective is to compare questioned soil present on the soles of a pair of shoes, surface soil samples, taken from a depth of 0 to 1 cm, should be taken. If the suspect footwear is heavily coated with mud on the uppers, the ground at the time was obviously very wet and soft, and control samples should be taken from a greater depth interval, for example, 0 to 10 cm. If the scene includes a shallow grave or other burial feature, soil samples should be taken at regular depth intervals (e.g., 0 to 5 cm, 5 to10 cm) corresponding to the full depth of the excavation. Samples should also be taken of any residual heaps of mixed earth. Useful information that will aid interpretation of the results can be obtained if additional soil pits are dug to a depth of c. 50 cm in order to expose the three-dimensional character of the soil (Figure 5.19).

For routine forensic soil and fine-grained sediment control sampling purposes, it is normally sufficient to take a sample measuring approximately $10 \times 10 \times 1$ cm, or $5 \times 5 \times 4$ cm (equivalent to a volume of c. 100 cc, or c. 160 g dry weight). However, larger samples should be taken from coarse grained or more heterogeneous materials (c. 500 g for coarse, poorly sorted sand and fine gravel, 1 to 2 kg for medium and coarse gravel). Samples of 1 cc (1.6 g) or less can provide enough material for several types of analysis, but the smaller the sample, the greater the potential for variation in results due to the irregular distribution of a few relatively large particles which may have an unusual composition (the "nugget" effect, Ramsay, 1997).

In the case of soils and sediments that are clearly layered, or taken from a grave, there is advantage in taking undisturbed soil samples which preserve the natural stratigraphy (Hanson, 2004). If the soil is cohesive, this can be done by cutting out a block of soil with a broad-bladed stainless steel knife, plasterer's trowel, or spade and placing it in a suitably packed plastic container. In some circumstances, it may be advantageous to remove large blocks of soil for detailed examination in the laboratory (Figure 5.20). These should be carefully cut and lifted using a stainless steel spade and placed in separate plastic crates for transport to the laboratory. In the case of unconsolidated granular materials, such as sands and friable soils, a box or a half-section of square pipe can be driven into the surface or wall of an excavated pit or trench and then excavated. Kubiena tins have been widely used in soil science for this purpose (Kubiena, 1938; Figure 5.21). However, if chemical analysis is to be undertaken, plastic boxes or pipes are preferable to metal ones.

Figure 5.19 A shallow pit dug to expose a soil profile in three dimensions.

(a)

(b)

Figure 5.20 (a) Area adjacent to a field entrance gateway near Ruan High Lanes, Cornwall, where a pair of glasses belonging to a murder victim were placed, possibly by the offender, some months after the murder; (b) a block of grass-covered soil on which the glasses were found, removed in one piece to the laboratory for detailed examination.

Figure 5.21 A Kubiena tin used for the collection of "undisturbed" soil samples.

5.5 Sampling Tools and Sample Containers

The most useful tools for taking control soil and sediment samples are a spade, small plasterer's trowel, and a selection of artist's pallette knives of varying thickness. All should be made of stainless steel. Plastic spades and trowels lack the rigidity required to deal with hard ground and are often prone to scratching or breakage by sharp stones. Similarly, disposable plastic or wooden spatulas lack the rigidity to take anything other than very soft surface sediment samples. Plastic sample tubes or other containers should

Figure 5.22 Basic equipment used for the recovery of soil samples in the field: stainless steel spade, small pointing trowel, and artist's pallette knife.

not be used to take surface scraping samples since this process inevitably involves contamination of the outside of the container with soil, often obscuring the label.

Stainless steel tools are resistant to scratching and corrosion and can be easily cleaned using distilled deionized water and a roll of paper towel (Figure 5.22). Experiments have shown that risks of cross-sample contamination using this method are minimal if the cleaning procedures are thorough. In some circumstances, for example, when dealing with oily or otherwise contaminated soils, additional cleaning using alcohol or a solvent such as acetone may also be required. As an additional precaution against cross-contamination, different sets of tools should be used for sampling at different locations (e.g., crime scene and suspect's home address).

Thin pallette knives are particularly useful for taking thin surface samples of mud or loose granular material that may be spread on a pavement or road surface. If it is necessary to take a thin sample from a grassed surface, a thin stainless steel slicing tool with a serrated edge will assist in cutting through the rootlets. In the case of noncohesive sandy or gravelly sediments, a stainless steel or plastic scoop provides better grain retention than a flat trowel or pallette knife. If only a very thin smear of soil or dust is present in the area of interest, it is often more efficient to recover it by way of an adhesive tape lift or micro-vacuum device (Figure 5.23). Where possible, wet recovery of

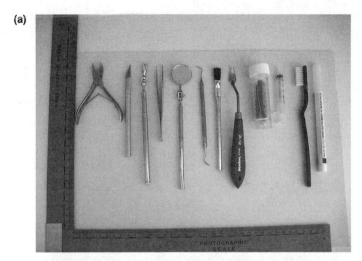

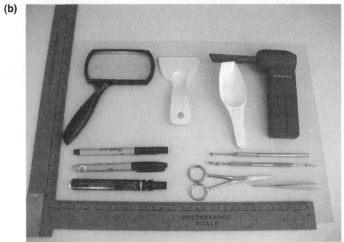

Figure 5.23 (a) and (b) A selection of implements used for the recovery of soil evidence.

the sample by washing it from the supporting surface should be avoided since some of the soil/dust constituents may be water soluble, and the original physical texture of the material may be lost as a result of wetting and subsequent drying.

Whether disturbed or undisturbed, control soil and sediment samples should preferably be placed in plastic containers rather than polythene bags or paper bags. In soil sampling for geochemical mapping and surveying purposes, samples have often been placed in special paper bags and left to air dry before sealing and transfer to the laboratory for analysis. In forensic work, this cannot be done due to the risks of contamination by airborne particles and the usual requirement for rapid sample analysis. Clip-seal polythene bags can be used, but they are prone to splitting or puncturing if the sample material contains sharp particles of angular gravel or woody matter. For these reasons, rigid plastic containers with air and water-tight lids are preferred.

In circumstances where the soils or sediments of interest might contain organic compounds or volatile substances of potential forensic significance, or which might react with plastic containers, the samples should be stored in glass containers with screw lids lined with a suitable nonreactive material. In general, however, glass should be avoided owing to the risk of breakage and sample loss or contamination.

Water samples should also be taken from any relevant water bodies (e.g., ponds, streams, puddles, and water tanks) using a plastic sampling device and used to completely fill a 500 ml polyethylene bottle. Both the sampling device and storage bottle should be rinsed three times using water from the water body from which the sample is to be taken, prior to filling. Such samples can be useful both for analysis of the water chemistry, suspended minerogenic matter, and the intrinsic bio-flora and fauna (diatoms, pollen, ostracods, and gastropods). If detailed chemical analysis is to be undertaken, it is preferable to filter samples in the field using a hand-pumped filtration system, followed by acidification using nitric acid. However, for most forensic purposes, these procedures can be performed in the laboratory if the interval between sampling and analysis is relatively short (hours to a few days). If water samples are to be used primarily for diatom and pollen extraction, they can be stabilized using a few drops of Lugol's iodine solution or commercially available tincture of iodine.

Samples of sub-aqueous sediment, for example, from the bottom of a pond or stream, can be obtained by pressing a plastic tube or container directly into the bed and removing it with a scooping action. The lid should be placed on the container before it is lifted from the pond/stream bed. The container should then be placed inside another, large container or polythene bag which is suitably labeled. In deeper water, samples can be obtained using specialized sampling devices, such as the Van-Veen-type grab sampler or Mackereth-type gravity corer (Mudroch and Azcue, 1995).

It may also be helpful to take control samples of the surrounding vegetation for later identification and comparison, although this task should generally be carried out by a forensic botanist.

5.6   Sample Labeling and Associated Information

All samples should be labeled in the field using an appropriate code number and supporting information (e.g., case name, person collecting, date, and time). In the author's experience, the best approach is to give each sample an identifying code that includes the initials of the collector and the sequential sample number for the case, for example, KP1, KP2, and KP3. This assists identification of any samples that may subsequently be used as exhibits in court. The use of slashes, hyphens, or underscores in the sample codes should be avoided to minimize the opportunities for confusion, for example, between KP1, KP11, and KP/1.

At least three photographs should be taken of each sampling location, one close-up to show the nature of the sampled surface, the other two from some distance away looking in perpendicular directions toward the sampling position, to provide information about the environmental context.

The geographical coordinates of the sampling location should be determined using a hand-held gps instrument (Figure 5.24). In circumstances

(a)

(b)

Figure 5.24 (a) and (b) Recording of sampling positions in the field using a digital camera and hand-held gps.

where the sampling positions are closer together than the error associated with the gps, direct measurements should be made relative to at least two fixed points, such as gate posts at a field entrance, or an easily identifiable point on a large tree, using a tape or laser distance measurer. This information should be recorded at the time in a notebook or hand-held computer, together with any relevant observations about ground conditions, weather, etc.

5.7 Sample Storage

The samples should be conveyed to the laboratory, unopened, as soon as possible after collection. If immediate delivery is not possible, the samples should be stored in the dark, in a refrigerator or cold room at c. 2°C. Soil samples should not be frozen unless very long-term storage is envisaged, since some physical properties of the soil and any contained organic matter may be changed or destroyed by the process of freezing. Similarly, it is preferable not to dry the samples until they have been delivered to the laboratory and subjected to preliminary examination and any necessary sub-sampling. If drying is unavoidable, for example, to reduce sample weight during an overseas sampling expedition, this should be done by removing the lids from the sample containers and placing them inside sealed paper bags to air dry. They should not be left uncovered or placed in an oven to accelerate drying. Rock samples should be handled in exactly the same way as soil and sediment samples. For medium-term storage, water samples are best left in a dark, cold room after appropriate fixation procedures have been applied.

Fresh vegetation samples are prone to rapid deterioration and should preferably be examined very soon after collection. If this is not possible, appropriate measures should be taken to preserve them in the best condition possible for later examination. While some plant tissues can withstand freezing without undue damage to the cell structure, others can be completely destroyed; it is often better to dry and press flowers and foliage (e.g., Knudsen, 1972). Freeze-drying is also an option in some circumstances. The advice of a professional botanist should be obtained to identify the best method of preservation for the particular vegetation matter concerned.

5.8 Questioned Soil Samples from Items Submitted for Forensic Examination

A large variety of different types of item may be submitted for forensic soil examination. Probably the most frequently encountered are items of footwear (Figure 5.25 and Figure 5.26) and clothing (Figure 5.27 and Figure 5.28).

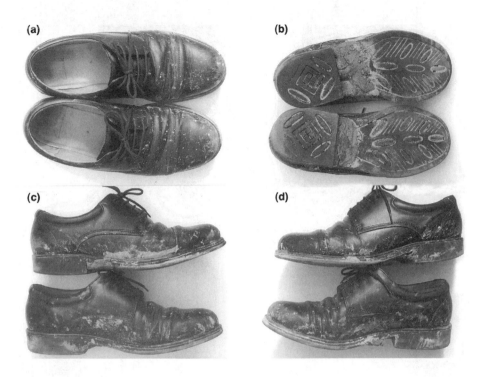

Figure 5.25 A pair of heavily mud-stained shoes photographed from four perspectives.

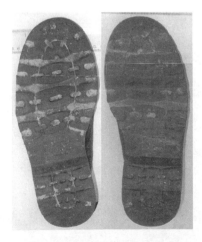

Figure 5.26 A pair of shoes worn by a defendant who was subsequently convicted of the murder of a young woman. Mud had been scraped off the sole of one of the shoes by the time they were recovered, but the other shoe bore significant traces of soil that showed a high degree of similarity to that present at the murder scene.

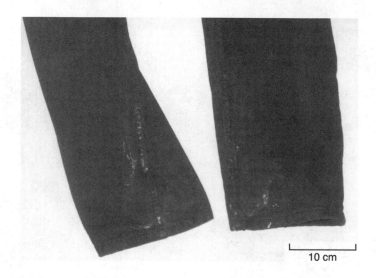

Figure 5.27 Mud stains on a pair of black Levi jeans.

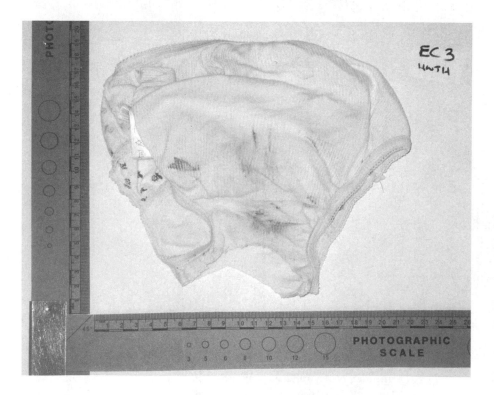

Figure 5.28 Soil staining on the underwear of a young woman who was raped in a London park.

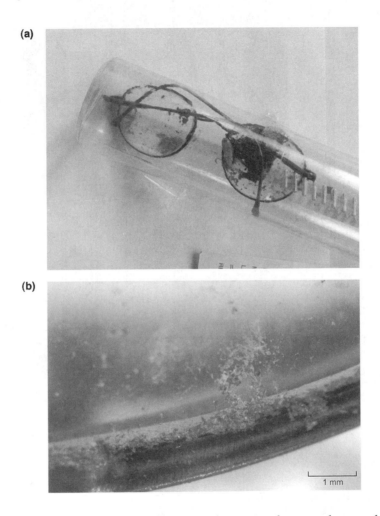

Figure 5.29 (a) Glasses belonging to a murder victim that were later replaced at the murder scene, possibly by the offender; (b) binocular microscope photograph showing soil traces and fine rootlets adhering to the lens and rim of the glasses.

Others include personal items such as spectacles (Figure 5.29) and fingernail clippings (Figure 5.30), spades (Figure 5.31), metal detectors (Figure 5.32), pieces of rope (Figure 5.33), and knives and firearms (Figure 5.34). Vehicles and their contents are also often the subject of soil examination. Traces of soil and sediment may be found on wheel arches, mud flaps, bodywork, suspension, footwell mats, carpets, and foot pedal rubbers (Figure 5.35, Figure 5.36, Figure 5.37, Figure 5.38, and Figure 5.39).

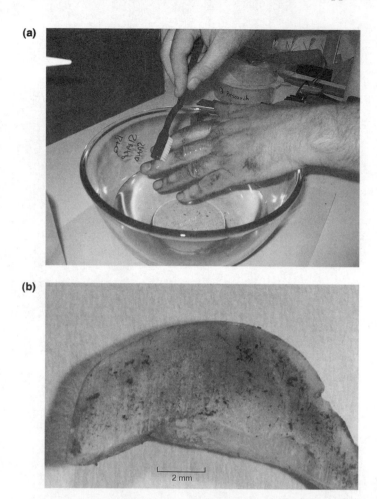

Figure 5.30 (a) Removal of soil from a pair of dirty hands using a nylon toothbrush and distilled water; (b) traces of soil on a fingernail clipping removed from a suspect in a drugs investigation.

The best results are usually obtained if the initial examination of the items of forensic interest is conducted by the soil examiner in person, perhaps jointly with other forensic scientists with biological or other trace evidence expertise. On initial examination, each item should be photographed in several different orientations to record its condition on receipt, prior to examination using a magnifying lens and high intensity light

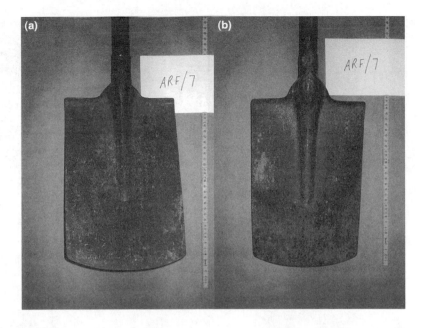

Figure 5.31 Front (a) and back (b) views of a spade seized from a man suspected of the murder of a teenage girl.

Figure 5.32 Soil-stained metal detector recovered from a vehicle owned by a suspected drugs dealer.

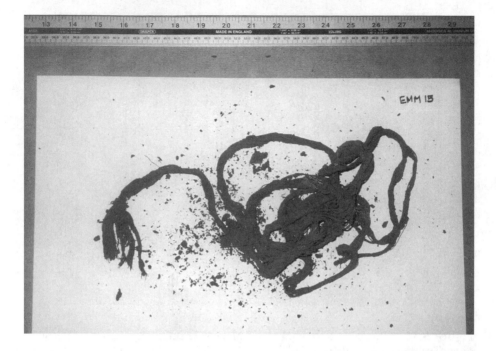

Figure 5.33 Heavily mud-stained rope recovered as part of a murder investigation in which the victim's body had been dismembered and the parts buried in several different locations.

source, or sub-sample removal. Written notes should be made at the time relating to the nature of the item, its condition, and any other relevant factors. Sketches should be made to show the distribution of any soil staining or vegetation debris present, with indications of the locations where samples are taken.

It is a great advantage to the soil examiner to see any soil or other debris *in situ*, rather than as a pot of mixed "debris" removed by another scientist. The nature (size, shape, thickness, and texture) and distribution of soil stains can provide useful information about the timing, sequence, and mechanism of mud acquisition, which assists interpretation of the analytical results obtained.

Where possible, photographic images should be obtained of the mud stains or trapped grains *in situ* using a close-focusing digital camera or stereo-microscope linked to a digital image capture system (Figure 5.39).

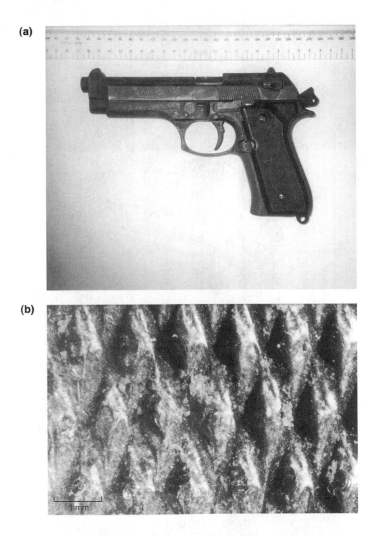

Figure 5.34 (a) A hand gun believed to have been concealed by burial in soil adjacent to a country lane; (b) binocular microscope picture showing traces of soil trapped in the hand grip.

The purpose of this close examination is to determine whether more than one type of soil, mud, or granular debris is present, and to allow separate sampling of each. Individual lumps of mud can be removed using dentists' tools and examined under the microscope in three dimensions to establish whether more than one layer of soil is present (Figure 5.40).

(a)

(b)

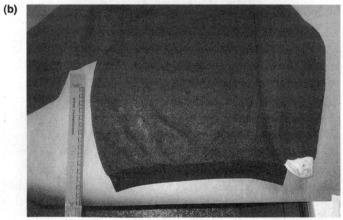

Figure 5.35 (a) A Fiat Ducato van used to abduct a young girl from a field in West Sussex; (b) a soil-stained jersey found inside the van.

Figure 5.36 A Ford Fiesta car used to transport the bodies of two young girls to a deposition site near Lakenheath, Suffolk.

If the item bearing the mud stains is too large to examine under a binocular microscope, pieces of mud-stained fabric can be cut out using a scalpel and examined separately. In this case, initial optical microscope examination can be followed by other types of *in situ* microscopic examination and micro-analysis, including UV-microscopy, SEM, EDXRA, and FTIR (Figure 5.41). The advantage of *in situ* analysis is that the original fabric of the soil can be identified and interpreted, information which is lost if the soil material is removed from its support.

If sufficient soil material is present, some can be removed for further processing and analysis using techniques such as color spectrophotometry, XRD, laser granulometry and ICP-AES/MS. In such cases, it is preferable to obtain several small samples and to analyze each individually, rather than to combine them and undertake a single analysis, since the stains may originate from more than one location.

During the transfer of soil on to clothing or other surfaces, particle fractionation may take place on the basis of size and shape, and further

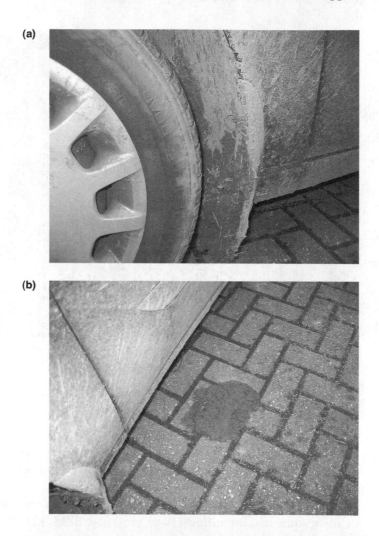

Figure 5.37 (a) Large quantities of mud and vegetation debris adhering to the wheel-arches of a car driven across a field; (b) a large lump of mud which had fallen onto the driveway after the car was parked, providing a significant opportunity for secondary transfer.

fractionation can occur during subsequent wear due to preferential retention of grains with a particular size range, shape, or surface texture. For this reason, the size distribution of the soil recovered from forensic exhibits should be noted, and preferably determined quantitatively by laser diffraction, so that an appropriate size range can be obtained from the control samples for

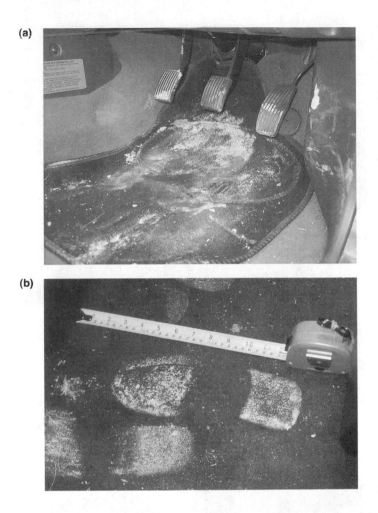

Figure 5.38 (a) A heavily mud-stained driver's footwell mat and pedal rubber covers; (b) a distinct shoe mark in mud in the front passenger footwell of the same car.

comparison. However, experience has shown that, for many routine applications, the <150 μm fraction of samples provides a suitable and convenient basis for comparison between many samples (Pye and Blott, 2004b; Pye et al., 2006c). If the available sample size is sufficient, additional useful information can be obtained by analyzing several different, more narrowly defined size fractions.

Figure 5.39 Binocular microscope picture showing particles of mud adhering to the brake pedal rubber of a suspect vehicle.

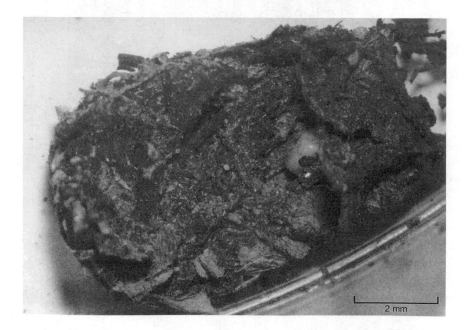

Figure 5.40 An intact "lump" of mud removed from the sole tread of a shoe worn by a murder suspect; detailed examination provided useful information about the sequence of particle acquisition onto the sole.

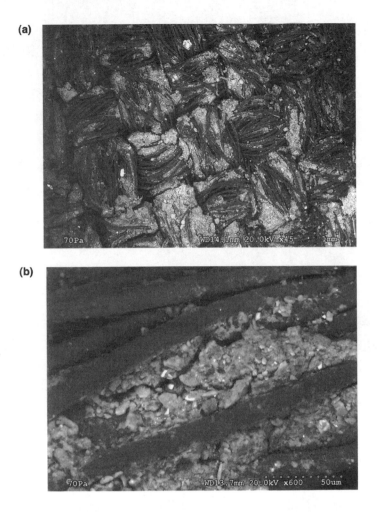

Figure 5.41 (a) and (b) Backscattered VP-SEM micrographs of phosphorus and nitrogen-rich mud trapped in the weave of clothing belonging to young woman who was allegedly subjected to a sexual attack in a chicken shed.

Whenever possible, more than one soil/mud sample should be taken for analysis from each item of footwear or clothing examined. The larger the number of samples analyzed, the better from a precision/representativeness point of view, and ideally at least three should be analyzed from each item. However, this must be considered alongside cost/time considerations and the need to preserve sample material for possible re-examination.

Evaluation of the Significance of Geological and Soil Evidence

6

6.1 General Procedures and Principles

After geological or soil evidence has been recovered and analyzed, the next step is to synthesize the data, make any necessary comparisons, and interpret the likely significance of the results. Several steps are normally involved in this process, including reduction of the data into summary forms suitable for comparison, interpretation and presentation, statistical testing of hypotheses relating to apparent differences and similarities between the samples, comparison of the data with available database information and experimental results, and formulation of an opinion. In many respects, the issues relating to the processing and interpretation of geological and soil evidence are also common to other forms of trace evidence, such as glass, paint, and fibers (Robertson and Vignaux, 1995; Robertson and Grieve, 2001; Curran et al., 2000; NicDaéid, 2001; Gallop and Stockdale, 2005), although certain specific factors must be taken into account.

The questions most frequently asked of the trace evidence examiner are (1) is there a "match," and (2) if so, what is its significance? In the author's opinion, the term match is often inappropriate in the context of geological and soil evidence comparison. A true match can only be said to exist in rare circumstances, for example, where pieces of a broken rock form a physical fit, or "jigsaw match." Where no such physical fit can be demonstrated, the best that can be said is that two soil or rock samples are of the same type and have very extremely similar, or perhaps indistinguishable, physical,

225

chemical and/or biological characteristics. The alternative possibilities that they originate from the same immediate source, from the same ultimate geological source but different immediate sources, or from different geological sources that contain indistinguishable material, often cannot be discriminated on the basis of examination of the trace evidence alone.

There are three possible opinions that can be reached after comparing a questioned sample with a possible source (control) sample. First, it may be possible to identify a firm positive association, for example, on the basis of a physical fit or similar criterion. Second, a confident exclusion may be made where two samples of soil or rock differ in terms of several major respects. Third, it may be concluded that there are no significant differences between the questioned sample and possible source sample, and therefore the possibility that the questioned sample and potential source sample are associated in some way cannot be excluded. The difficulty arises in attempting to provide a measure of the likely significance of the observed similarity; that is, to assess the likelihood that the two samples actually are associated.

Most forms of trace evidence, unlike DNA evidence, generally cannot be assessed in strict statistical probability terms. There are several reasons for this, one being that measured parameters for most forms of trace evidence vary between sub-samples of an object or material, whereas nuclear DNA is invariant among samples from a given person (Koons, 1999). Whereas there are a finite number of chromosomes and possible genetic combinations associated with the human genome, the number of soil or rock samples, and number of combinations of soil properties, are not fixed. The number of potential soil and sediment samples in the world is impossible to quantify and constantly changing. Soils are inherently heterogeneous at a variety of scales, and further heterogeneity may arise as a result of selective particle transfer onto an item, and selective retention after the transfer has taken place.

The inability to place a specific statistical probability estimate on the likelihood of a chance association has led to a widespread view that trace evidence is much weaker than DNA evidence, a problem which has been referred to as "the tyranny of numbers" (Mayr, 1982; Houck, 1999). However, as noted by Houck (1999, p. 3), "the tyranny of numbers is a consequence of an over-reliance on deduction and mathematics, and these ultimately limit a discipline by requiring it to fit into a pre-ordained model. Equating quantification with science to justify and validate its 'science-ness' indicates that a 'faulty notion of science, or no notion at all,' is at the heart of the tyranny."

Quantitative methods, including formal hypothesis testing, clearly have an important role to play in the assessment of all forms of trace evidence, but they will never be able to provide the whole story. Issues of "uniqueness," "rarity," "randomness," and "representativeness" in relation to trace evidence are difficult to deal with, and usually impossible to demonstrate in an exact mathematical or meaningful statistical way. Statistical estimates of frequency of occurrence can only ever be context-dependent, based on the size, spatial extent, and timing of any sampling exercise carried out, and the methods used for sample collection and analysis. However, this context-dependency of trace evidence can also be regarded as a strength, as well as a weakness. As pointed out by Houck (1999, p. 6), "context, is in fact, the crucial component to a proper grasp of the significance of trace evidence. Without context, we are communicating mere facts with no foundation of meaning, much in the way Poincare's pile of stones is not a house."

Availability of a suitable context within which to place the findings of a comparative trace evidence examination depends partly on the experience/knowledge of the forensic trace evidence examiner, with regard to the materials under examination, and the environmental contexts in which they occur, and partly on the provision to the examiner of sufficient relevant information relating to the circumstances of the case. While the former is substantially under the control of the examiner himself, unfortunately the latter is not. Frequently, the forensic examiner is not provided with detailed information relating to the circumstances of a case, for example, about the employment or geographical movements of suspects, which may have an important bearing on the interpretation of the *evidential value* of the findings. For this reason, the examiner should, so far as possible, restrict his or her assessment of possible significance purely to "scientific" matters. The wider issue of evidential significance is more properly one for a court to decide, and any opinion on this aspect which a forensic examiner or expert witness is asked to give must always be regarded within the context of the background information with which he/she has been provided.

6.2 Exploratory Data Analysis

In simple situations, a forensic geological comparison may simply involve a visual comparison of two objects, such as two pieces of rock or mineral gemstones, and there may be few, if any, numerical data produced. Photographic evidence may be all that is required to demonstrate a physical fit or high degree of similarity in terms of shape, size, surface texture, and color. However, in most situations, a combination of several different physical and chemical tests is also employed, resulting in the production of a large quantity of numerical data. In order to facilitate comparison of these data, it is standard practice to summarize them using *descriptive statistics*. An essential initial step is that of *exploratory data analysis*, which involves visual examination of the data sets, displayed in the form of tables, histograms, and scatter plots (Smith and Prentice, 1993; Johnson and Ehrlich, 2001; Webster, 2001). Numerous textbooks are available which describe these procedures, in the specific context of geology and soil science (Webster and Oliver, 1990;

Rollinson, 1993; Swan and Sandilands, 1995; Davis, 2002), forensic science (Lucy, 2005), and more generally (e.g., Tukey, 1977; Hollander and Wolfe, 1999; Kirk, 1999; Miller and Miller, 2005; Buccianti et al., 2006).

The first step in interrogating any data set should be to plot the values to gain a visual impression of the nature of the distribution of values. This can be done by using a variety of methods, including scatter plots, frequency histograms, cumulative frequency, curves, or box and whisker-plots (Figure 6.1, Figure 6.2, and Figure 6.3). These procedures allow the

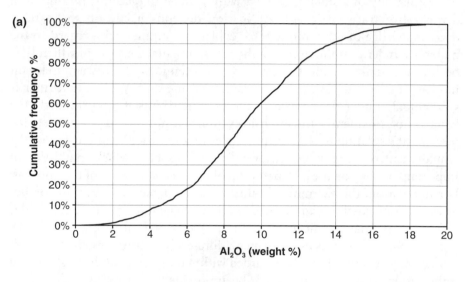

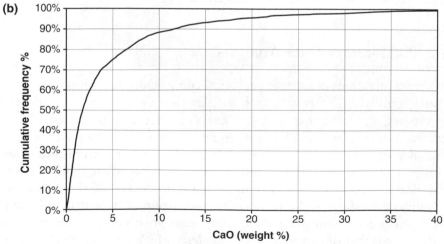

Figure 6.1 Cumulative frequency distribution curves for concentrations (a) of %Al_2O_3 and (b) %CaO in the KPAL soil chemistry database (1895 samples included).

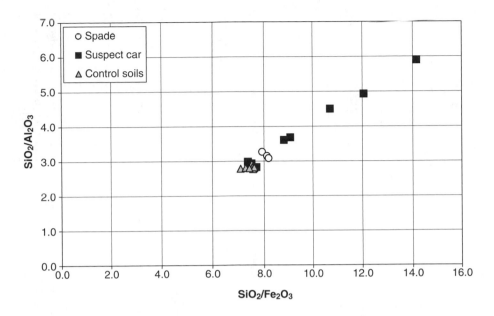

Figure 6.2 Bivariate plot of SiO_2/Al_2O_3 ratio against SiO_2/Fe_2O_3 ratio for three groups of samples taken in a murder investigation. The suspect claimed to have used the spade to dig his car out of the mud at the location from which the control samples were taken. Only some of the mud samples taken from the car were found to be similar to those at the "control" site.

investigator to identify any outlier values, which in some cases may indicate problems with the analyses or data entry process, and to establish whether the values approximate a *normal distribution* or not. The latter issue is of considerable significance since it determines which further statistical procedures should most appropriately be employed (see Sutherland, 2001, for a useful illustration). If the distribution of values approximates a normal distribution, parametric statistical tests can be used, but if the distribution deviates markedly from normal, it is better to use a non-parametric test that makes no assumptions about the distribution of values, or to transform the original distribution before applying a parametric test. Since a normal distribution is essentially symmetrical, that is, approximates the shape of a bell-shaped curve, an indication of the degree of non-normality in a distribution is provided by a measure of the skewness, which can be calculated using the method of moments, together with values for the mean, variance, and kurtosis (peakedness). The measure of skewness is provided by the third moment of the data about the mean. A skewness value of 0 indicates a symmetrical distribution, values of >1 indicate positive skewness (a long tail of larger values), whereas a value <1 indicates negative skewness (a tail of smaller values). If the skewness is less than 0.5, transformation should not be

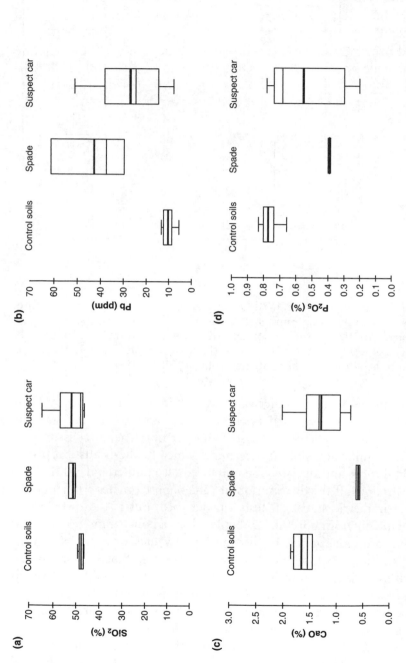

Figure 6.3 Box and whisker plots for %SiO$_2$, %CaO, %P$_2$O$_5$, and Pb concentration in the three groups of samples referred to in Figure 6.2. The thick horizontal line represents the mean, the central thin horizontal line represents the median, the lower edge of each box represents the lower quartile value, the upper edge of the box represents the upper quartile, and the outermost pair of shorter horizontal lines represent the maximum and minimum values.

necessary; if it is between 0.5 and 1.0, transformation using square roots may be sufficient, but if it is >1.0, transformation to logarithms should be applied (Webster, 2001).

A rapid assessment of the likely normality/log-normality of a distribution can also be made by a visual inspection of box and whisker plot of the type proposed by Tukey (1977) and Emerson and Strenio (2000), or of a histogram of values or cumulative frequency curve. In the case of a box plot (Figure 6.3), if the median value lies close to the lower or upper limits of the box, the distribution is markedly skewed. In the case of cumulative frequency curves, a near-normal distribution has an even "S" shaped form (e.g., the distribution of soil database Al_2O_3 values shown in Figure 6.1a), whereas a markedly non-normal distribution has a steeply rising, asymptotic form (e.g., the distribution of database soil CaO values shown in Figure 6.1b).

Bivariate scatter plots, or "dotplots" (Sasieni and Royston, 1996), allow a rapid visual comparison to be made between samples in terms of different pairs of attributes, for example, elemental concentration values or ratios (Figure 6.2). This may allow identification of possible associations between variables, which may warrant more detailed investigation using statistical tests. Using programs such as Microsoft Excel, Excel Stat, or more sophisticated statistical software packages, plots of numerous combinations of parameters can be obtained in a relatively short time. Details of several suitable software programs are provided by Sutherland (2001).

6.3 "Classical" Hypothesis Testing

In a simple case, graphical comparison may suggest that there is a difference in some specified attribute (e.g., mean particle size) between two groups of samples. However, the distribution of values for the mean of individual samples in the two groups of samples may overlap, and there may be a relatively small difference in the calculated average value of the mean for each group. The question then arises whether the observed difference between the sample means is statistically significant. The null hypothesis, H_0, that the observed difference is not statistically significant, and the alternative hypothesis, H_1, that the difference is significant, are conventionally evaluated using *inferential statistics*. A variety of tests can be used for hypothesis testing of this type, selection being dependent on the type of data and what is known about the nature of the distribution (i.e., whether it is normally distributed or not). The tests most commonly used to compare two samples in terms of a single variable are Student's *t* test and its non-parametric equivalent, the Mann-Whitney U-test. If more than two samples are to be compared, one-way analysis of variance (ANOVA), sometimes known as the F-Test, or the non-parametric equivalent, the Kruskal-Wallis test, are commonly used.

Where comparisons between samples are required on the basis of more than one variable, the multivariate equivalent of Student's t test, Hotelling's T^2 test, or multivariate analysis of variance (MANOVA) can be used (for further information, see Kirk, 1999; Davis, 2002; Rencher, 2002).

By way of a simple example, Table 6.1 shows the results of two-tailed t tests carried out to compare the particle size parameters in four groups of soil samples taken during a murder investigation. Two of the groups relate to two distinct types of mud deposit which were removed from the car, one group relates to control samples taken at a location where the car was abandoned, and the fourth group relates to soil recovered from a spade claimed by a suspect to have been used to attempt to dig the car out of the mud at the location where it was abandoned. In terms of the particle size parameters considered, only mud samples in one of the sample groups from the car showed statistically significant differences to the others at the 95% confidence level.

It is important to recognize that the outcome of any such statistical test need not necessarily be correct; that is, H_o may be rejected when it is true (a Type I error), or accepted when it is false (a Type II error), even at relatively high levels of specified statistical significance ($p > 0.05$ or $p > 0.01$; Koons and Buscaglia, 2002; Miller and Miller, 2005). The likelihood of this occurring is dependent on a number of factors, including the size of the sample (number of data points) and the *power* of the test being used; risks of a false rejection or acceptance increase markedly with very small sample sizes (Cohen, 1988; Sutherland, 2001).

Student's t test and the F-test (ANOVA) are probably the most widely used in science. Although these are parametric tests, they are relatively insensitive to non-normality in the data distributions (Webster, 2001). The use of

Table 6.1 Probabilities Resulting from Two-Tailed t-tests Comparing the Particle Size Parameters Measured in Four Groups of Samples (Soil from a Spade and a Car Associated with a Murder Suspect, and Soil from the Location where the Car was Abandoned, Data Shown in Table 3.6)

Sample Groups Compared	Mean (μm)	Mode (μm)	D10 (μm)	D50 (μm)	D90 (μm)	%Sand	%Silt	%Clay
Spade and suspect car (group 1)	18.1	11.6	12.9	15.2	84.1	14.9	14.7	23.1
Spade and suspect car (group 2)	43.8	5.1	61.4	73.4	11.7	69.9	54.7	11.9
Spade and control soils	95.5	81.5	74.1	71.8	12.3	98.4	95.8	64.8
Suspect car (group 1) and suspect car (group 2)	18.9	21.5	70.3	23.6	13.1	12.8	8.4	27.1
Suspect car (group 1) and control soils	1.3	29.8	2.2	0.9	14.4	0.6	0.6	4.1
Suspect car (group 2) and control soils	43.2	8.1	70.8	57.4	27.1	63.9	45.7	8.8

Note: Values lower than 5% indicate a significant difference between the groups at the 95% confidence level.

non-parametric equivalents, which effectively "degrade" the data by converting interval data to ranks, may be regarded as a "conservative" approach that reduces the chances that the differences between the means of two data sets will be shown to be statistically significant.

Dudley and Smalldon (1978a) compared the particle size distributions of 18 duplicate soil samples from Berkshire, England, using conventional two-sample statistical tests. They found that soils from different locations could not always be distinguished, and more worryingly, that some of the duplicate soil samples were identified as being different at conventional levels of statistical significance. They concluded that such tests were too unreliable for routine use and suggested an alternative "elliptical contour model" which utilized two new parameters derived from the particle size distribution data for duplicate samples.

Wanogho et al. (1985) developed a method for comparing closely spaced soils based on an experimental investigation of particle size properties in a ploughed field near Glasgow, which had been divided into one hundred 3 × 3 m cells. Two-way analysis of variance was performed on three parameters which were chosen to characterize the samples: the median (D_{50} value), the percentage mass in the modal size class of 90 to 250 μm, and the percentage of organic matter determined as weight loss following hydrogen peroxide treatment. A method of comparing the similarity of different cells in terms of the three parameters, termed the *Similarity Probability Model*, was developed and found to be capable of accurately predicting the identity of "unknown" samples in a high proportion of cases. This method, like that developed by Dudley and Smalldon (1978a), appeared to offer some potential, but neither has been subjected to extensive testing under operational conditions (Gettinby, 1991).

6.4   Correlation and Regression Analysis

The degree of association between two variables where scores for one variable are paired with the scores for the other variable can be quantified by correlation analysis and regression analysis. The two procedures have much in common and are frequently confused. Correlation analysis measures the degree of association or relationship between variables, without making any assumption about dependency in the relationship. By contrast, in regression analysis, one variable is identified as the independent variable and the other as the dependent variable. For example, soil moisture content may be expected to be at least partly dependent on rainfall amount, but rainfall is unlikely to be affected by soil moisture content; rainfall can therefore be identified as the independent variable and soil moisture content as the dependent variable. However, in many situations, the nature of the relationship between two variables may be unknown, and no assumptions can be made about causal association. In such circumstances, correlation analysis is applicable. Regression procedures are most commonly employed in situations where the dependent/independent variable relationship is known and where there is a desire to predict the effects of changes in the independent variable on the dependent variable (Cook and Weisberg, 1994), but there have been many instances of its misuse in geological and soil science research (Mark and Church, 1977; Webster, 1997).

6.4.1   Correlation

Two variables may be said to be positively correlated if an increase in values of one variable is associated with a general increase in values of the other variable. Conversely, the variables are said to be negatively correlated if an increase in the values of one variable are associated with a decrease in the values of the other variable. If the data points plot approximately as a straight line on a bivariate diagram, the correlation is said to be linear. However, correlations may also be non-linear (e.g., curvilinear).

The degree of association or strength of a relationship between two variables can be represented by a correlation coefficient. Probably the most widely used measure of degree of correlation is *Pearson's product-moment correlation coefficient*, r, values of which can range from -1 to $+1$. A value of $+1$ indicates a perfect positive relationship (perfect correlation), whereas a value of -1 indicates a perfect negative relationship. A value of 0 indicates no association. Two derivative statistics of association may also be employed, the coefficient of determination, r^2, which is equal to the square of the correlation coefficient, and the coefficient of non-determination, k^2, which is equal to $1 - r^2$.

Pearson's correlation analysis is a parametric test which utilizes interval data (infinitely variable values), rather than ranks (ordinal data). The non-parametric equivalent is *Spearman's rank correlation coefficient*, r_s, which also has values ranging from -1 to $+1$. In geological and soil studies, the nature of the distribution of background population values is often uncertain, or varies from one soil attribute to another (e.g., compare the Al_2O_3 and CaO concentrations in database soils shown in Figure 6.1), and it can be argued that the use of parametric tests is frequently unjustified. However, the conversion of measured values or percentages to ranks involves degrading the data. Spearman's rank correlation coefficient simply provides a measure of the monotonic relationship between two sets of ranks. As such it may underestimate or overestimate the degree of similarity in circumstances where measured values are converted to ranks. For example, many samples of soil taken within an area may have the same rank sequence of major oxide concentrations, making them effectively indistinguishable using Spearman's rank correlation coefficient, even though the actual percentage concentration values of particular oxides may differ considerably. In such circumstances, there can be an advantage in performing both Pearson's and Spearman's correlations and comparing the results, all of which should be treated with caution. In the case of Pearson's correlation coefficient, spuriously high linear relationships between two variables or samples may be indicated due to the effect of "outlier" values (Rong, 2000; Sutherland, 2001). However, despite its limitations, experience has shown that Pearson's correlation analysis can often accurately group soil samples of known origin on the basis of similarities and differences in elemental composition (Pye and Blott, 2004b).

Table 6.2 shows an example of Pearson's correlation matrices for the <150 μm fractions of soils from two 1 × 1 m experimental grids at two locations in Berkshire, U.K. Separate matrices are shown for correlations carried out using all 49 elements determined, 10 major elements only, 25 trace elements only, and 14 rare earth elements only. Two levels of correlation, 0.99 and 0.95, have been selected and highlighted to help identify similarities and differences. The two sites can be differentiated using three of the four matrices, the exception being the rare earths, although the differences in r values between the two groups are small, using the major element data only. The matrices for all 49 elements, and for the 25 trace elements only, show that within the Simon's Wood grid, three samples, located along one side, are clearly distinguishable from others in the grid. The differences are statistically significant at the 99% confidence level.

Table 6.3 compares Pearson's and Spearman's rank correlation values for five soil samples taken from the boots of a suspect in an aggravated burglary with nine samples taken from a nearby location where his car was found abandoned, allegedly having been stolen. The question of investigative interest was whether the suspect had in fact driven and abandoned the car at the location where it was found. Pearson's correlation using data for 49 elements indicated a great majority of r values higher than 0.99, suggesting a high degree of similarity between the samples from the boots and the location where the car was abandoned. However, Spearman's rank correlation gave significantly lower values of r_s, higher than 0.95 but lower than 0.99, providing less support for a possible association between the questioned samples and the controls.

6.4.2   Regression

In situations where there is a requirement to extrapolate beyond the limits of a measured data set, a best fit regression line may be fitted to the data and an index of "best fit," or *regression coefficient*, obtained. There are several different types of regression, the most commonly used being ordinary linear least squares regression (OLLS), which is only truly applicable if the data plot approximately

Table 6.2 Matrices of Pearson's Correlation Values Obtained by Comparing <150 μm Soil Fractions from Two Distinctly Different Locations (Simon's Wood and Arborfield Bridge, Both in Berkshire, U.K.) in Terms of Elemental Concentrations Determined by ICP-AES and ICP-MS (a) All 49 Elements

		Simon's Wood									Arborfield Bridge								
		A1	A2	A3	A4	A5	A6	A7	A8	A9	B1	B2	B3	B4	B5	B6	B7	B8	B9
Simon's Wood	A1	1.000																	
	A2	0.998	1.000																
	A3	0.997	0.999	1.000															
	A4	0.999	0.998	0.998	1.000														
	A5	0.998	0.999	0.998	0.999	1.000													
	A6	0.992	0.998	0.997	0.992	0.996	1.000												
	A7	0.982	0.992	0.990	0.983	0.989	0.997	1.000											
	A8	0.951	0.932	0.935	0.951	0.938	0.910	0.881	1.000										
	A9	0.931	0.908	0.910	0.930	0.915	0.881	0.850	0.997	1.000									
Arborfield Bridge	B1	0.751	0.730	0.743	0.757	0.738	0.690	0.652	0.894	0.901	1.000								
	B2	0.707	0.688	0.703	0.715	0.697	0.647	0.612	0.856	0.862	0.995	1.000							
	B3	0.746	0.725	0.739	0.752	0.733	0.686	0.650	0.888	0.895	0.999	0.997	1.000						
	B4	0.684	0.659	0.674	0.690	0.669	0.616	0.575	0.850	0.862	0.995	0.996	0.995	1.000					
	B5	0.836	0.824	0.836	0.843	0.830	0.792	0.764	0.929	0.926	0.983	0.976	0.984	0.963	1.000				
	B6	0.788	0.771	0.783	0.795	0.778	0.733	0.700	0.910	0.914	0.995	0.990	0.995	0.984	0.993	1.000			
	B7	0.762	0.750	0.764	0.770	0.756	0.715	0.687	0.878	0.876	0.988	0.992	0.991	0.980	0.991	0.994	1.000		
	B8	0.812	0.792	0.803	0.818	0.800	0.754	0.720	0.932	0.938	0.993	0.982	0.991	0.979	0.991	0.997	0.984	1.000	
	B9	0.740	0.713	0.725	0.745	0.723	0.670	0.630	0.897	0.911	0.995	0.986	0.993	0.992	0.969	0.987	0.971	0.990	1.000

Legend: 0.950 – 0.990 ; 0.990 – 1.000

(b) 10 Major Oxides Only

Legend: ▢ 0.950 – 0.990   ▢ 0.990 – 1.000

		A1	A2	A3	A4	A5	A6	A7	A8	A9	B1	B2	B3	B4	B5	B6	B7	B8	B9
Simon's Wood	A1	1.000																	
	A2	1.000	1.000																
	A3	1.000	1.000	1.000															
	A4	1.000	1.000	1.000	1.000														
	A5	1.000	1.000	1.000	1.000	1.000													
	A6	1.000	1.000	1.000	1.000	1.000	1.000												
	A7	1.000	1.000	1.000	1.000	1.000	1.000	1.000											
	A8	1.000	1.000	1.000	1.000	1.000	1.000	1.000	1.000										
	A9	1.000	1.000	1.000	1.000	1.000	1.000	1.000	1.000	1.000									
Arborfield Bridge	B1	0.995	0.995	0.995	0.995	0.995	0.995	0.995	0.996	0.996	1.000								
	B2	0.992	0.992	0.993	0.992	0.993	0.992	0.992	0.993	0.994	0.999	1.000							
	B3	0.994	0.994	0.995	0.994	0.995	0.994	0.994	0.995	0.996	1.000	1.000	1.000						
	B4	0.992	0.992	0.993	0.992	0.993	0.992	0.992	0.993	0.994	1.000	1.000	1.000	1.000					
	B5	0.995	0.995	0.996	0.996	0.996	0.996	0.996	0.996	0.997	1.000	0.999	0.999	0.999	1.000				
	B6	0.995	0.995	0.996	0.995	0.995	0.995	0.995	0.996	0.996	1.000	0.999	1.000	1.000	1.000	1.000			
	B7	0.995	0.995	0.996	0.995	0.995	0.995	0.995	0.996	0.997	1.000	0.999	0.999	0.999	1.000	1.000	1.000		
	B8	0.996	0.996	0.996	0.996	0.996	0.996	0.996	0.997	0.997	1.000	0.999	1.000	0.999	1.000	1.000	1.000	1.000	
	B9	0.995	0.995	0.996	0.996	0.996	0.995	0.995	0.996	0.997	1.000	1.000	1.000	0.999	0.999	1.000	1.000	1.000	1.000

(*continued*)

Table 6.2 (c) 25 trace elements only (Continued)

		Simon's Wood									Arborfield Bridge								
		A1	A2	A3	A4	A5	A6	A7	A8	A9	B1	B2	B3	B4	B5	B6	B7	B8	B9
Simon's Wood	A1	1.000																	
	A2	0.998	1.000																
	A3	0.997	0.999	1.000															
	A4	1.000	0.998	0.998	1.000														
	A5	0.998	0.999	0.999	0.999	1.000													
	A6	0.992	0.998	0.997	0.992	0.995	1.000												
	A7	0.981	0.991	0.989	0.982	0.988	0.997	1.000											
	A8	0.946	0.925	0.928	0.945	0.931	0.900	0.867	1.000										
	A9	0.923	0.897	0.900	0.922	0.905	0.868	0.832	0.996	1.000									
Arborfield Bridge	B1	0.729	0.703	0.716	0.733	0.713	0.662	0.616	0.888	0.894	1.000								
	B2	0.679	0.655	0.670	0.684	0.665	0.613	0.570	0.845	0.849	0.995	1.000							
	B3	0.721	0.696	0.711	0.726	0.706	0.656	0.613	0.880	0.886	0.999	0.997	1.000						
	B4	0.653	0.624	0.639	0.657	0.635	0.579	0.531	0.839	0.850	0.994	0.996	0.995	1.000					
	B5	0.825	0.809	0.821	0.830	0.816	0.777	0.741	0.929	0.922	0.981	0.972	0.981	0.958	1.000				
	B6	0.771	0.749	0.762	0.775	0.758	0.711	0.670	0.908	0.909	0.996	0.990	0.996	0.984	0.993	1.000			
	B7	0.741	0.725	0.740	0.747	0.732	0.690	0.654	0.870	0.866	0.987	0.991	0.991	0.978	0.990	0.993	1.000		
	B8	0.798	0.773	0.785	0.801	0.783	0.735	0.693	0.933	0.937	0.993	0.980	0.991	0.977	0.990	0.997	0.981	1.000	
	B9	0.717	0.685	0.697	0.719	0.696	0.640	0.592	0.892	0.906	0.994	0.984	0.992	0.991	0.965	0.987	0.968	0.990	1.000

0.950 – 0.990

0.990 – 1.000

(d) 14 Rare Earth Elements Only

		Simon's Wood									Arborfield Bridge								
		A1	A2	A3	A4	A5	A6	A7	A8	A9	B1	B2	B3	B4	B5	B6	B7	B8	B9
Simon's Wood	A1	1.000																	
	A2	0.975	1.000																
	A3	0.993	0.994	1.000															
	A4	0.982	0.999	0.997	1.000														
	A5	0.941	0.991	0.974	0.987	1.000													
	A6	0.998	0.987	0.998	0.992	0.960	1.000												
	A7	0.954	0.996	0.983	0.994	0.998	0.971	1.000											
	A8	1.000	0.980	0.995	0.986	0.948	0.999	0.961	1.000										
	A9	0.999	0.985	0.998	0.990	0.957	1.000	0.968	1.000	1.000									
Arborfield Bridge	B1	0.988	0.997	0.999	0.999	0.980	0.995	0.988	0.991	0.995	1.000								
	B2	0.985	0.998	0.998	1.000	0.983	0.994	0.991	0.989	0.993	1.000	1.000							
	B3	0.982	0.999	0.997	1.000	0.985	0.992	0.993	0.987	0.991	0.999	1.000	1.000						
	B4	0.980	0.999	0.996	1.000	0.987	0.991	0.994	0.985	0.989	0.999	1.000	1.000	1.000					
	B5	0.978	1.000	0.995	1.000	0.989	0.989	0.995	0.983	0.987	0.998	0.999	1.000	1.000	1.000				
	B6	0.993	0.994	1.000	0.997	0.972	0.998	0.981	0.996	0.998	0.999	0.998	0.997	0.996	0.995	1.000			
	B7	0.988	0.997	0.999	0.999	0.979	0.996	0.988	0.992	0.995	1.000	1.000	0.999	0.999	0.998	0.999	1.000		
	B8	0.992	0.995	0.999	0.998	0.974	0.998	0.984	0.994	0.997	1.000	0.999	0.998	0.998	0.997	1.000	1.000	1.000	
	B9	0.986	0.998	0.999	0.999	0.983	0.994	0.990	0.989	0.993	1.000	1.000	1.000	1.000	0.999	0.998	1.000	0.999	1.000

Legend:
- 0.950 – 0.990
- 0.990 – 1.000

Note: In each case, the highest correlation values are highlighted.

Table 6.3 Matrices of Correlation Values Obtained by Comparing Soil Samples Taken from a Suspect's Boots with Control Soil Samples in Terms of 49 Elemental Concentrations Determined by ICP-AES and ICP-MS (a) Parametric Pearson's Correlation Coefficients

Legend: 0.950 – 0.990 ; 0.990 – 1.000

		Boot Mud					Control Soils								
		KP 1	KP 2	KP 3	KP 4	KP 5	KP 17	KP 18	KP 19	KP 20	KP 21	KP 22	KP 23	KP 24	KP 25
Boot mud	KP 1	1.000													
	KP 2	0.998	1.000												
	KP 3	1.000	0.998	1.000											
	KP 4	0.997	0.999	0.997	1.000										
	KP 5	1.000	0.998	1.000	0.997	1.000									
Control soils	KP 17	0.993	0.988	0.994	0.984	0.994	1.000								
	KP 18	0.997	0.993	0.998	0.991	0.998	0.999	1.000							
	KP 19	0.999	0.997	0.999	0.996	0.999	0.996	0.999	1.000						
	KP 20	0.997	0.996	0.998	0.994	0.998	0.997	0.999	0.999	1.000					
	KP 21	0.999	0.997	0.999	0.995	0.999	0.996	0.999	1.000	0.999	1.000				
	KP 22	0.996	0.992	0.997	0.989	0.997	1.000	1.000	0.998	0.999	0.998	1.000			
	KP 23	0.996	0.992	0.997	0.989	0.997	1.000	1.000	0.998	0.999	0.998	1.000	1.000		
	KP 24	0.998	0.999	0.998	0.998	0.998	0.989	0.995	0.998	0.997	0.998	0.993	0.993	1.000	
	KP 25	0.996	0.992	0.997	0.989	0.997	0.999	1.000	0.998	0.999	0.998	1.000	1.000	0.993	1.000

(b) Nonparametric Spearman's Rank Correlation Coefficients

	Boot Mud					Control Soils								
	KP 1	KP 2	KP 3	KP 4	KP 5	KP 17	KP 18	KP 19	KP 20	KP 21	KP 22	KP 23	KP 24	KP 25
Boot mud														
KP 1	1.000													
KP 2	0.995	1.000												
KP 3	0.993	0.997	1.000											
KP 4	0.992	0.993	0.995	1.000										
KP 5	0.993	0.996	0.998	0.997	1.000									
Control soils														
KP 17	0.950	0.962	0.974	0.975	0.975	1.000								
KP 18	0.951	0.963	0.975	0.975	0.976	0.998	1.000							
KP 19	0.943	0.957	0.970	0.968	0.969	0.997	0.998	1.000						
KP 20	0.949	0.961	0.974	0.973	0.974	0.998	0.998	0.998	1.000					
KP 21	0.959	0.969	0.981	0.981	0.981	0.997	0.997	0.997	0.998	1.000				
KP 22	0.952	0.965	0.976	0.976	0.977	0.998	0.999	0.998	0.999	0.999	1.000			
KP 23	0.950	0.962	0.975	0.975	0.975	0.999	0.999	0.998	0.998	0.997	0.999	1.000		
KP 24	0.966	0.976	0.984	0.985	0.985	0.995	0.995	0.993	0.994	0.996	0.995	0.996	1.000	
KP 25	0.951	0.963	0.975	0.974	0.975	0.998	0.999	0.999	0.999	0.999	0.999	0.998	0.994	1.000

Legend:
0.950 – 0.990
0.990 – 1.000

Note: In each case, the highest correlation values are highlighted.

is a straight line (Clark and Hosking, 1986). If the data plot is exactly on a straight line, the regression coefficient will have a value of +1. However, in most cases, the data points will be scattered around the line. The difference between the measured and predicted value of any given data point is known as a *prediction error*, or *residual*. The line of best fit through the data is one which minimized the sum of the squared prediction errors. Both linear and non-linear lines of best fit can be determined using linear and non-linear regression models. The degree to which one variable (Y) is predicted to vary as a function of the other (X) is reflected by the intercept of the regression line with the X axis and by the slope of the regression line, which are used to calculate the *coefficient of determination*, r^2. A similar relationship can be determined to quantify the dependency of X on Y. Although regression analysis is widely used in geological and soil research, it is a rare requirement in environmental and criminal forensic investigations which are primarily compared with comparisons between samples or groups of samples; if prediction is not a major goal of the investigation, correlation should be used instead because it makes fewer assumptions about the data (Sutherland, 2001).

6.5 Multivariate Analysis

Multivariate analysis methods can be used when several measurements are made on each individual or object, such as a soil sample, and are usually employed when considerably more than two samples are to be compared. The measurements may relate to different aspects of the same attribute (e.g., concentrations of different elements determined in a single chemical analysis), or to distinctly different attributes measured on different scales (e.g., chemical data, particle size data, and color parameters) and considered in combination. Multivariate techniques are essentially exploratory, searching for groupings with data sets and helping to generate hypotheses, rather than testing them (Webster and Oliver, 1990; Rencher, 2002).

Many different methods of multivariate analysis are available, including multiple correlation, multiple regression, multivariate analysis of variance, discriminant analysis, canonical analysis, principal component analysis (PCA), factor analysis, and hierarchical cluster analysis (HCA). A full discussion is beyond the scope of this book, and the interested reader is referred to texts such as Rencher (2002) and Miller and Miller (2005) for further information. In forensic geological work, PCA and HCA are among the most widely used. PCA has been used in soils research since the 1960s, and has proved to be a highly useful tool which has contributed to many fundamental insights (Webster, 2001). However, PCA is really only useful where the variables concerned are correlated, for example, the determinands in a chemical analysis, the percentage frequencies in a particle size distribution analysis, or the reflectance values at different wavelengths in a quantitative color analysis. The technique can be performed with as few as two variables, but most frequently analyses involve many more.

6.5.1 Principal Component Analysis

A series of concentration values for a number of different elements (n values) relating to a suite of N soil samples can be envisaged as a distribution of N points in a Euclidean space of n dimensions, rather like a scatter graph with n orthogonal axes. PCA seeks to reduce the number of dimensions to a smaller number which allow most of the variance in the data set to be explained. The analysis seeks new axes in the multidimensional space so that the first axis lies in the direction of maximum variance. The second axis, orthogonal to the first, lies in the direction of maximum variance in the residuals from the first, whereas the second lies in the direction of maximum variance in residuals from the second, and so on. There are several variants of PCA, but most contain three sets of values in the output: eigenvalues, eigenvectors, and component scores. The component scores are most commonly plotted on scatter diagrams in two or three dimensions with an indication on the axes of the percentage of the variance which is "explained" by each component. In many situations, a high proportion of the total variance (60 to 95%) is explained by the first two or three components. A plot of the first two components may adequately distinguish between groups, but in some instances the third, fourth, or higher components may need to be considered to adequately resolve different groups (Johnson and Ehrlich, 2002).

A simple example of a two-dimensional PCA plot is illustrated in Figure 6.4. The data relate to the same soil samples shown in Figure 6.2 and Figure 6.3. From this plot alone it is evident that samples from the suspect car fall into two groups, one being tightly clustered and overlapping with control soil samples taken from the location where the car was found abandoned, the other being more variable and representing soil derived from more than one location, possibly over a significant period of time. The samples from the suspect's spade form a group which can be distinguished both from the control sample group from the scene of interest and from both of the car sample groups.

6.5.2 Hierarchical Cluster Analysis

Although PCA will often identify groups of like samples or objects, it is not always successful in doing so. A complementary technique that specifically attempts to identify groups is cluster analysis. This procedure searches for objects (or samples) that are close together in the n-dimensional space. The distance, d, between two points in n-dimensional space with coordinates

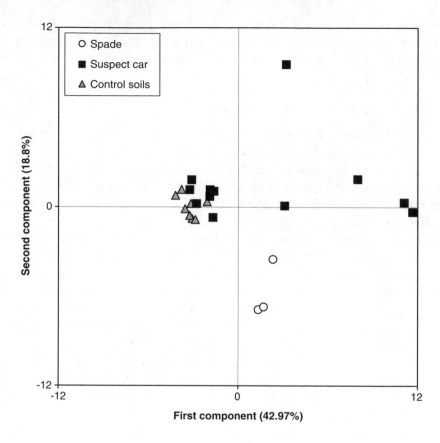

Figure 6.4 A plot of the first two principal components, taking into account data for 49 major and trace elements, for the samples referred to in Figure 6.2. The questioned samples taken from the spade plot well outside the field of the control samples.

$(x_1, x_2...x_n)$ and $(y_1, y_2...y_n)$ is termed the Euclidean distance. The successive stages of grouping according to the Euclidean distance can be shown diagrammatically on a dendrogram. Figure 6.5 illustrates a hierarchical cluster dendrogram for the same samples shown in Figure 6.4. The figure again clearly shows that the soil samples from the car fall into two distinct groups, one of which is very similar to the control soil samples and one of which is very dissimilar to all the other samples analyzed. The soil samples from the spade form a discrete group, which can be distinguished from the control soil sample group and both groups of samples from the car.

Both PCA and HCA make no assumptions about the nature of the distribution of data being analyzed, and consequently no prior transformations are necessary. Equally, these techniques provide no information about the statistical significance of any of the indicated groupings.

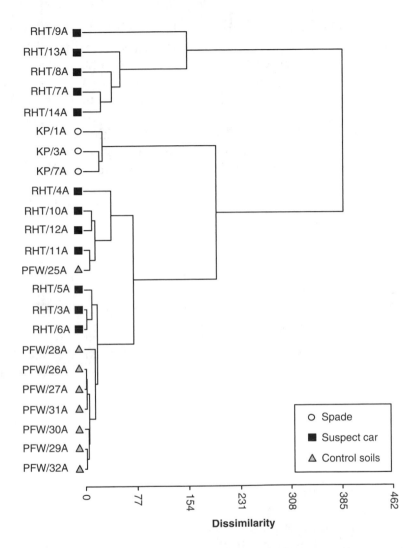

Figure 6.5 Hierarchical cluster dendrogram calculated using Ward's method based on chemical data for the samples referred to in Figure 6.2. The two groups of mud samples on the car are clearly differentiated, while the spade mud samples form a discrete group which is distinguishable from the control group and both groups from the car.

6.6 Combined Approaches

Rawlins and Cave (2004) used a combination of univariate and multivariate analysis to determine whether soils samples developed on the same parent material in part of eastern England could be discriminated on the basis of both individual and multi-element geochemistry. Discrimination was based on estimates of measurement uncertainty, which was calculated from analysis of duplicates and sub-samples. The methods used were *robust ANOVA* and nested analysis of covariance using a residual maximum likelihood (REML) algorithm. Results of the univariate analysis showed that for 13 individual elements, it was possible on average to discriminate between 80% of the samples within parent material groups. CaO and MgO were found to be the most discriminatory and U and Mo the least so. With regard to sample discrimination within parent material groups, between 15 and 17 of the 19 elements discriminated individual samples, taking into account the measurement uncertainty. The results of the multivariate analysis confirmed that multi-element geochemical signatures could be established within most parent material groups.

6.7 Assessment of Coincidence Probabilities and Likelihood Ratios

6.7.1 Chance Coincidence Probabilities

A meaningful assessment of the statistical probability of a chance coincidence is extremely difficult, and perhaps impossible, to make with any type of trace evidence. However, some form of assessment can be made if a significant amount of relevant information exists in the form of a relevant database. However, all databases suffer limitations imposed by the number of samples they contain, the degree to which those samples are representative of the "population" at large, the number and quality of the criteria which form a basis for inter-sample comparison, and the fact that nothing stands still, and some types of database may be rendered out of date very quickly. In the case of trace evidence such as glass and textile fibers, changes in production methods, delivery systems, and disposal of manufactured products over time effectively negates the value of use of database information for the preparation of probability statements (Koons, 1999). Such databases can, however, provide useful sources of contextual information (Houck, 1999).

Classical probability theory requires that the number of possible outcomes can be calculated precisely based on accurate knowledge of the number of individuals or items and the number of attributes involved. This is very rarely the case when considering geological or soil evidence, although exceptional circumstances may arise when working, for example, with gemstones or artefacts which have been made in relatively small, determinable numbers. The number of potential soil samples which could be taken from around the world is impossible to define precisely, since a potential soil sample has no defined physical size, mass, or character, and the same body of soil material could be sampled, re-mixed, and re-sampled in many different ways.

Moreover, any results relating to the physical and chemical properties of soils and sediments are both spatial scale and time-scale dependent.

6.7.2   Bayesian Statistical Approaches

Particularly over the past 20 years, Bayesian statistical approaches have become popular in many branches of science, including forensic science (Aitken and Stoney, 1991; Aitken, 1985; Cook et al., 1998; Aitken and Taroni, 2004). Bayesian approaches have been applied to the interpretation of DNA (Buckleton et al., 2005), glass (Evett, 1986; Evett and Buckleton, 1989; Curran et al., 2000), fibers (Taroni and Aitken, 1998; Champod and Taroni, 1999), paint (Willis et al., 2001), and bullet lead analyses (Finkelstein and Levin, 2005). Within geosciences, Bayesian approaches have been applied to the issue of sediment fingerprinting and discrimination of alternative sediment sources in fluvial systems (Franks and Rowan, 2000; Small et al., 2004), but there have so far been very few applications in forensic casework.

A detailed discussion of the Bayesian philosophy and its applications to forensic evidence is beyond the scope of this book, and the interested reader should consult the sources mentioned above for further information. However, in essence, *Bayes' theorem*, which was initially developed by Thomas Bayes in the 18th century (Bayes, 1763), simply relates a set of unconditional probabilities (or *prior probabilities*) to a set of conditional probabilities (or *posterior probabilities*).

The ratio of the probability of evidence, or information relating to one proposed scenario (proposition) to that of the probability of evidence/information relating to an alternative proposition, is defined as the *likelihood ratio*, LR (Taroni and Aitken, 1998). This ratio has been used to provide a quantitative measure of the value, V, of a piece of evidence. Values of LR theoretically range from 0 to infinity. A qualitative verbal terminology scale equating to various ranges of LR (or V) has been proposed and widely used in the context of DNA and related evidence (Evett et al., 2000; Table 6.12). In the case of DNA, for which very large likelihood ratios can be calculated, it has become common practice to apply the phrase *extremely strong evidence* for likelihood ratios of 1 million or more.

Advocates of the Bayesian approach maintain that one of its main advantages is that it provides a flexible framework for the evaluation of probabilities, in which different scenarios can be tested using different prior information (Lucy, 2005). However, critics point to the frequently arbitrary and unjustified adoption of prior probability information. The quality of this information remains the critical limiting factor, and in the case of many types of trace evidence, including soil, the information can only be provided in a qualitative, rather than a quantitative, form (Koons, 1999).

6.8 Direct Data Comparison: Deciding if Two Samples Are Indistinguishable, Similar, or Different

No two samples of rock, sediment, or soil can be regarded as identical in every respect. Indeed, if any given sample is divided into two parts, each part will differ slightly from the other in detailed respects. In undertaking forensic comparison of two unknown samples, or one unknown and a reference sample, the key question is therefore to decide what level of difference is significant in order to draw the conclusion that the two samples are "different."

If there are no evident differences, the conclusion can be drawn that the two samples are *indistinguishable*. This is preferable to the statement that the two samples are "the same" or "identical," since it might be possible to demonstrate significant differences if additional techniques of comparison were employed. If there are slight evident differences between the samples but they are judged to be insignificant, for example, two rocks are of different size but they have a similar texture and evident mineral composition, it can be concluded that the samples are *indistinguishable in terms of texture and mineralogy*. If there are slight differences in the texture or apparent mineralogical composition, the samples can be described as *very similar but distinguishable*. In such situations, the challenge is then to assess the likely significance of the evident differences and similarities and to come to a conclusion about the likelihood that they originate from the same source, or are otherwise associated.

There is a widely held misconception that if a questioned soil sample is derived from a particular location, such as a crime scene, it should be indistinguishable, or at least very similar to, control samples collected at that location. There have been many instances in the author's forensic casework experience where expert witnesses have sought to *exclude* a questioned sample as having originated from a particular location because the two samples do not "match" in every respect. Reasons for such suggested exclusion have included, for example, the fact that one or more elemental concentrations in a questioned sample lies outside the range of that found in the control samples taken from the scene, even when allowance is made for analytical and sub-sampling "error." Other reasons suggested are the fact that one or more particle types found in the crime scene samples was not recorded in the questioned sample, or vice versa, or that the measured particle size frequency distributions and summary parameters do not exactly coincide. However, such suggestions are often unfounded, for a number of reasons. First, soils at most crime scenes and other locations show a significant degree of heterogeneity which may not be fully represented by the number of control samples collected. Second, the soil transfer process may be particle-selective, resulting in a "biased" sample on the questioned item (e.g., shoe). Third, the questioned sample may be subsequently modified by loss of some particles (selective retention) and acquisition of other particles from different sources (i.e., it represents a mixture, part of which may originate from the scene). Fourth, samples of soil that are removed from a suspect item for forensic analysis are often very small in size and few in number. Apparent differences in composition may arise due simply to the irregular distribution of particles within the source material, and to effective sub-sampling variation which arises during the transfer process from soil source to item (e.g., footwear); this effect is known in geochemistry as the "nugget" effect (Ramsay, 1997), which becomes potentially more pronounced with decreasing sample size. Analytical and laboratory weighing errors also become relatively more significant with smaller samples.

The reality of these effects has been demonstrated by experimental work. A simple example is shown in Table 6.4, in which major element oxide and trace element data are compared for a sample of aluminosilicate-rich dust, taken from a 5 mm thick layer on a concrete surface, and a sample of dust removed from the sole of a shoe which had been in contact with the dust-covered surface. Since the shoe was removed from the foot of the wearer immediately after contact with the dust, selective particle loss and/or mixing with extraneous material from other locations during subsequent contacts can be ruled out. Only a single contact between the shoe and the dust-covered surface was made, and all of the dust transferred to the sole of the shoe was recovered, mixed and ground before sub-sampling for ICP analysis. The sample of the source dust was taken by carefully removing the surface dust from beneath the shoe impression and processing it in exactly the same way as the shoe dust sample.

Table 6.4 shows the measured elemental and oxide concentrations together with the numerical differences and percentage differences between the measured values for the two samples. The percentage differences for the major oxides ranged from 1.1% for aluminum to 40.2% for magnesium, and in the case of the trace elements from 2.0% for gallium to 83.9% for cobalt. The percentage difference for silicon, the most abundant element, was 5.7%. To some extent the differences reflect instrumental measurement and laboratory sub-sampling precision (Pye et al., 2006a), but the predominant factors contributing to the observed differences are likely to be heterogeneity in the source material and the potentially particle-selective nature of the transfer mechanism.

Particle size distribution curves and selected grain size summary parameters for sub-samples of the two dust samples, determined by laser granulometry, are shown in Figure 3.12. Although there are differences between

Table 6.4 Comparison of the Elemental Composition of Tile Dust Taken from a Floor (Sample KP890) with Dust Collected from the Sole of a Shoe which had made a single contact with the floor (Sample KP901) (a) Raw Elemental Values (Major Oxides in Weight Percent, Other Elements in ppm)

	SiO₂	Al₂O₃	Fe₂O₃	MgO	CaO	Na₂O	K₂O	TiO₂	P₂O₅	MnO	Ba	Co	Cr	Cu	Ni	Pb	Sc	Sr	V	Zn	Zr	Ga	Rb	Nb
KP890	65.7	15.5	2.96	1.34	6.20	0.85	2.64	0.66	0.11	0.03	569	47.1	108	611	30.3	188	12.3	142	80.7	3940	2558	18.9	114	13.5
KP901	62.0	15.3	3.71	1.88	6.45	0.74	3.14	0.64	0.15	0.04	581	86.6	172	568	45.3	175	12.0	155	94.7	3682	2503	19.3	137	12.5
Numerical difference	3.8	0.2	0.75	0.54	0.25	0.11	0.50	0.01	0.03	0.01	13	39.5	65	43	15.0	13	0.3	13	14.1	258	55	0.4	23	1.0
Percentage difference	5.7	1.1	25.4	40.2	4.0	12.7	18.8	2.2	29.2	34.5	2.2	83.9	59.9	7.1	49.4	6.9	2.5	8.9	17.4	6.5	2.1	2.0	20.2	7.1

	Mo	Sn	Cs	Hf	Ta	W	Tl	Y	Be	U	Th	La	Ce	Pr	Nd	Sm	Eu	Gd	Tb	Dy	Ho	Er	Tm	Yb	Lu
KP890	1.01	155	8.79	61.2	1.13	1.84	0.60	25.6	2.49	4.50	13.0	32.4	61.2	7.32	26.1	5.05	1.07	4.36	0.71	3.98	1.02	3.03	0.48	3.44	0.59
KP901	1.42	141	11.9	59.8	1.19	2.78	0.63	26.7	3.16	4.62	13.2	35.0	65.9	8.11	28.3	5.60	1.10	4.70	0.79	4.28	1.10	3.26	0.50	3.54	0.62
Numerical difference	0.41	14	3.12	1.4	0.06	0.94	0.03	1.1	0.67	0.12	0.2	2.7	4.7	0.79	2.2	0.54	0.03	0.35	0.08	0.30	0.09	0.23	0.02	0.10	0.02
Percentage difference	40.7	9.1	35.5	2.3	4.9	50.7	5.2	4.3	27.0	2.6	1.6	8.2	7.6	10.8	8.4	10.7	2.8	8.0	11.2	7.6	8.5	7.6	4.0	2.9	3.9

(b) Values Normalized to the Concentration of Al₂O₃

	SiO₂	Fe₂O₃	MgO	CaO	Na₂O	K₂O	TiO₂	P₂O₅	MnO	Ba	Co	Cr	Cu	Ni	Pb	Sc	Sr	V	Zn	Zr	Ga	Rb	Nb
KP890	424	19.1	8.66	40.0	5.48	17.0	424	0.73	0.19	3668	304	696	3941	195	1215	79.5	917	520	25419	16501	122	734	86.9
KP901	404	24.2	12.3	42.1	4.84	20.5	420	0.96	0.26	3790	565	1125	3704	295	1144	78.3	1010	618	24018	16327	126	892	81.6
Numerical difference	20	5.1	3.62	2.1	0.64	3.4	4	0.22	0.07	122	261	429	238	100	71	1.1	92	97	1401	173	4	158	5.3
Percentage difference	4.7	26.8	41.8	5.1	11.7	20.2	1.1	30.6	36.0	3.3	86.0	61.6	6.0	51.1	5.9	1.4	10.1	18.7	5.5	1.0	3.1	21.6	6.1

	Mo	Sn	Cs	Hf	Ta	W	Tl	Y	Be	U	Th	La	Ce	Pr	Nd	Sm	Eu	Gd	Tb	Dy	Ho	Er	Tm	Yb	Lu
KP890	6.52	998	56.7	395	7.31	11.9	3.89	165	16.1	29.1	83.8	209	395	47.2	168	32.6	6.91	28.1	4.56	25.7	6.57	19.6	3.13	22.2	3.84
KP901	9.28	918	77.7	390	7.75	18.1	4.13	174	20.6	30.2	86.1	229	430	52.9	184	36.5	7.18	30.7	5.12	27.9	7.21	21.3	3.29	23.1	4.03
Numerical difference	2.76	81	21.0	5	0.44	6.2	0.25	9	4.6	1.1	2.3	20	35	5.7	16	3.9	0.27	2.6	0.57	2.3	0.63	1.7	0.16	0.9	0.19
Percentage difference	42.3	8.1	37.0	1.2	6.0	52.4	6.3	5.5	28.4	3.8	2.8	9.4	8.8	12.1	9.6	12.0	3.9	9.2	12.4	8.8	9.7	8.7	5.1	4.0	5.1

(c) Selected Elemental Ratios

	SiO₂/Al₂O₃	SiO₂/Fe₂O₃	Al₂O₃/Fe₂O₃	Al₂O₃/K₂O	SiO₂/CaO	CaO/MgO	K₂O/Na₂O	Cu/Pb	Zn/Pb	Rb/Sr	Ta/W	Ce/La	Nd/Sm	U/Th
KP890	4.24	22.2	5.24	5.88	10.6	4.62	3.11	3.24	20.9	0.80	0.61	1.89	5.16	0.35
KP901	4.04	16.7	4.13	4.89	9.61	3.42	4.23	3.24	21.0	0.88	0.43	1.88	5.05	0.35
Numerical difference	0.20	5.5	1.11	0.99	1.0	1.19	1.12	0.01	0.1	0.08	0.19	0.01	0.11	0.00
Percentage difference	4.7	24.8	21.1	16.8	9.3	25.8	36.1	0.2	0.4	10.4	30.4	0.5	2.1	1.0

the samples (e.g., the two modes are reversed), the general shapes of the distribution curves are similar. Both samples are very fine grained, but the sample from the source is slightly coarser (mean size 3.1 µm compared with 2.5 µm, with a more prominent coarser mode). Although the difference between the means is quite large in percentage terms (19.35%), it is very small in numerical terms (0.6 µm) and in terms of the range of particle sizes found in soils and sediments. The very fine mean size of this material, compared with those found in most soils and sediments, and the similar, markedly non-normal shapes of the particle size frequency distribution curves, can be regarded as potentially distinguishing features supporting the propositions of an association and possibly a common source.

The issue of soil heterogeneity at various scales has been explored by the author and colleagues in a number of other studies. For example, Pye et al. (2006b) examined the variation in particle size, color, elemental composition, and stable isotope ratios within two 1 m grids on contrasting soil types in Berkshire, U.K. A summary of the elemental data is provided in Table 6.5. In this instance, the % coefficient of variation (%CV) for the nine samples analyzed from each grid was taken as a measure of the within-grid variation (also incorporating analytical and sub-sampling variation). At the Simon's Wood site, %CV values of >40% were recorded for three major and five trace elements, whereas the Arborfield Bridge grid showed a lower degree of variability (%CV >20% for one major element and five trace elements).

Table 6.6 presents ICP chemical data for eight control samples taken in an approximate linear array from a section of a Devon dirt track measuring c. 30 × 5 m, and for 15 samples collected from an area of woodland near Gatwick Airport, Sussex, measuring c. 100 × 100 m. In the former case, the %CV exceeded 10% for only one major oxide (CaO, 32.7%), and six trace elements. In the second case, the %CV exceeded 20% for 8 major oxides and 14 trace elements. Clearly, the degree of heterogeneity varies from location to location, and can be large over quite small distances in some circumstances. Of the major elements, CaO (present mainly in calcium carbonate minerals) shows the general highest incidence of small-scale variability.

The important point to be drawn from this illustration is that the possibility that a questioned sample and control samples have a common source should not be excluded purely on the basis that their elemental concentration values, particle size parameters, or other measured values do not exactly coincide or fall within the same range. The general nature of the chemical profiles, or shapes of the particle size distribution curves, color reflectance curves or other distribution features, are often of greater interpretative significance than strict numerical values.

Table 6.5 Local-Scale Chemical Variation Across 1 m² Grids at Two Locations in Berkshire, U.K: A Woodland Site in Simon's Wood, and a Riverside Waste-Ground Location Near Arborfield Bridge

Element	Simon's Wood (*n*=9)					Arborfield Bridge (*n*=9)				
	Mean	Max	Min	SD	CV %	Mean	Max	Min	SD	CV %
				ICP-AES Elements						
SiO_2	78.07	86.98	66.97	6.97	8.9	50.64	55.08	45.08	3.00	5.9
Al_2O_3	1.93	2.22	1.73	0.15	7.7	5.52	5.98	5.12	0.32	5.8
Fe_2O_3	0.41	0.83	0.28	0.17	41.6	3.72	3.98	3.45	0.20	5.4
MgO	0.07	0.13	0.03	0.03	50.9	1.29	1.56	1.00	0.20	15.5
CaO	0.88	1.48	0.34	0.38	43.6	3.75	4.81	2.92	0.75	20.0
Na_2O	0.09	0.10	0.08	0.01	8.1	0.26	0.32	0.24	0.02	8.8
K_2O	0.96	1.05	0.91	0.04	4.5	1.30	1.36	1.26	0.03	2.4
TiO_2	0.33	0.39	0.27	0.04	11.3	0.45	0.48	0.41	0.02	4.8
P_2O_5	0.08	0.11	0.05	0.02	27.1	0.50	0.56	0.44	0.04	8.0
MnO	0.01	0.02	0.01	0.00	30.5	0.08	0.10	0.07	0.01	13.0
Be	<LLQ	<LLQ	<LLQ	<LLQ	<LLQ	1.09	1.23	1.01	0.07	6.3
Sc	<LLQ	<LLQ	<LLQ	<LLQ	<LLQ	5.09	5.61	4.58	0.37	7.3
Ni	<LLQ	<LLQ	<LLQ	<LLQ	<LLQ	25.83	28.07	21.55	2.29	8.9
Zn	56.10	82.32	32.86	16.09	28.7	365.84	401.10	333.80	22.32	6.1
Y	9.31	10.24	7.87	0.73	7.9	19.49	20.47	18.10	0.77	3.9
Ba	279.61	313.20	240.30	24.32	8.7	479.18	597.50	427.30	57.71	12.0
				ICP-MS Elements						
V	17.96	25.98	8.38	6.23	34.7	61.43	83.59	43.25	14.69	23.9
Cr	38.47	52.23	24.81	9.13	23.7	74.15	86.41	60.30	10.52	14.2
Co	0.98	2.08	0.58	0.46	46.5	9.57	10.84	7.77	1.14	11.9
Cu	12.24	23.09	6.99	5.10	41.7	78.70	93.04	66.01	9.14	11.6
Ga	2.17	2.80	1.77	0.30	13.7	6.76	7.92	5.63	0.82	12.1
Rb	26.37	28.57	23.92	1.58	6.0	60.84	72.34	50.42	8.05	13.2
Sr	33.09	37.17	29.38	2.48	7.5	72.09	81.57	56.16	8.37	11.6
Zr	371.82	474.11	203.67	95.00	25.5	237.80	311.98	159.77	49.69	20.9
Nb	4.95	5.66	4.04	0.53	10.8	7.78	9.48	6.30	1.15	14.7
Mo	0.57	1.15	0.06	0.39	67.8	2.85	3.91	2.14	0.54	19.0
Sn	1.66	2.51	0.88	0.55	33.3	7.09	9.02	5.14	1.25	17.7
Cs	1.06	1.30	0.88	0.16	15.0	2.87	3.34	2.37	0.36	12.5
La	6.92	9.48	2.76	2.01	29.0	19.30	23.58	14.51	3.55	18.4
Ce	14.42	18.16	10.05	2.39	16.6	42.51	48.41	33.84	6.05	14.2
Pr	1.59	1.93	0.83	0.32	20.1	4.61	5.42	3.57	0.74	16.1
Nd	6.06	7.82	3.77	1.11	18.3	17.86	20.33	14.46	2.49	13.9
Sm	1.15	1.52	0.81	0.24	20.9	3.45	3.97	2.84	0.46	13.2
Eu	0.23	0.28	0.19	0.03	12.0	0.78	0.89	0.64	0.10	12.8
Gd	1.02	1.22	0.79	0.14	13.6	3.08	3.55	2.53	0.45	14.5
Tb	0.17	0.20	0.14	0.02	12.9	0.49	0.56	0.42	0.07	13.2
Dy	1.09	1.27	0.87	0.14	12.4	2.87	3.32	2.37	0.40	13.9
Ho	0.26	0.30	0.20	0.04	13.7	0.64	0.74	0.52	0.09	14.5
Er	0.79	0.89	0.60	0.10	13.2	1.77	2.05	1.46	0.26	14.6
Tm	0.13	0.15	0.09	0.02	15.4	0.26	0.31	0.22	0.04	15.3

Table 6.5 Local-Scale Chemical Variation Across 1 m² Grids at Two Locations in Berkshire, U.K: A Woodland Site in Simon's Wood, and a Riverside Waste-Ground Location Near Arborfield Bridge (Continued)

Element	Simon's Wood (n=9)					Arborfield Bridge (n=9)				
	Mean	Max	Min	SD	CV %	Mean	Max	Min	SD	CV %
Yb	0.92	1.11	0.65	0.16	17.1	1.71	2.11	1.37	0.26	15.4
Lu	0.15	0.17	0.10	0.03	17.0	0.25	0.30	0.21	0.04	15.5
Hf	8.60	10.63	4.71	2.18	25.4	5.95	7.86	4.01	1.31	22.0
Ta	0.40	0.48	0.32	0.05	12.9	0.61	0.76	0.49	0.10	16.7
W	1.32	3.20	0.32	1.63	123.0	0.89	1.74	0.44	0.41	45.5
Tl	0.16	0.37	0.07	0.09	56.8	0.22	0.42	0.08	0.12	53.6
Pb	18.40	25.82	8.74	6.36	34.6	30.42	40.37	20.14	6.04	19.8
Th	2.26	2.98	1.62	0.44	19.7	5.34	6.59	4.29	0.83	15.5
U	0.90	1.02	0.75	0.09	10.5	1.52	1.78	1.24	0.21	14.1

Note: Values for the 10 major oxides are given in weight percent, all other values are in ppm (μg g^{-1}). <LLQ indicates that values were below the lower limit of quantitation.

Table 6.6 (a) Chemical Variation in Eight Control Soil Samples Taken in a Linear Array Along a Trackway, Covering an Area of Approximately 30×5 m, near Ilfracombe, Devon, U.K. (b) Chemical Variation in 15 Control Soil Samples Taken within a Woodland Area Measuring Approximately 100×100 m, Near Crawley, Sussex, U.K.

Element	(a)					(b)				
	Mean	Max	Min	SD	CV %	Mean	Max	Min	SD	CV %
				ICP-AES Elements						
SiO$_2$	47.69	49.17	46.51	1.03	2.2	73.16	82.74	58.57	6.08	8.3
Al$_2$O$_3$	16.90	17.23	16.47	0.30	1.8	7.88	10.94	4.80	1.94	24.6
Fe$_2$O$_3$	6.41	6.65	6.21	0.13	2.0	2.51	3.52	2.04	0.44	17.3
MgO	1.19	1.25	1.12	0.04	3.3	0.38	0.57	0.23	0.09	23.4
CaO	1.66	2.79	0.87	0.54	32.7	0.68	1.80	0.20	0.49	72.3
Na$_2$O	0.54	0.60	0.51	0.03	5.1	0.24	0.42	0.17	0.06	25.0
K$_2$O	3.18	3.27	3.09	0.07	2.3	1.23	1.54	0.86	0.22	17.9
TiO$_2$	0.83	0.85	0.81	0.02	1.9	1.16	1.36	0.98	0.10	8.7
P$_2$O$_5$	0.77	0.90	0.66	0.07	9.3	0.13	0.23	0.05	0.04	34.0
MnO	0.21	0.22	0.19	0.01	5.5	0.17	0.28	0.01	0.07	43.8
Be	2.84	2.92	2.76	0.06	2.1	1.25	1.80	0.76	0.32	25.5
Sc	16.23	16.71	15.76	0.38	2.3	8.24	11.24	5.54	1.79	21.7
Ni	38.84	43.30	36.31	2.59	6.7	22.54	32.23	13.43	5.46	24.2
Zn	154.56	182.00	121.20	17.80	11.5	78.38	172.60	48.99	31.32	40.0
Y	30.04	31.19	29.11	0.72	2.4	42.61	51.86	36.70	3.85	9.0
Ba	479.30	496.90	459.70	13.73	2.9	239.57	317.80	173.70	47.65	19.9

(continued)

Table 6.6 (a) Chemical Variation in Eight Control Soil Samples Taken in a Linear Array Along a Trackway, Covering an Area of Approximately 30 × 5 m, near Ilfracombe, Devon, U.K. (b) Chemical Variation in 15 Control Soil Samples Taken within a Woodland Area Measuring Approximately 100 × 100 m, Near Crawley, Sussex, U.K. (Continued)

	(a)					(b)				
Element	Mean	Max	Min	SD	CV %	Mean	Max	Min	SD	CV %
				ICP-MS Elements						
V	104.71	117.10	93.44	8.56	8.2	77.39	100.75	45.94	17.72	22.9
Cr	90.66	92.60	88.84	1.27	1.4	64.68	76.72	45.05	9.43	14.6
Co	15.02	15.76	14.75	0.33	2.2	10.08	14.61	7.32	1.78	17.7
Cu	29.64	40.18	25.33	4.62	15.6	44.03	154.56	22.56	32.57	74.0
Ga	22.42	22.79	21.38	0.46	2.1	10.03	13.69	6.06	2.53	25.3
Rb	160.06	163.84	150.90	4.17	2.6	68.58	89.58	44.71	13.10	19.1
Sr	140.34	159.38	120.14	12.41	8.8	48.15	81.54	31.69	14.54	30.2
Zr	150.23	153.27	146.43	2.44	1.6	419.45	555.83	307.57	64.95	15.5
Nb	14.06	14.61	13.10	0.44	3.1	20.34	23.59	16.08	2.03	10.0
Mo	0.98	1.50	0.56	0.33	33.9	1.12	1.68	0.68	0.26	23.6
Sn	3.65	4.17	3.28	0.28	7.7	4.63	20.31	2.67	4.40	95.0
Cs	11.25	11.68	10.52	0.34	3.0	5.84	8.39	3.08	1.72	29.5
La	29.28	31.94	26.40	1.76	6.0	36.10	45.21	27.66	4.30	11.9
Ce	66.19	69.14	64.54	1.48	2.2	76.78	97.81	60.69	9.03	11.8
Pr	7.08	7.51	6.70	0.23	3.3	8.25	10.33	6.42	0.98	11.9
Nd	27.61	28.76	26.64	0.60	2.2	30.75	38.75	24.02	3.69	12.0
Sm	5.49	5.64	5.27	0.12	2.2	5.78	7.32	4.54	0.68	11.8
Eu	1.28	1.30	1.24	0.02	1.7	1.23	1.55	0.93	0.16	13.0
Gd	4.95	5.13	4.81	0.10	2.0	5.64	6.97	4.45	0.60	10.7
Tb	0.82	0.84	0.79	0.02	1.9	1.02	1.26	0.84	0.10	10.0
Dy	4.84	5.03	4.65	0.12	2.4	6.59	8.01	5.43	0.61	9.3
Ho	1.07	1.10	1.03	0.03	2.5	1.51	1.82	1.27	0.14	9.0
Er	2.93	3.03	2.84	0.06	2.2	4.17	5.03	3.56	0.37	8.9
Tm	0.43	0.44	0.42	0.01	2.1	0.62	0.73	0.53	0.05	8.0
Yb	2.80	2.89	2.69	0.06	2.2	3.95	4.60	3.43	0.31	7.7
Lu	0.40	0.42	0.39	0.01	2.9	0.59	0.68	0.51	0.04	7.5
Hf	4.13	4.23	3.90	0.11	2.6	10.66	13.61	8.09	1.43	13.4
Ta	1.15	1.17	1.11	0.02	1.6	1.68	1.96	1.33	0.17	10.0
W	1.39	3.08	0.92	0.72	52.1	2.21	3.94	1.35	0.55	24.9
Tl	0.61	0.92	0.37	0.16	26.6	0.31	0.70	<LLQ	0.20	66.1
Pb	10.38	13.31	5.75	2.58	24.9	42.38	105.84	18.64	21.43	50.6
Th	11.74	12.14	11.40	0.21	1.8	10.86	14.75	8.63	1.49	13.7
U	2.41	2.64	2.30	0.11	4.6	3.50	4.43	2.93	0.36	10.4

Note: Values for the 10 major oxides are given in weight percent, all other values are in ppm ($\mu g\,g^{-1}$).
     <LLQ indicates that values were below the lower limit of quantitation.

6.9   Use of Multi-Technique Comparison Data for Exclusion/Inclusion

A decision about whether to exclude or include a questioned soil or other geological sample as having possibly originated from the same source as a control or reference sample should always rest on an assessment of several independent lines of evidence acquired using a range of analytical techniques. In the case of soils and sediments, most frequently this will include the results from analysis of some or all of the following: color, particle size distribution, elemental composition, particle type assemblage, including mineralogical composition, stable isotope composition, pollen assemblage, and diatom assemblage. The discriminatory value of each type of data will vary from situation to situation, and prescriptive guidelines regarding the best technique or type of evidence to use in every situation cannot be provided. Often, the selection of the most appropriate techniques can only be made after a consideration of the environment at the crime scene and the nature of the questioned samples submitted for examination. If, for example, a crime has been committed at a location where there are exotic plants, botanical examination techniques, including pollen analysis, may provide highly discriminatory, and even diagnostic, information. On the other hand, pollen analysis is usually destructive and in many circumstances samples turn out to be barren, or to contain only common pollen types which are of very limited discriminatory value. Consequently, for routine work, it is often better to employ initial techniques which are either non-destructive or which are more or less guaranteed to yield information of comparative, and often highly discriminatory, value. Examples include geochemical and mineralogical analysis. The inorganic geochemical and mineralogical composition of soils and sediments is also generally less prone to significant change over time, and during sample storage, than many of the organic properties of soils.

In general, as many different techniques as possible should be used to characterize a sample, where possible non-destructively or with minimal destructive effect. Wherever sample size and type permit, at least three, and preferably more, independent techniques should be used to provide data for comparison. The results from each should be evaluated independently and then combined to arrive at an overall conclusion.

In the case of rock samples, the strongest evidence of an association between two samples is provided by a physical fit combined with indistinguishability in terms of properties such as texture, mineralogy, and chemical composition. However, in making comparisons it should be borne in mind that, in some circumstances, certain properties of either the questioned samples or control samples may have been changed (e.g., color changes during exposure to fire, or prolonged sub-aerial exposure). In the case of soil samples, the strongest evidence of an association is often provided by the presence of a number of unusual particle types in common, supported by a very high degree of similarity in terms of bulk physical, chemical and/or biological characteristics. Occasionally, samples of interest from a scene may contain one particle type that is extremely unusual and may even be regarded as unique; for example, fragments from a single piece of home-made, hand-painted pottery with a complex design. More commonly, however, the samples may contain a small number of particle types with shapes, textures, and colors which can be regarded as distinctive but whose precise nature, origin, and distribution in the environment is unknown. The assemblages of such particle types are likely to be very rare on a global or regional scale, but the possibility that they occur in more than one location at a local scale often cannot be discounted (e.g., in several gardens of houses on the site of a former waste tip, or which have been supplied with topsoil from a common source). Analysis of comparison samples from various control locations unrelated to the crime, and reference to available local database information, may assist in assessing such possibilities.

6.10 Geological and Soil Databases and Database Interrogation

Relatively detailed geological maps, in many cases supported by descriptive memoirs, are now available for many parts of the Earth's surface, although the accuracy and date of the information varies considerably from country to country, and even between different parts of the same country. In the U.K., the British Geological Survey holds an extensive collection of maps, borehole records, and data sets of geophysical, physical, and chemical information. In the U.S., similar information is held in the archives of the U.S. Geological Survey and many state geological surveys. There is also a large amount of information held by commercial companies concerned with mining and mineral exploration. Local authorities in many countries also hold information relating to the geology and soils within their area which is relevant to planning and environmental management issues, although not always in a form which is readily accessible to the public. However, all such avenues should be explored as sources of relevant contextual information. Frequently, useful initial leads to the existence of information can be gained from the published scientific literature and from Internet searches.

A considerable number of soil databases also exist, but in the main they do not contain data in a form which is suitable for direct comparison with forensic samples (Batjes, 2000; Nachtergaele, 2000; van Engelen, 2000). Most of these databases have been developed for the purposes of agricultural assessment, regional geochemical mapping, and mineral exploration. In the simplest case, a soil database can be defined as a structured set of data that can be used for purposes of reference, comparison, or interrogation.

Most modern databases are hosted on a computer, although there are still some which consist only of written records and hard copy illustrative material. Most include information about location (the place where the sample was taken and/or soil profile described), environmental setting (terrain type, soil parent material, local climatic, and hydrological regime), soil properties (physical and chemical), vegetation type, land-use, and possibly soil management regime. The information derived from systematic soil surveys is often displayed in thematic map form. Physical and chemical properties that are included often include soil texture, particle size distribution, soil moisture, color, pH, electrical conductivity, and cation exchange capacity.

Two relatively large-scale soil maps of the world have been produced, a 1:10 million scale map prepared by Kovda (1977) and a 1:5 million scale map produced in nine volumes by the Food and Agriculture Organization (FAO) and United Nations Environment and Scientific Cooperation Organization (UNESCO) (FAO 1971–1981). There are also a number of simplified and modified versions of the latter map such as those produced by the US Department of Agriculture using the USDA Soil Taxonomy system of soil classification (Soil Survey Staff, 1990). The FAO–UNESCO 1:5 million map is widely regarded as the most authoritative at continental and global scales, and a digitized version in Arc-Info format has been available for some years (FAO, 1996; Nachtergaele, 2000). Another useful source of information at the global scale, although not yet providing complete global coverage, is the SOTER Soil–Terrain Database. This initiative, sponsored by the United Nations Environment Programme, aimed to map areas having a distinctive, and often repetitive, pattern of landforms, parent materials and soils at a scale of 1:1 million (van Engelen, 2000).

Another source of global information is the WISE (World Inventory of Soil Emission Potentials) database developed by the International Soil Reference and Information Centre (ISRIC). Information in the World Soil Reference Collection was computerized as early as the mid-1980s (Batjes et al., 1994). The WISE project was intended to integrate this information within the spatial framework defined by the FAO–UNESCO digital soil map and SCOTER program.

Soil maps have been produced at the scale of individual countries by numerous national and international organizations. The maps range generally in scale from 1:1 million to 1:50,000, although few countries have complete national coverage at scales larger than 1:250,000. For example, the Soil Survey of England and Wales (1983) published 1:250,000 maps, with an explanatory soil key, for the whole country. Maps at 1:63,360 (1" to 1 mile), 1:50,000 scale and 1:25,000, some with supporting explanatory memoirs, are available for less than half of the country.

About 94% of the private land in the U.S. is covered by soil surveys, mostly at 1:12,000 or 1:24,000 scale, and the information is summarized in approximately 2200 country level soil survey reports and databases. Digital information at 1:50,000 scale is available for the coterminous United States and at 1:1 million scale for Alaska. A program is in hand to digitize the data from individual country surveys. The National Cooperative Soil Survey (NCSS) is responsible for developing and maintaining the Soil Survey Databases. In Canada, a National Soil Database has been developed within a GIS framework (Coote and MacDonald, 2000). It includes a soil map of the whole of Canada at 1:5 million scale, a soil landscapes map at 1:1 million scale and detailed soil survey information and maps for some areas at scales ranging from 1:20,000 to 1:250,000. In addition, Soil Attribute Files provide information about soil properties which vary in a vertical direction (i.e., down the soil profile).

Uncertainty in databases, especially global ones, arises from several different sources, including errors in the information used for database compilation, inconsistencies in translating the original survey data into a standardized format for mapping and data processing purposes, and the scaling up of site-specific information to larger areas (Batjes, 2000). The information relating to soil attributes, especially detailed physical and chemical properties, varies greatly from area to area, and there is considerable variation in the degree of standardization in the methods used to obtain the data.

In addition to soil classification and soil property maps, a range of geological and geochemical maps is available at local, national, regional, and global scales. From a forensic investigation point of view, the utility of the information contained on such maps varies greatly, depending on the scale of the map, the date and reliability of survey, whether the maps indicate geological age or lithological types, and whether solid geology (bedrock), drift geology (surface deposits), or both are indicated. In general, maps showing the distribution of surface (drift) deposits, or solid geology and drift, are of greatest use. Most forensic investigations relate to local, regional, and national scales, so maps produced at scales of 1:50,000, 1:100,000 or 1:250,000 are, in general, of greatest utility. In the U.K., for example, all of the country is covered at 1:250,000 scale and more than 95% at 1:63,360 or 1:50,000 scales. However, the dates of survey vary considerably, and some map sheets still rely on surveys carried out more than 100 years ago. Drift, or solid and drift editions, are not available for all areas, and for a few areas no map sheets at all are available, or are available only as provisional editions. Explanatory memoirs are available for some map sheet areas or wider regional areas, and original survey and borehole data can be obtained from the British Geological Survey, some now in digital format.

While such sources often provide background information about the spatial distribution of soil, sediment, or rock types, which is useful for intelligence or sampling design purposes, they usually do not contain sufficient detailed information about physical and chemical properties for direct forensic comparison. Geochemical atlases, and the underlying databases on which they rest, are sometimes of greater assistance in this regard. In the U.K., several geochemical atlases have been published for soil research and geological/environmental purposes. A soil geochemical atlas of England and Wales was published in 1992 based on sampling at approximately 5 km intervals across most of the country, with the exception of urban areas (McGrath and Loveland, 1992). A statistical and geostatistical appraisal of this data set is also available (Oliver et al., 2002). A similar atlas relating to soils in Northern Ireland was published in 2000 (Jordan et al., 2000). Geochemical atlases of England and Wales, and of Northern Ireland, based on analyses of stream sediments, were first published in the 1970s (Applied Geochemistry Research Group, Imperial College London, 1972, 1978). Subsequently, the Institute of Geological Sciences, later known as the British Geological Survey, has published a series of regional atlases based on analysis of stream sediments, or a combination of stream sediments, water, and soils (Institute of Geological Sciences, 1978a, 1978b, 1978c, 1982, 1983; British Geological Survey, 1987, 1990, 1991, 1992, 1993, 1996, 1997, 1999, 2000). However, at present, there is no published coverage for eastern and southern England, where most of the population lives. The most recently published volumes contain summaries of the regional geology, the nature of mineralization and any mining activities in the area, maps showing the spatial variation in recorded concentrations of determined elements, and brief summaries of the relationship between the geochemical patterns and underlying geology.

A geochemical atlas containing geochemical data for soils in 10 countries in northern Europe is available (Reimann et al., 2003). Samples were taken from different soil horizons at approximately 750 sites over a 180,000 km^2 area, giving an average sampling density of one site per 2500 km^2. Although the spatial resolution of such information is too coarse for direct forensic comparison, such sources do provide useful contextual information about, for example, the frequency with which particular elemental concentration values are encountered.

Only relatively recently have attempts been made to build sediment and soil databases that can be used for forensic or similar purposes, and the geographical coverage of data is still patchy. A database containing chemical and particle size information for coastal dune sediments around the entire coast of England and Wales was described by Saye and Pye (2004). A preliminary database for coastal dune sediments along the west coast of Denmark has also been developed (Saye and Pye, 2006; Saye et al., 2006). Pye and Blott (2007) describe a computer-searchable database that contains chemical information for 1896 soil samples from known locations in England and Wales. This database contains concentrations for up to 49 major and trace elements in a standardized <150 µm size fraction of the soils, analyzed by inductively coupled plasma spectrometry-atomic emission spectrometry (ICP-AES) and inductively coupled plasma spectrometry-mass spectrometry (ICP-MS). Although the geographical coverage of sampling points is not even across the country, it is informative to examine the frequency distribution of concentration values for each element (Figure 6.1), and to compare values obtained from questioned forensic samples against the mean, maximum, and minimum values recorded in the database (Table 6.7). Two methods of

Table 6.7 Chemical Variation in 1896 Control Soil Samples Contained in the KPAL Database of U.K. Soils

Element	n	Mean	Max	Min	SD	CV %
		ICP-AES Elements				
SiO_2	1011	59.35	96.11	2.56	16.00	26.9
Al_2O_3	1895	9.16	29.98	0.33	3.50	38.2
Fe_2O_3	1895	4.63	40.72	0.16	2.14	46.2
MgO	1895	0.91	9.24	0.01	0.70	77.0
CaO	1891	4.42	54.04	<LLQ	6.80	153.9
Na_2O	1894	0.48	4.31	<LLQ	0.32	65.9
K_2O	1895	1.68	4.58	0.02	0.63	37.7
TiO_2	1895	0.56	2.14	0.02	0.20	36.6
P_2O_5	1895	0.34	2.44	0.01	0.24	69.8
MnO	1890	0.10	6.01	0.00	0.19	180.4
Be	399	1.75	4.46	<LLQ	0.80	45.7
Sc	1874	9.25	50.00	<LLQ	4.14	44.8
Ni	1860	37.25	381.00	<LLQ	23.82	63.9
Zn	1870	226.80	4040.00	11.00	308.85	136.2
Y	1887	24.97	172.00	<LLQ	11.49	46.0
Ba	1895	411.31	4049.00	<LLQ	276.06	67.1
		ICP-MS Elements				
V	1875	82.70	744.00	<LLQ	40.14	48.5
Cr	1875	86.89	9492.00	1.00	298.62	343.7
Co	1872	11.53	52.00	<LLQ	5.70	49.4
Cu	1860	65.15	3002.00	<LLQ	107.84	165.5
Ga	400	11.45	27.59	1.35	5.27	46.0
Rb	1003	78.18	704.61	1.05	38.29	49.0
Sr	1895	122.83	985.50	7.00	88.22	71.8
Zr	1896	216.50	1188.43	1.42	153.14	70.7
Nb	1022	12.69	144.09	0.23	6.94	54.7
Mo	400	2.51	80.73	<LLQ	4.97	198.1
Sn	400	7.37	205.67	<LLQ	12.08	164.0
Cs	1003	4.98	89.09	0.11	4.74	95.1
La	1887	30.68	142.00	1.00	11.38	37.1
Ce	1887	60.50	378.00	5.00	22.25	36.8
Pr	1003	6.86	32.06	0.70	2.91	42.4
Nd	1887	29.45	204.26	0.84	12.68	43.0
Sm	1887	5.31	31.84	0.09	2.30	43.3
Eu	1886	1.09	6.40	0.13	0.52	47.6
Gd	1003	4.68	22.90	0.66	1.94	41.5
Tb	400	0.79	2.99	0.10	0.33	41.7
Dy	1887	3.91	28.50	0.31	1.72	44.1
Ho	1003	0.93	3.60	0.12	0.35	36.9
Er	1003	2.46	9.08	0.32	0.91	36.8
Tm	337	0.41	1.23	0.05	0.16	38.1
Yb	1887	2.24	14.10	<LLQ	0.88	39.3
Lu	1001	0.39	1.10	0.05	0.13	32.3

Table 6.7 Chemical Variation in 1896 Control Soil Samples Contained in the KPAL Database of U.K. Soils (Continued)

Element	n	Mean	Max	Min	SD	CV %
Hf	1003	7.98	30.39	<LLQ	3.57	44.7
Ta	1002	0.95	23.33	<LLQ	0.97	102.6
W	400	2.16	26.26	<LLQ	2.50	115.8
Tl	640	0.43	2.36	<LLQ	0.40	93.0
Pb	1269	184.30	13533.26	<LLQ	734.86	398.7
Th	1003	8.17	26.83	0.69	2.88	35.2
U	1003	2.12	29.15	0.17	1.11	52.5
Li	1337	51.35	365.00	0.00	28.98	56.4

Note: Values for the 10 major oxides are given in weight percent, all other values are in ppm ($\mu g\, g^{-1}$). <LLQ indicates values below the lower limit of quantitation.

comparing questioned samples with those in the database have been developed. The first uses the calculated Euclidean distance, as described in relation to HCA in Section 6.6.2), between a questioned sample and each sample in the database. The database samples, which can include control samples and other questioned samples in a given case, are then ranked in terms of their overall chemical similarity to the questioned sample under consideration (e.g., Table 6.8). The second method involves comparing a questioned sample with the database samples in terms of a series of chosen elemental or oxide concentrations and ratios, after incorporating a margin of "error", which is based on the "Type 5 Precision" determinations for each element reported by Pye et al. (2006a). The example shown in Table 6.9 uses the Al_2O_3/Fe_2O_3 and Al_2O_3/K_2O ratios as a screening tool, followed by the Nd/Sm ratio and CaO content. Experience has shown that these oxide and elemental ratios provide a useful tool for discrimination between different mineral assemblages in the fine sand, silt and clay-size fractions of soils and sediments (Pye and Blott, 2004b). The use of ratios also gets round a problem which sometimes arises in comparing samples which have varying contents of organic matter or calcium carbonate. The choice of other potentially discriminatory elemental ratios and concentrations can be varied according to the chemical nature of the questioned and/or crime scene samples under consideration; suitable elements and ratios can be identified during exploratory data analysis.

In the example shown in Tables 6.8 and 6.9, three questioned mud samples taken from a pair of trainers allegedly worn by a murder suspect were compared with the database. On the basis of the two major element ratios alone, the vast majority of the samples in the database could be excluded as providing a possible "match" for the three questioned samples. When the Nd/Sm ratio was also included, only two samples in the database could not be excluded as providing a "match" with the first questioned sample (KP26A), both samples having been taken at the murder scene, a river bank

Table 6.8 Comparison of the Elemental Composition of Soil Samples Taken from a Suspect Shoe (Samples KP26A and KP29A) with a Database of 1896 Soil Samples from Known Locations Across England and Wales

Rank	Subset of 989 Samples with 40 Elemental Analyses			Subset of 1876 Samples with 27 Elemental Analyses		
	Case Name	Sample ID	Euclidean Distance from KP26A	Case Name	Sample ID	Euclidean Distance from KP26A
1	COMET	KP26A	0.000	COMET	KP26A	0.000
2	COMET	KP29A	1.201	COMET	KP29A	1.016
3	COMET	KP 03A	1.529	COMET	KP 03A	1.356
4	COMET	KP 01A	1.591	AVON	KP102	1.360
5	AVON	KP102	1.593	COMET	KP 01A	1.409
6	COMET	KP182A	1.634	COMET	KP182A	1.418
7	COMET	KP05A	1.700	COMET	KP181A	1.448
8	COMET	KP170A	1.714	MAROON	KP 10A	1.457
9	AVON	KP1036	1.730	COMET	DC 12	1.463
10	COMET	KP174A	1.737	COMET	KP174A	1.470
11	COMET	KP 02A	1.766	COMET	KP05A	1.488
12	AVON	KP108	1.774	AVON	KP108	1.490
13	COMET	DC 12	1.793	COMET	KP170A	1.506
14	COMET	KP06A	1.825	COMET	KP175A	1.606
15	COMET	KP181A	1.850	AVON	KP1036	1.621
16	COMET	KP 04A	1.947	COMET	KP06A	1.667
17	COMET	KP171A	1.953	COMET	DC 09	1.676
18	COMET	DC 09	1.975	COMET	KP 04A	1.692
19	AVON	KP902	1.993	COMET	DC 10	1.703
20	HUNTLEY	KP/81A	1.994	AVON	KP105	1.716
21	AVON	KP1038	2.005	COMET	KP 02A	1.716
22	AVON	KP111	2.009	AVON	KP1038	1.737
23	AVON	KP105	2.032	AVON	KP133	1.750
24	COMET	DC 10	2.097	HUNTLEY	KP/81A	1.776
25	COMET	KP175A	2.098	COMET	DC 13	1.824
26	AVON	KP920	2.129	AVON	KP103	1.835
27	AVON	KP123	2.137	COMET	KP 170A	1.837
28	AVON	KP129	2.148	GOLDFINCH	DIJ23	1.838
29	TRITTON	KP19A	2.173	DAMASCUS	KP88A	1.849
30	COMET	KP 170A	2.192	COMET	DC 11	1.853
31	AVON	KP1037	2.219	AVON	SFC22	1.855
32	AVON	KP103	2.225	AVON	KP1015	1.856
33	COMET	KP184A	2.240	NEWTON	KP 105A	1.857
34	AVON	KP2018	2.253	COMET	KP171A	1.858
35	AVON	KP905	2.264	AVON	KP129	1.862
36	AVON	KP1712R	2.264	MAROON	KP122	1.868
37	TRITTON	KP21A	2.273	AVON	KP902	1.870
38	MAROON	KP 10A	2.279	COMET	KP184A	1.875
39	COMET	KP316A	2.287	AVON	KP123	1.896
40	AVON	KP901	2.296	AVON	KP24	1.900
41	DAMASCUS	KP65A	2.306	AVON	KP111	1.901
42	DAMASCUS	KP64A	2.323	AVON	KP2018	1.910

Table 6.8 Comparison of the Elemental Composition of Soil Samples Taken from a Suspect Shoe (Samples KP26A and KP29A) with a Database of 1896 Soil Samples from Known Locations Across England and Wales (Continued)

	Subset of 989 Samples with 40 Elemental Analyses			Subset of 1876 Samples with 27 Elemental Analyses		
Rank	Case Name	Sample ID	Euclidean Distance from KP26A	Case Name	Sample ID	Euclidean Distance from KP26A
43	AVON	KP1027	2.331	COMET	KP 182A	1.910
44	DAMASCUS	KP 29A	2.336	DAMASCUS	KP 29A	1.936
45	AVON	KP1708R	2.367	GOLDFINCH	DIJ30	1.936
46	COMET	DC 11	2.371	GOLDFINCH	DIJ31	1.937
47	COMET	KP185A	2.372	AVON	KP107	1.950
48	COMET	KP 182A	2.377	AVON	KP901	1.959
49	SHERSTON	RJF213A	2.383	RAIL/NAVY	KP36	1.962
50	HUNTLEY	KP/82A	2.386	DAMASCUS	KP65A	1.970

Note: For comparison purposes, two sample subsets were used: the first comprising 989 samples with 40 elemental analyses and the second comprising 1876 samples with 27 elemental analyses. For each subset, a principal component analysis was performed on the whole dataset, including samples from the suspect shoe, and the Euclidean distance between control samples and KP26A calculated in terms of all components. In each case, the two samples taken from the shoes were found to be most similar to each other, followed by samples taken from the alleged scene of the crime at Bradford-on-Avon, Wiltshire, U.K. (KP01A and KP03A).

Table 6.9 Example of a Search of an Elemental Composition Database Containing 1896 Soil Samples Collected from Known Locations Across England and Wales

		Database Searched in Terms of:			
Soil Samples from Suspect Trainers		Al_2O_3/Fe_2O_3	Al_2O_3/Fe_2O_3 Al_2O_3/K_2O	Al_2O_3/Fe_2O_3 Al_2O_3/K_2O Nd/Sm	Al_2O_3/Fe_2O_3 Al_2O_3/K_2O Nd/Sm CaO
KP26A	Total number of hits:	261	18	2	2
	Number of hits from scene of interest:	39	4	2	2
KP28A	Total number of hits:	288	77	8	4
	Number of hits from scene of interest:	16	2	1	1
KP29A	Total number of hits:	261	56	8	5
	Number of hits from scene of interest:	40	14	6	5

Note: Three soil samples from suspect trainers were compared to each sample in the database in terms of three elemental ratios (Al_2O_3/Fe_2O_3, Al_2O_3/K_2O, and Nd/Sm) and the concentration of CaO. The application of successive criteria narrows down the number of possible "hits." The search was conducted after allowing an "error" margin for each element equivalent to the "Type 5" precision error defined by Pye et al. (2006a).

near Bradford-on-Avon, Wiltshire. Only eight samples provided a possible chemical "match" with the second questioned mud sample (KP28A), one of which was from the murder scene, and eight samples provided a possible "match" with the third questioned sample, six of which were from the murder scene. When a further criterion, %CaO, was introduced, the number of "hits" for sample KP28A was reduced to four, and for sample KP29A was reduced to five samples, all of which were from the murder scene. In the author's operational casework, both approaches have used, alongside other chemical data comparison methods, including PCA, HCA, and correlation analysis, as part of the data evaluation process.

6.11 Evaluation of the Overall Strength of Geological and Soil Evidence

In undertaking any forensic geology or soil investigation, several stages are usually involved. The sequence and number of stages may vary slightly depending on whether the work is undertaken at the request of the prosecution (e.g., Police, Customs, or other Government agencies) or the defense, but the essential approaches to sample collection, sampling analysis, and data evaluation share much in common (Table 6.10 and Table 6.11). In both cases, the ultimate tasks are to evaluate the data and formulate an opinion regarding the significance of the results obtained.

In undertaking any evaluation, it is of crucial importance to understand the limitations of the data concerned, the techniques used to analyze and compare them, and the constraints placed on the evaluation by the extent of background information provided. In ideal circumstances, the forensic geologist will have been involved in the investigation for the prosecution from an early stage, and will have been provided with sufficient information relating to the circumstances of the case to allow a sufficiently large and appropriate set

Table 6.10 Stages in a Typical Forensic Soil Investigation for the Prosecution

1. Attend briefing session and read relevant background information
2. Carry out laboratory examination of suspect items and obtain questioned soil samples
3. Visit scene and other relevant locations to collect control samples
4. Undertake laboratory analysis
5. Compare results/data using graphical and statistical methods
6. Compare data with reference database information
7. If necessary, collect control samples from additional locations
8. Undertake further laboratory analysis on additional samples
9. Compare data with those for questioned samples and initial control samples from scene
10. Synthesise results, formulate opinion and write report/witness statement
11. If required, attend court to give oral evidence

Table 6.11 Stages in a Typical Forensic Soil Investigation for the Defense

1. Read reports/statements by prosecution experts and other relevant background information
2. Read/discuss any additional information provided by the defendant or their solicitor
3. Visit prosecution expert's laboratory to examine case file; obtain copies of data and photographs for further analysis; examine exhibits of primary importance and take sub-samples as necessary for further analysis
4. Visit crime scene and other relevant locations; collect additional samples as necessary
5. Undertake laboratory analysis of prosecution samples and additional control samples
6. Compare data obtained with those produced by the prosecution and relevant database information
7. Synthesise results, formulate opinion, and write report
8. If required, attend court to give oral evidence

of control samples to be collected for comparison with the questioned samples. In designing the sampling strategy, the forensic geologist should have borne in mind a number of alternative propositions relating to the possible significance of the questioned samples. In essence this is the method of *multiple working hypotheses*, which has been used by many geologists to guide their investigations for over a hundred years (Chamberlin, 1890, 1897; Blewett, 1993). However, there are, unfortunately, many casework situations where the forensic geologist is not approached by the investigative team at an early stage and where the sampling has been undertaken by others who may not have adopted the multiple working hypotheses approach. The amount of background information about the circumstances of the case provided to the geologist may be very limited, and the "appropriateness" of the samples collected may not be clear. In such circumstances, assessments of the significance of the results obtained from any comparisons should be made with great caution.

In cases where the forensic geologist is instructed by the defense, the samples and data available for evaluation are inevitably constrained by the practices which were adopted by the scientists instructed by the prosecution, and the extent to which the defence examiner can work along the lines he would wish will be determined by such factors as the amount and condition of questioned and control sample material remaining, the willingness or otherwise of the accused to provide relevant and accurate information, the time available before a trial, and the extent of approved funding available.

An assessment of the overall strength of geological and soil evidence must take into account a number of factors. The first is the degree of similarity observed between the questioned samples and that of the control samples of interest (e.g., from the crime scene); this, in turn, is dependent on the number and type of characteristics in common, which may be limited by the size and type of samples available for examination, and the number and types of analytical techniques which could be applied. The second concerns the number of samples for which comparative data are available, and the extent to which they are considered to be adequately representative of the crime scene and any other locations which may be relevant to the particular enquiry. Thirdly, the availability or otherwise of suitable database information, or other reference information, which determines the context within which the results can be placed; in particular, the ability to comment on the known or expected frequency of occurrence of samples with similar characteristics.

At the end of the evaluation process, the examiner must usually (1) state the observed degree of similarity between the questioned sample and the control samples (i.e., whether they are considered to be *different, similar,* or *indistinguishable* in terms of the comparison criteria used), and (2) give an opinion as to the *level of overall scientific support* which has been obtained for the proposition that there is an association between the questioned sample and the comparison samples from the scene of crime, and any alternative propositions that may have been formulated.

The fact that two samples are indistinguishable in all considered respects is not in itself evidence that the questioned sample was derived from the crime scene; the alternative possibility must always be considered that the questioned sample was derived from another location with soil/sediment/ rock properties which are very similar to, or indistinguishable from, those at the crime scene, and which could be derived from the same ultimate geo- logical source.

There are different views regarding the best way to express weight of opinion in relation to the second issue. Some geological and soil expert witnesses in the U.K. have used the qualitative eight category verbal scheme commonly used by members of the Forensic Science Service to convey weight of evidence relating to footwear and other forms of trace evidence (Table 6.12b), whereas others have preferred their own systems of assessment. One such system, with 11 verbal categories, which has been used by the author, is shown in Table 6.12c. For the purposes of convenient shorthand, each verbal category is assigned a numeral rank ranging from 0 (no scientific evidence) to 10 (conclusive). No statistical significance of the ranks is implied.

The use of ranks or other numerical indices has been criticized by some lawyers and scientists involved in forensic soil investigation on the grounds that it gives a false sense of numerical precision. However, there is a long history of the use of numerical scales in the context of evidential and legal matters. For example, Aitken and Taroni (2004, p. 109), following Fienberg (1989), quote from the 19th century jurist, Jeremy Bentham (1827): "The scale being understood to be composed of ten degrees — in the language applied by the French philosophers to thermometers, a decigrade scale — a man says, My persuasion is at 10 or 9 etc. *affirmative*, or at least 10 etc. *negative*…." Numerical category scales also have a long history of use in the

Table 6.12 Suggested Schemes for Interpretation of Weight of Evidence (a) Scheme Proposed by Evett et al. (2000) for Classification of DNA Evidence

Verbal Terminology	Equivalent Likelihood Ratio
Limited evidence to support	>1 to 10
Moderate evidence to support	10 to 100
Moderately strong evidence to support	100 to 1000
Strong evidence to support	1,000 to 10,000
Very strong evidence to support	>10,000

(b) Scheme Often Used by Employees of the U.K. Forensic Science Service to Assess Evidence Relating to Footwear Marks and Various forms of Trace Evidence Other than DNA

Verbal Terminology
None
Limited
Moderate
Moderately strong
Strong
Very strong
Extremely strong
Conclusive

(c) Scheme Used by the Author

Verbal Terminology	Relative Ranking	Examples of Type of Evidence
None	0	Different in virtually all respects
Extremely limited	1	
Very limited	2	Some general similarity in terms of color, texture and/or relatively common particle types present
Limited	3	
Limited to moderate	4	General similarity in terms of color and/or texture, similar assemblage of relatively common particle types in common, some of which may have distinctive textural or chemical features
Moderate	5	
Moderately strong	6	Fairly high degree of physical, chemical, mineralogical, and/or biological similarity, including relatively unusual particle types in common
Strong	7	
Very strong	8	High degree of physical, chemical, mineralogical, and/or biological similarity, including several relatively unusual particle types present
Extremely strong	9	
Conclusive	10	Physical fit (rocks) and high degree of physical, chemical, and/or biological similarity; one or more very unusual particle types present

earth and environmental sciences, having been applied to a wide variety of phenomena, including mineral hardness (e.g., Table 2.5), and the relative severity of hurricanes and earthquakes.

The assessment of the strength of scientific evidence to support the proposition of an association between two samples should be considered separately from any evaluation of the *evidential significance* of any such association. The latter is dependent on a whole set of contextual circumstances about which the soil examiner may know little or nothing. For example, the demonstration that soil on a pair of shoes seized from a suspect is highly likely to originate from the scene of a crime, or another unidentified location with indistinguishable soil properties, in itself says nothing about who committed the crime. For example, the shoes could have been worn at the time by someone else other than the accused, or the shoes could have been cross-contaminated at some stage, for example, by *secondary transfer* from a source other than crime scene. It may be possible to eliminate these possibilities through other lines of enquiry, but these tasks are not normally the responsibility of the soil examiner.

It may, however, be possible for the examiner to provide an opinion on specific questions such as the relative likelihood that transfer of soil onto the shoes occurred by primary transfer at the crime scene, rather than by or secondary transfer at some other location, based on his/her own observations relating to the distribution and quantities of soil present on the shoes, or an observed textural association between the questioned soil and "contaminants" which are not present at the crime scene but which may be environmentally or locationally diagnostic. An illustration of an opportunity for secondary transfer of mud onto the footwear of an innocent party is shown in Figure 5.37b.

The examiner may also be able to provide an opinion about the potential significance of negative findings in his examination. In some circumstances, the existence of "negative evidence," or more correctly the absence of "positive evidence," may provide a sound basis on which to draw conclusions about the likelihood of a suggested set of events. For example, if it is suggested that a man was killed on a sandy beach but his body and clothing, when found elsewhere, bear no traces of sand, it is likely that the suggestion is incorrect, since even the most thorough attempts at cleaning will fail to remove all traces of such contact from hair, nails, ears, mouth, trachea, etc., even if the clothing has been changed and the body washed after death. In other circumstances, however, no conclusions can be drawn on the basis of an absence of positive evidence. For example, when a man is arrested on suspicion of murder several days after a crime has been committed, and forensic examination of his footwear and clothing fails to identify any traces of soil similar to that at a location where the victim's body was found, the finding cannot be taken as evidence of innocence. Various alternative possibilities exist which cannot be ruled out without additional evidence: the suspect may have been present at the crime scene but no soil was transferred, soil may have been transferred but has since fallen off or been cleaned up, the footwear/clothing worn at the time may have been destroyed, or the suspect may have committed the murder elsewhere and someone else transported the body to the deposition site.

As general practice, the geological and soil expert witness should resist all attempts to draw from him/her an opinion which goes beyond the limits of his specialist expertise and knowledge regarding the specific items/samples which he has examined. Useful general guidance regarding the nature of the desired relationships between an "expert" and those instructing him/her, on the preparation of experts' scientific reports, an on the presentation of expert evidence in court, can be obtained from sources such as Bond et al. (1999), Townley and Ede (2004), and White (2004).

6.12 The Future

Forensic geology and soil analysis has advanced a long way since Murray and Tedrow (1975) published the first edition of their book. As yet, however, the potential value of this type of evidence has not been fully explored. There are numerous particle and material types in the environment which potentially can be useful as trace evidence, and sophisticated techniques are now widely available for their analysis and comparison. At present, the chief factors limiting the use of this type of evidence in a forensic context are lack of awareness among the investigative and legal communities, the inadequacy of resources to develop relevant databases and bodies of experimental data, and the relatively small number of suitably trained and experienced personnel able to undertake the work. Many of these problems could be overcome through better communication between members of the earth and environmental science research communities, practising forensic scientists, representatives of the legal community, and government officials in the relevant departments who have responsibility for funding. It is hoped that this book will contribute in some small way toward that objective.

References

Abbott, D.M., Jr. (2005) Investigating mining frauds. *Geotimes* January 2005, 1–5.

Adams, A.E. and MacKenzie, W.S. (1998) *A Colour Atlas of Carbonate Sediments and Rocks Under the Microscope*. Manson Publishing, London, 180pp.

Adams, A.E., MacKenzie, W.S., and Guilford, C. (1984) *Atlas of Sedimentary Rocks Under the Microscope*. Longman Scientific and Technical, Harlow, 104pp.

Aitken, C.G.G. and Stoney, D.A. (eds.) (1991) *The Use of Statistics in Forensic Science*. Ellis Horwood, Chichester, 242pp.

Aitken, C.G.G. and Taroni, F.A. (1998) A verbal scale for the interpretation of evidence. *Science and Justice* 38, 279–281.

Aitken, C.G.G. and Taroni, F. (2004) *Statistics and the Evaluation of Evidence for Forensic Scientists*, 2nd ed. John Wiley & Sons Ltd., Chichester, 509pp.

Aitken, M.J. (1985) *Thermoluminescence Dating*. Academic Press, London, 359pp.

Aleon, J., Chaussidon, M., Marty, B., Schutz, L., and Jaenicke, R. (2002) Oxygen isotopes in single micrometer-sized quartz grains: tracing the source of Saharan dust over long-distance atmospheric transport. *Geochimica et Cosmochimica Acta* 66, 3351–3365.

Allen, T. (1997a) *Particle Size Measurement. Volume 1. Powder Sampling and Particle Size Measurement*, 5th ed. Chapman & Hall, London, 525pp.

Allen, T. (1997b) *Particle Size Measurement. Volume 2. Surface Area and Pore Size Distribution*. Chapman & Hall, London, 251pp.

Allmars, R.R. and Kempthorne, O. (2002) Errors, variability, and precision. In Dane, J.H. and Topp, G.G. (eds.) *Methods of Soil Analysis. Part 4. Physical Methods*. Soil Science Society of America, Madison, Wisconsin, 15–44.

Alloway, B.J. (ed.) (1995) *Heavy Metals in Soils*, 2nd ed. Blackie Academic and Professional, Glasgow & London, 368pp.

Ambach, E., Trubutsch, W. and Henn, R. (1991) Fatal accidents on glaciers: forensic, criminological, and glaciological conclusions. *Journal of Forensic Sciences* 36, 1469–1473.

Amthauer, G., Grodzicki, M., Lottermoser, W., and Redhammer, G. (2004) Mossbauer spectroscopy: basic principles. In Beran, A. and Libowitzky, E. (eds.) *Spectroscopic Methods in Mineralogy*. Eotvos University Press, Budapest, 345–367.

Andrasko, J. (1979) *An Analysis of Polycyclic Aromatic Hydrocarbons in Soils and Its Application to Forensic Science.* Linkoping University Report 4, 15pp.

Antoci, P.R. and Petraco, N. (1993) A technique for comparing soil colors in the forensic laboratory. *Journal of Forensic Sciences* 38, 437–441.

Applied Geochemistry Research Group, Imperial College London (1972) *A Provisional Geochemical Atlas of Northern Ireland.* Imperial College, London.

Applied Geochemistry Research Group, Imperial College London (1978) *The Wolfson Geochemical Atlas of England and Wales.* Clarendon Press, Oxford, 14pp plus maps.

Armstrong, H.A. and Brasier, M.D. (2005) *Microfossils,* 2nd ed. Blackwell Publishing, Oxford, 296pp.

Ashenbrenner, B.C. (1956) A new method of expressing particle sphericity. *Journal of Sedimentary Petrology* 26, 15–31.

Avery, B.W. and Bascomb, C.L. (1974) *Soil Survey Laboratory Methods.* Soil Survey Technical Monograph No. 6, Soil Survey of England and Wales, Rothamsted Experimental Station, Harpenden, 83pp.

Barker, C.E. and Kopp, O.C. (eds.) (1991) *Luminescence Microscopy and Spectroscopy. Qualitative and Quantitative Applications.* SEPM Short Course 25, Dallas, Texas 1991. SEPM (Society for Sedimentary Geology), Tulsa, Oklahoma, 195pp.

Barksdale, R.D. and Irani, S.Y. (1994) Influence of aggregate shape on base behaviour. *Transportation Research Record* 1227, 171–182.

Barrett, P.J. (1980) The shape of rock particles, a critical review. *Sedimentology* 27, 291–303.

Bates, D.M., Anderson, G.J., and Lee, R.D. (1997) Forensic botany: trichome evidence. *Journal of Forensic Sciences* 42, 380–386.

Bathurst, R.G.C. (1975) *Carbonate Sediments and Their Diagenesis. Developments in Sedimentology 12.* Elsevier, Amsterdam, 658pp.

Batjes, N.H. (2000) Development of a 0.5° by 0.5° resolution global database. In Sumner, M.E. (ed.) *Handbook of Soil Science.* CRC Press, Boca Raton, H-29 to H-40.

Batjes, N.H., van Engelen, V.W.P., Kauffman, J.H., and Oldeman, L.R. (1994) Development of soil databases for global modeling. *Transactions of the 15th World Congress on Soil Science* 6a, 40–57.

Bayes, T. (1763) An essay towards solving a problem in the doctrine of chances. *Philosophical Transactions of the Royal Society of London* 53, 370–418.

Beall, G.H. (1981) Glass, devitrification of volcanic. In Frye, J.C. (ed.) *The Encyclopedia of Mineralogy.* Encyclopedia of Earth Sciences Series Volume IVB, Hutchinson Ross Publishing Company, Stroudsberg, Pennsylvania, 187–s189.

Becker, J.S., Pickhardt, C., and Dietze, J.J. (2000) Laser ablation inductively coupled plasma mass-spectrometry for determination of trace elements in geological glasses. *Mikromchimica Acta* 135, 71–80.

Benko, J.J. (2000) Why is my house so dirty? *The Microscope* 48(3), 167–171.

Benn, D.I. (2004) Clast morphology. In Evans, D.J.A. and Benn, D.I. (eds.) *A Practical Guide to the Study of Glacial Sediments*. Arnold, London, 78–92.

Benn, D.I. and Ballantyne, C.K. (1993) The description and representation of particle shape. *Earth Surface Processes and Landforms* 18, 665–672.

Bennett, K.D., Fossitt, J.A., Sharp, M.J., and Switsur, V.R. (1990) Holocene vegetational and environmental history at Loch Lang, South Uist, Western Isles, Scotland. *New Phytologist* 114, 281–298.

Bennett, R.H., Bryant, W.R., and Hulbert, M.H. (eds.) (1991) *Microstructure of Fine-Grained Sediments*. Springer Verlag, New York, 582pp.

Bentham, J. (1827) In Mill, J.S. (ed.) *Rationale of Judicial Evidence, Specially Applied to English Practice*. Hunt and Clarke, London.

Benton, M. and Harper, D. (1997) *Basic Palaeontology*. Prentice Hall, Harlow, 342pp.

Beran, A. and Libowitzky, E. (eds.) (2004) *Spectroscopic Methods in Mineralogy*. Eotvos University Press, Budapest, 661pp.

Beran, A, Voll, D. and Schneider, H. (2004) IR spectroscopy as a tool for the characterization of ceramic precursor phases. In: Beran, A. and Libowitzky (eds.) *Spectroscopic Methods in Mineralogy*. EMU Notes in Mineralogy No. 6, Eotvos University Press, Budapest, 189–226.

Bernet, M. and Bassett, K. (2005) Provenance analysis by single-quartz-grain SEM-CL/optical microscopy. *Journal of Sedimentary Research* 75, 492–500.

Bevan, B.W. (1991) The search for graves. *Geophysics* 56, 1310–1319.

Bever, R.A. (2004) Plant Identification by DNA. Part I: identification of plant species from particulate material using molecular methods. In Coyle, H.M. (ed.) *Forensic Botany: Principles and Applications to Criminal Casework*. CRC Press, Boca Raton, 151–158.

Bigham, J.M. and Ciolkosz, E.J. (eds.) (1993) *Soil Color*. Soil Science Society of America Special Publication 31, Soil Science Society of America, Madison, Wisconsin, 159pp.

Birks, H.J.B. and Birks, H.H. (1980) *Quaternary Palaeoecology*. Edward Arnold, London, 289pp.

Bisbing, R.E. (2001) Finding trace evidence. In Houck, M.M. (ed.) *Mute Witness. Trace Evidence Analysis*. Academic Press, San Diego, 87–115.

Bisdom, E.B.A. (1981) A review of the applications of submicroscopic techniques in soil micromorphology. II. Electron microprobe analyzer (EMA), scanning electron microscope-energy-dispersive X-ray analyzer (SEM-EDXRA), laser microprobe mass-analyzer (LAMMA 500), electron spectroscopy for chemical analysis (ESCA), ion microprobe mass analyzer (IMMA), and the secondary ion microscope. In Bisdom, E.B.A. (ed.) *Sub-microscopy of Soils and Weathered Rocks*. Centre for Agricultural Publishing and Documentation, Wageningen, 117–162.

Bisdom, E.B.A., Henstra, S., Werner, H.W., Boudewijn, P.R., Knippenberg, W.F., de Grefte, H.A.M., and Migeon, H.N. (1983) Quantitative analysis of trace and major elements in thin sections of soils with the secondary ion microscope (CAMECA). *Geoderma* 30, 117–134.

Black, C.A., Evans, J.L., White, L.E., Ensminger, L.E., and Clark, F.E. (eds.) (1965) *Methods of Soil Analysis. Parts 1 and 2*. Agronomy Monograph No. 9. American Society of Agronomy, Madison, Wisconsin, 1572pp.

Black, R.M. (1989) *The Elements of Palaeontology*, 2nd ed. Cambridge University Press, Cambridge, 404pp.

Blatt, H. (1967) Original characteristics of clastic quartz grains. *Journal of Sedimentary Petrology* 37, 401–424.

Blatt, H., Middleton, G., and Murray, R. (1980) *Origin of Sedimentary Rocks*, 2nd ed. Perentice-Hall Inc., Englewood Cliffs, New Jersey, 782pp.

Blewett, W.L. (1993) Description, analysis and critique of the method of multiple working hypotheses. *Journal of Geological Education* 41, 254–259.

Blott, S.J. (2002) *Morphological and Sedimentological Changes on Artificially Nourished Beaches, Lincolnshire, UK*. PhD Thesis, University of London, 422pp.

Blott, S.J. and Pye, K. (2001) GRADISTAT: a grain size distribution and statistics package for the analysis of unconsolidated sediments. *Earth Surface Processes and Landforms* 26, 1237–1248.

Blott, S.J. and Pye, K. (2006) Particle size distribution analysis of sand-sized particles by laser diffraction: an experimental investigation of instrumental sensitivity and the effects of particle shape. *Sedimentology* 53, 671–685.

Blott, S.J. and Pye, K. (2007) A review of methods for the characterization of particle shape and some suggested new methods. *Earth Science Reviews* (in press)

Blott, S.J., Al-Dousari, A.M., Pye, K., and Saye, S.E. (2004a) Three-dimensional characterization of sand grain shape and surface texture using a nitrogen gas adsorption technique. *Journal of Sedimentary Research* 74, 156–159.

Blott, S.J., Croft, D.J., Pye, K., Saye, S.E., and Wilson, H.E. (2004b) Particle size analysis by laser diffraction. In Pye, K. and Croft, D.J. (eds.) *Forensic Geoscience: Principles, Techniques and Applications*. Geological Society, London, Special Publications 232, 63–73.

Bock, J.H. and Norris, D.O. (1997) Forensic botany: an under-utilized resource. *Journal of Forensic Sciences* 42, 364–367.

Boer de, D.H. and Crosby, G. (1995) Evaluating the potential of SEM/EDS analysis for fingerprinting suspended sediment derived from two contrasting topsoils. *Catena* 24, 243–258.

Boggs, S. and Krinsley, D.H. (2006) *Application of Cathodoluminescence Imaging to the Study of Sedimentary Rocks*. Cambridge University Press, Cambridge, 180pp.

Boggs, S., Krinsley, D.H., Goles, G.G., Seyedolali, A., and Dypvik, H. (2001) Identification of shocked quartz by scanning cathodoluminescence imaging. *Meteoritics and Planetary Science* 36, 783–778.

Bond, C., Solon, M., and Harper, P. (1999) *The Expert Witness in Court: A Practical Guide*, 2nd ed. Shaw & Sons Ltd, Crayford, Kent, 207pp.

Borg, L.E. and Banner, J.L. (1996) Neodymium and strontium isotopic constraints on soil sources in Barbados, West Indies. *Geochimica et Cosmochimica Acta* 60, 4193–4206.

Borgia, G.C., Bortolotti, V., Brancolini, A., Brown, R.J.S., and Tantazzini, P. (1996) Developments in core analysis by NMR measurements. *Magnetic Resonance Imaging* 14, 751–760.

Boughriet, A., Laurenyns, J., Recourt, P., Sobanska, S., Billon, G., Ouddane, B., and Bemard, C. (2004) Raman and SEM/EDS microanalysis of an environmental lead ore. *Microscopy and Microanalysis* November, 13–14.

Boutton, T.W. (1996) Stable carbon isotope ratios of soil organic matter and their use as indicators of vegetation and climate change. In Boutton, T.W. and Yamasaki, S. (eds.) *Mass Spectrometry of Soils*. Marcel Dekker, New York, 47–82.

Bowen, G.J., Winter, D.A., Spero, H.J., Zierenberg, R.A., Reeder, M.D., Cerling, T.E., and Ehleringer, J.R. (2005) Stable hydrogen and oxygen isotope ratios of bottled waters of the world. *Rapid Communications in Mass Spectrometry* 19, 3442–3450.

Bowie, S.H.U. and Simpson, P.R. (1977) Microscopy: reflected light. In Zussman, J. (ed.) *Physical Methods in Determinative Mineralogy*, 2nd ed. Academic Press, London, 109–165.

Bowles, F.A. (1968) Microstructure of sediments: investigation with ultra-thin sections. *Science* 159, 1236–1237.

Bowman, E.T., Soga, K., and Drummond, T.W. (2001) Particle shape characterization using Fourier analysis. *Geotechnique* 51, 545–554.

Brady, N.C. and Weil, R.R. (1999) *The Nature and Properties of Soils*, 12th ed. Prentice Hall, Upper Saddle River, New Jersey, 881pp.

Bremner, J.M. (1996) Total nitrogen. In: Sparks, D.L., Page, A.L., Helmake, P.A., Loeppert, R.H., Soltanpour, P.N., Tabatabal, M.A., Johnston, C.T. and Summer, M.E. (eds.) *Methods of Soil Analysis. Part 3 Chemical Methods*. Soil Science Society of America Book Series No. 5, Madison, Wisconsin, 1085–1122.

Brewer, R. (1976) *Fabric and Mineral Analysis of Soils*, 2nd ed. R.E. Krieger Publishing Co., Huntingdon, New York, 482pp.

Brick Development Association (1974) *Bricks. Their Properties and Use*. The Construction Press, Ltd., Lancaster, 251pp.

Bridgland, D.R. (ed.) (1986) *Clast Lithological Analysis*. Quaternary Research Association Technical Guide No. 3, Quaternary Research Association, Cambridge, 207pp.

Brillis, G.M., Gerlach, C.L., and van Waasbergen, R.J. (2000) Remote sensing tools assist in environmental forensics. Part I. Traditional methods. *Journal of Environmental Forensics* 1, 63–67.

Brindley, G.W. and Brown, G. (1980) *Crystal Structures of Clay Minerals and Their X-ray Identification*. Mineralogical Society Monograph No. 5, Mineralogical Society, London, 495pp.

Bristow, C.S. and Jol, H.M. (eds.) (2003) *Ground Penetrating Radar in Sediments.* Geological Society, London, Special Publications 211, 330pp.

British Geological Survey (1987) *Regional Geochemical Atlas: Great Glen.* British Geological Survey, Keyworth, Nottingham.

British Geological Survey (1990) *Regional Geochemical Atlas: Argyll.* British Geological Survey, Keyworth, Nottingham.

British Geological Survey (1991) *Regional Geochemistry of the East Grampians Area.* British Geological Survey, Keyworth, Nottingham.

British Geological Survey (1992) *Regional Geochemistry of the Lake District and Adjacent Areas.* British Geological Survey, Keyworth, Nottingham, 98pp.

British Geological Survey (1993) *Regional Geochemistry of Southern Scotland and Parts of Northern England.* British Geological Survey, Keyworth, Nottingham.

British Geological Survey (1996) *Regional Geochemistry of North-east England. British Geological Survey,* Keyworth, Nottingham, 100pp.

British Geological Survey (1997) *Regional Geochemistry of Parts of North-west England and North Wales.* British Geological Survey, Keyworth, Nottingham, 128pp.

British Geological Survey (1999) *Regional Geochemistry of Wales and Part of West-Central England: Stream Water.* British Geological Survey, Keyworth, Nottingham.

British Geological Survey (2000) *Regional Geochemistry of Wales and Part of West-Central England: Stream Sediment and Soil.* British Geological Survey, Keyworth, Nottingham, 156pp.

Brooks, M. and Newton, K. (1969) Forensic pedology. *Police Journal (London)* 42, 107–112.

Brown, A.G. (2000) Going to ground. *Police Review* 4 February 2000, 18–20.

Brown, A.G. (2006) The use of forensic botany and geology in war crime investigation in NE Bosnia. *Forensic Science International* 163, 204–210.

Brown, A.G., Smith, A., and Elmhurst, O. (2002) The combined use of pollen and soil analysis in a search and subsequent murder investigation. *Journal of Forensic Sciences* 47, 614–618.

Brown, A.T., Nelson, D.E., Matthews, R.W., Vogel, J.S., and Southon, J.R. (1989) Radiocarbon dating of pollen by accelerator mass spectrometry. *Quaternary Research* 32, 205–212.

Brown, A.T., Farrell, G.W., Grootes, P.M., and Schmidt, F.H. (1992) Radiocarbon dating of pollen extracted from peat samples. *Radiocarbon* 34, 550–556.

Bruce, R.G. and Dettmann, M.E. (1996) Palynological analyses of Australian surface soils and the potential in forensic science. *Forensic Science International* 81, 77–94.

Bruce, L.G. & Schmidt, G.W. (1994) Hydrocarbon fingerprinting for application in forensic geology: review with case studies. *Bulletin of the Association of American Petroleum Geologists* 78, 1692–1710.

Bryant, V.M. and Jones, G.D. (2006) Forensic palynology: current status of a rarely used technique in the United States of America. *Forensic Science International* 163, 183–197.

Bryant, V.M., Jr., Jones, J.G., and Mildenhall, D.C. (1990) Forensic palynology in the United States of America. *Palynology* 14, 193–208.

Bryant, V.M. Jr. and Mildenhall, D.C. (1998) Forensic palynology: a new way to catch crooks. In Bryant, V.M., Jr., and Wren, J.W. (eds.) *New Developments in Palynomorph Sampling, Extraction and Analysis*. American Association of Stratigraphic Palynologists Foundation, Contribution Series 33, 145–155.

Buccianti, A., Mateu-Figueras, G. and Pawlowsky-Glahn, V. (eds.) (2006) *Compositional Data Analysis in the Geosciences*. Geological Society, London, Special Publications 264, 212pp.

Buck, S.C. (2003) Searching for graves using geophysical technology: field tests with ground penetrating radar, magnetometry and electrical resistivity. *Journal of Forensic Sciences* 48, 1–7.

Buckleton, J., Triggs, C.M., and Walsh, S.J. (eds.) (2005) *Forensic DNA Evidence Interpretation*. CRC Press, Boca Raton, 534pp.

Bull, P.A. and Morgan, R.M. (2006) Sediment fingerprints: a forensic technique using quartz sand grains. *Science and Justice* 46, 107–124.

Bull, P.A., Parker, A. & Morgan, R.M. (2006) The forensic analysis of soils and sediment taken from the cast of a footprint. *Forensic Science International* 162, 6–12.

Bullock, P. and Murphy, C.P. (1980) Towards the quantification of soil structure. *Journal of Microscopy* 120, 317–328.

Bullock, P., Federoff, N., Jongerius, A., Stops, G., and Tursina, T. (1985) *Handbook for Soil Thin Section Description*. Waine Research Publications, Wolverhampton, 152pp.

Caddy, B. (ed.) (2001) *Forensic Examination of Glass and Paint: Analysis and Interpretation*. Taylor & Francis, New York, 292pp.

Cady, J.G., Wilding, L.P., and Drees, L.R. (1986) Petrographic microscopic techniques. In Klute, A. (ed.) *Methods of Soil Analysis, Part 1. Physical and Mineralogical Methods*. Agronomy Monography No. 9, American Society of Agronomy, Madison, Wisconsin, 185–218.

Cameron, N.G. (2004) The use of diatom analysis in forensic geoscience. In Pye, K. and Croft, D.J. (eds.) *Forensic Geoscience: Principles, Techniques and Applications*. Geological Society, London, Special Publications 232, 277–280.

Camps, F.E. (1962) Soil—some medico-legal aspects. In Neil, M.W. and Warren, F.L. (eds.) *Soil*. British Academy of Forensic Sciences Teaching Symposium No. 1. Sweet and Maxwell Ltd., London, 47–51.

Canti, M.G. (2005) *Forensic Micromorphology of the Crow Down Hoard*. English Heritage, Centre for Archaeology, Report 52/2005, English Heritage, Portsmouth, 10pp.

Carcaillet, C. (2001) Soil particles reworking evidences by AMS 14C dating of charcoal. *Comptes Rendues Academie de Sciences Paries, Sciences de la Terre et des Planetes* 332, 21–28.

Carlson, W.D., Rowe, T., Ketcham, R.A., and Colbert, M.W. (2003) Applications of high-resolution X-ray computed tomography in petrology, meteoritics and palaeontology. In Mees, F., Swennen, R., Van Geet, M., and Jacobs, P. (eds.) *Applications of X-ray Computed Tomography in the Geosciences.* Geological Society, London, Special Publications 215, 7–22.

Carita, E.J. (2004) Classical and future DNA typing technologies for plants. In Coyle, H.M. (ed.) *Forensic Botany: Principles And Applications to Criminal Casework.* CRC Press, Boca Raton, 253–274.

Carver, R.E. (1971) Heavy mineral separation. In Carver, R.E. (ed.) *Procedures in Sedimentary Petrology.* Wiley Interscience, New York, 427–452.

Cerling, T.E. and Yang, W. (1996) Stable carbon and oxygen isotopes in soil CO_2 and soil carbonate: theory, practice and application to some Prairie soils of Upper Midwestern North America. In Boutton, T.W. and Yamasaki, S. (eds.) *Mass Spectrometry of Soils.* Marcel Dekker, New York, 113–132.

Chamberlin, T.C. (1890) The method of multiple working hypotheses. *Science* 15, 92–96.

Chamberlin, T.C. (1897) The method of multiple working hypotheses. *Journal of Geology* 5, 837–848.

Champod, C. and Taroni, F. (1999) The Bayesian approach. In Robertson, J. and Grieve, M. (eds.) *Forensic Examination of Fibres*, 2nd ed. Taylor & Francis, London, 379–398.

Chaperlin, K. (1981) Lead content and soil discrimination in forensic science. *Forensic Science International* 18, 79–84.

Chaperlin, K. and Howarth, P.S. (1983) Soil comparison by the density gradient method — a review and evaluation. *Forensic Science International* 23, 161–177.

Charman, D.J., Hendon, D., and Woodland, W.A. (2000) *The Identification of Testate Amoebae (Protozoa: Rhizopoda) in Peats.* Quaternary Research Association Technical Guide No. 9, Quaternary Research Association, London, 145pp.

Chayes, F. (1949) A simple point counter for thin section analysis. *American Mineralogist* 34, 1–11.

Chazottes, V., Brocard, C. and Peyrot, B. (2002) Particle size analysis of soils under simulated scene of crime conditions: the interest of multivariate analyses. *Forensic Science International* 140, 159–166.

Chen, Q., Kinzelbach, W., and Oswald, S. (2002) Nuclear magnetic resonance imaging for studies of flow and transport in porous media. *Journal of Environmental Quality* 31, 477–486.

Christensen, J.N., Halliday, A.N., Lee, C.D., and Hall, C.M. (1995) *In situ* Sr isotopic analysis by laser ablation. *Earth and Planetary Science Letters* 136, 79–85.

Clark, A. (1996) *Seeing Beneath the Soil: Prospecting Methods in Archaeology.* Routledge, London, 192pp.

Clark, J.S. (1988a) Stratigraphic charcoal analysis on petrographic thin sections: application to fire history in Northwestern Minnesota. *Quaternary Research* 30, 81–91.

Clark, J.S. (1988b) Particle motion and the theory of charcoal analysis: source area, transport, deposition, and sampling. *Quaternary Research* 30, 67–80.

Clark, M. (1981) Quantitative shape analysis: a review. *Mathematical Geology* 13, 303–319.

Clark, R.L. (1982) Point count estimation of charcoal in pollen preparations and thin sections of sediments. *Pollen et Spores* 24, 523–535.

Clark, R.L. (1984) Effects on charcoal of pollen preparation procedures. *Pollen et Spores* 26, 559–576.

Clark, W.A.V. and Hosking, P.L. (1986) *Statistical Methods for Geographers.* John Wiley and Sons, New York.

Cleveland, G.B. (1973) CDMG (California Division of Mines and Geology) helps find kidnapper. *California Geology* 26(10), 240–241.

Cline, M.G. (1944) Principles of soil sampling. *Soil Science* 58, 275–288.

Cohen, J. (1988) *Statistical Power Analysis for the Behavioural Sciences,* 2nd ed. Lawrence Erlbaum Associates, Hillside, New Jersey, 415pp.

Commision Internationale de l'Eclairage (1931) *Proceedings of the Eighth Session, Cambridge, England, 1931.* Bureau Central de la CIE, Paris.

Commision Internationale de l'Eclairage (1978) *Recommendations on Uniform Colour Spaces, Colour Differences and Psychometric Colour Terms.* Supplement No 2 to Publication No 15, *Colorimetry,* CIE, Paris.

Cook, R., Evett, I.W., Jackson, G., Jones, P.J., and Lambert, J.A. (1998a) A model for case assessment and interpretation. *Science and Justice* 38, 151–156.

Cook, R., Evett, I.W., Jackson, G., Jones, P.J., and Lambert, J.A. (1998b) A hierarchy of propositions: deciding which to address in forensic casework. *Science and Justice* 38, 231–239.

Cook, R.D. and Weisberg, S. (1994) *An Introduction to Regression Graphics.* John Wiley and Sons, New York.

Coote, D.R. and MacDonald, K.B. (2000) The Canadian Soil database. In Sumner, M.E. (ed.) *Handbook of Soil Science.* CRC Press, Boca Raton, H-41 to H-51.

Corey, A.T. (1949) *Influence of Shape on the Fall Velocity of Sand Grains.* MS Thesis, A & M. College, Colorado.

Cox, E.A. (1927) A method of assigning numerical and percentage values to the degree of roundness of sand grains. *Journal of Palaeontology* 1, 179–183.

Cox, R.A. (2004) Cathodoluminescence and laser-ablation ICP-MS analysis of topaz crystals. *Microscopy and Analysis* March, 5–7.

Cox, R.J., Peterson, H.J., Young, J., Cusik, C., and Espinoza, E.O. (2000) The forensic analysis of soil organic matter by FTIR. *Forensic Science International* 108, 107–116.

Coyle, H.M. (ed.) (2004) *Forensic Botany: Principles and Applications to Criminal Casework*. CRC Press, Boca Raton, 318pp.

Croft, D.J. (2003) *Forensic Geoscience: Development of Techniques for Soil Analysis*. PhD Thesis, University of London, 426pp.

Croft, D.J. and Pye, K. (2003) The potential use of continuous flow isotope ratio mass spectrometry as a tool in forensic soil analysis: a preliminary report. *Rapid Communications in Mass Spectrometry* 17, 2581–2584.

Croft, D.J. and Pye, K. (2004a) Colour theory and the evaluation of an instrumental method of measurement using geological samples for forensic applications. In Pye, K. and Croft, D.J. (eds.) *Forensic Geoscience: Principles, Techniques and Applications*. Geological Society, London, Special Publications 232, 49–62.

Croft, D.J. and Pye, K. (2004b) Stable carbon and nitrogen isotope variations in soils: forensic applications. In Pye, K. and Croft, D.J. (eds.) *Forensic Geoscience: Principles, Techniques and Applications*. Geological Society, London, Special Publications 232, 257–268.

Croft, D.J. and Pye, K. (2004c) Multi-technique comparison of source and primary transfer soil samples: an experimental investigation. *Science and Justice* 44, 21–28.

CSIRO (1983) Soils — An Australian Viewpoint. CSIRO Division of Soils and Academic Press, Melbourne and London, 928pp.

Culver, S.J., Bull, P.A., Campbell, S., Shakesby, R.A., and Whalley, W.B. (1983) Environmental discrimination based on quartz grain surface textures: a statistical evaluation. *Sedimentology* 30, 129–136.

Curran, J.M, Hicks, T.N., and Buckleton, J.S. (2000) *Forensic Interpretation of Glass Evidence*. CRC Press, Boca Raton, 178pp.

Daugherty, L.A. (1997) Soil science contribution to an airplane crash investigation, Ruidoso, New Mexico. *Journal of Forensic Sciences* 42, 401–405.

Davis, A. and Vastola, P.J. (1977) Developments in automated reflectance microscopy of coal. *Journal of Microscopy* 109, 3–12.

Davis, J.C. (2002) *Statistics and Data Analysis in Geology*, 3rd ed. John Wiley & Sons, New York, 638pp.

Dawson, L.A., Towers, W., Mayes, R.W., Craig, J., Vaisanen, R.K., and Waterhouse, E.C. (2004) The use of plant hydrocarbon signatures in characterizing soil organic matter. In Pye, K. and Croft, D.J. (eds.) *Forensic Geoscience: Principles, Techniques and Applications*. Geological Society, London, Special Publications 232, 269–276.

Day, R.W. (1998) *Forensic Geotechnical and Foundation Engineering*, Blackwell Publishing, Oxford, 296pp.

Dearing, J.A. (1999) Magnetic susceptibility. In Walden, J., Oldfield, F., and Smith, J. (eds.) *Environmental Magnetism: A Practical Guide*. Quaternary Research

Association Technical Guide, No. 6, Quaternary Research Association, London, 35–62.

Dearing, J.A. (2000) Natural magnetic traces in fluvial geomorphology. In Foster, I.D.L. (ed.) *Tracers in Geomorphology*. Wiley, Chichester, 57–82.

De Boer, D.H. and Crosby, G. (1997) Evaluating the potential of SEM/EDS analysis for fingerprinting suspended sediment derived from two contrasting topsoils. *Catena* 24, 243–258.

De Gruijter, J.J. (2002) Sampling. In Dane, J.H. and Topp, G.G. (eds.) *Methods of Soil Analysis. Part 4. Physical Methods*. Soil Science Society of America Book Series No.5. Soil Science Society of America Inc., Madison, Wisconsin, 45–80.

Deer, W.A., Howie, R.A., and Zussman, J. (1992) *An Introduction to the Rock-Forming Minerals*, 2nd ed. Longman, Harlow, 696pp.

Demmelmeyer, H. and Adam, J. (1995) Forensic investigation of soil and vegetable materials. *Forensic Science Review* 7, 119–142.

Dia, A., Chauvel, C., Belourde, M., and Gerard, M. (2006) Eolian contribution to soils on Mount Cameroon: isotopic and trace element records. *Chemical Geology* 226, 232–252.

Dickin, A.P. (2005) *Radiogenic Isotope Geology*, 2nd ed. Cambridge University Press, Cambridge, 508pp.

Dixon, J.B. and Weed, S.B. (eds.) (1977) *Minerals in Soil Environments*. Soil Science Society of America, Madison, Wisconsin, 948pp.

Dixon, J.B. and Schulze, D.G. (eds.) (2002) *Soil Mineralogy with Environmental Applications*. Soil Science Society of America Book Series No. 7. Soil Science Society of America Inc., Madison, Wisconsin, 866pp.

Dobkins, J.E., Jr. and Folk, R.L. (1970) Pebble shape development on Tahiti-Nui. *Journal of Sedimentary Petrology* 40, 1167.

Doran, D.K. (ed.) (1992) *Construction Materials Reference Book*. Butterworth Heinemann, Oxford.

Dowdeswell, J.A. (1982) Scanning electron micrographs of quartz sand grains from cold environments examined using Fourier shape analysis. *Journal of Sedimentary Petrology* 52, 1315–1323.

Dowdeswell, J.A., Osterman, L.E., and Andrews, J.T. (1985) Quartz sand grain shape and other criteria used to distinguish glacial and non-glacial events in a marine core from Frobisher Bay, Baffin Island, N.W.T., Canada. *Sedimentology* 32, 119–132.

Dryden, A.L. (1931) Accuracy in percentage representation of heavy mineral frequencies. *Proceedings of the National Academy of Sciences* 17, 233–238.

Dudley, R.J. (1975) The use of color in the discrimination between soils. *Journal of the Forensic Science Society* 15, 209–218.

Dudley, R.J. (1976a) A simple method for determining the pH of small soil samples and its use in forensic science. *Journal of the Forensic Science Society* 16, 21–27.

Dudley, R.J. (1976b) A colorimetric method for the determination of soil saccharide content and its application in forensic sciences. *Medicine Science and the Law* 16, 226–231.

Dudley, R.J. (1976c) The use of cathodoluminescence in the identification of soil minerals. *Journal of Soil Science* 27, 487–494.

Dudley, R.J. (1977) The particle size analysis of soils and its use in forensic science — the determination of particle size distributions within silt and sand fractions. *Journal of the Forensic Science Society* 16, 219–229.

Dudley, R.J. (1979) The use of density gradient columns in the forensic comparison of soils. *Medicine Science and the Law* 19, 39–48.

Dudley, R.J. and Smalldon, K.W. (1978a) The comparison of distributional shapes with particular reference to a problem in forensic science. *International Statistical Review* 46, 53–63.

Dudley, R.J. and Smalldon, K.W. (1978b) The evaluation of methods for soil analysis under simulated scenes of crime conditions. *Forensic Science International* 12, 49–60.

Dudley, R.J. and Smalldon, K.W. (1978c) The objective comparison of the particle size distribution in soils with particular reference to the sand fraction. *Medicine, Science and The Law* 18, 278–282.

Duller, G.A.T. (1997) Behavioural studies of stimulated luminescence from feldspars. *Radiation Measurements* 27, 663–694.

Edwards, H.G.M. (2001) Raman spectroscopic applications to archaeological biomaterials. In Lewis, I.R. and Edwards, H.G.M. (eds.) *Handbook of Raman Spectroscopy: From the Research Laboratory to the Process Line*. Marcel Dekker, New York, 1011–1044.

Edwards, H.G.M. (2004) Forensic applications of Raman spectroscopy to the non-destructive analysis of biomaterials and their degradation. In Pye, K. and Croft, D.J. (eds.) *Forensic Geoscience: Principles, Techniques and Applications*. Geological Society, London, Special Publications 232, 159–170.

Edwards, H.G.M. and Hassan, N.F.N. (2004) Evaluation of Raman spectroscopy for the non-destructive differentiation between elephant and mammoth ivories. *Asian Chemistry Letters* 7, 185–196.

Eglinton, G., Parkes, R.J., and Zhao, M. (1993) Lipid biomarkers in biogeochemistry: future roles? *Marine Geology* 113, 141–145.

Ehlers, E.G. and Blatt, H. (1982) *Petrology: Igneous, Sedimentary and Metamorphic*. W.H. Freeman and Company, San Francisco, 732pp.

Ehrlich, R. and Weinberg, B. (1970) An exact method for characterization of sand shape. *Journal of Sedimentary Petrology* 40, 205–212.

Ehrlich, R., Brown, P.J., Yarus, J.M., and Przygocki, R.S. (1980) The origin of shape frequency distributions and the relationships between size and shape. *Journal of Sedimentary Petrology* 50, 475–483.

Elzenga, W., Schwan, J., Baumfalk, T.A., Vandenbergh, J., and Krook, L. (1987) Grain surface characteristics of periglacial aeolian sand and fluvial sands. *Geologie en Mijnbouw* 65, 273–286.

Emerson, J.D. and Strenio, J. (2000) Box plots and batch comparison. In Hoaglin, D.C., Mosteller, F., and Tukey, J.W. (eds.) *Understanding Robust and Exploratory Data Analysis*. Wiley, New York, 58–96.

Engelen, van, V.W.P. (2000) SOTER: The World Soils and Terrain Database. In Sumner, M.E. (ed.) *Handbook of Soil Science*. CRC Press, Boca Raton, H-19 to H28.

Erdtman, G. (1986) *Handbook of Palynology*. Hafner, New York, 486pp.

Evans, J.E. and Tokar, F.J., Jr. (2000) Use of SEM. EDS and X-ray diffraction analyses for sand transport studies, Lake Erie, Ontario. *Journal of Coastal Research* 16, 926–933.

Evett, I.W. and Buckleton, J.S. (1989) Some aspects of the Bayesian approach to evidence evaluation. *Journal of the Forensic Science Society* 29, 317–324.

Evett, I.W., Jackson, G., and Lambert, J.A. (2000) More on the hierarchy of propositions: exploring the distinction between explanations and propositions. *Science and Justice* 40, 3–10.

Evett, I.W., Jackson, G., Lambert, J.A., and McCrossan, S. (2000) The impact of the principles of evidence interpretation on the structure and content of statements. *Science and Justice* 40, 233–239.

Eyring, M.B. (1996) Soil pollen analysis from a forensic point of view. *The Microscope* 44(2), 81–97.

Faegri, K., Iversen, J., Kaland, P.E., and Krzywinski, K. (2000) *Textbook of Pollen Analysis*, 5th ed. The Blackburn Press, Caldwell, New Jersey, 340pp.

FAO (1996) Guidelines for Soil Profile Description. Fourth Edition. United Nations Food and Agriculture Organization, Paris.

FAO–ISRIC (1989) Soil Database (SDB). *World Soil Resources*. Report No. 64. Food and Agriculture Organization, Rome, 89pp.

FAO–ISRIC (1990) *Guidelines for Soil Description*, 3rd ed. Food and Agriculture Organization of the United Nations, Rome, 69pp.

FAO–UNESCO (1974) *Soil map of the World 1: 5 000 000. Volume 1: Legend*. UNESCO, Paris.

FAO–UNESCO–ISRIC (1988) *Revised Legend of the FAO–Unesco Soil Map of the World*. World Soil Resources Reports No. 60, FAO, Rome.

Farmer, V.C. (ed.) (1974) *The Infrared Spectra of Minerals*. Mineralogical Society Monograph 4, Mineralogical Society, London, 539pp.

Fattahi, M. and Stokes, S. (2003) Dating volcanic and related sediments by luminescence methods: a review. *Earth Science Reviews* 62, 229–264.

Faure, G. (1986) *Principles of Isotope Geology*, 2nd ed. John Wiley & Sons, Chichester, 589pp.

Faure, G. and Mensing, T.M. (2004) *Isotopes*, 3rd ed. Cambridge University Press, Cambridge, 928pp.

Fedorowich, J.S., Jain, J.C., and Kerrich, R. (1995) Trace element analysis of garnet by laser ablation microprobe ICP-MS. *Canadian Mineralogist* 33, 469–480.

Fenning, P.J. and Donnelly, L.J. (2004) Geophysical techniques for forensic investigation. In Pye, K. and Croft, D.J. (eds.) *Forensic Geoscience: Principles, Techniques and Applications*. Geological Society, London, Special Publications 232, 11–20.

Fienberg, S.E. (1989) *The Evolving Role of Statistical Assessment as Evidence in the Courts*. Springer Verlag, New York.

Figueral, I. and Mosbrugger, V. (2000) A review of charcoal analysis as a tool for assessing Quaternary and Tertiary environments: achievements and limits. *Palaeogeography, Palaeoclimatology, Palaeoecology* 164, 397–407.

Finkelstein, M.O. and Levin, B. (2005) Compositional analysis of bullet lead as forensic evidence. *Journal of Law and Policy* 13, 119–142.

Fisher, B.A.J. (2000) *Techniques of Crime Scene Investigation*, 6th ed. CRC Press, Boca Raton, 552pp.

Fitzpatrick, E.A. (1984) *Micromorphology of Soils*. Chapman and Hall, London, 433pp.

Fitzpatrixk, E.A. (1993) *Soil Microscopy and Micromorphology*. John Wiley & Sons, New York.

Fleet, W.F. (1926) Petrological notes on the Old Red Sandstone of the West Midlands. *Geological Magazine* 63, 505–516.

Flemming, N.C. (1965) Form and function of sedimentary particles. *Journal of Sedimentary Petrology* 35, 381–390.

Flint, A.L. and Flint, L.E. (2002a) Particle density. In Dane, J.H. and Topp, G.G. (eds.) *Methods of Soil Analysis. Part 4. Physical Methods*. Soil Science Society of America Book Series No.5. Soil Science Society of America Inc., Madison, Wisconsin, 241–254.

Flint, A.L. and Flint, L.E. (2002b) Porosity. In Dane, J.H. and Topp, G.G. (eds.) *Methods of Soil Analysis. Part 4. Physical Methods*. Soil Science Society of America Book Series No.5. Soil Science Society of America Inc., Madison, Wisconsin, 241–254.

Folk, R.L. (1954) The distinction between grain size and mineral composition in sedimentary rock nomenclature. *Journal of Geology* 62, 344–359.

Folk, R.L. (1955) Student operator error in determination of roundness, sphericity and grain size. *Journal of Sedimentary Petrology* 25, 297–301.

Folk, R.L. (1974) *Petrology of Sedimentary Rocks*, 2nd ed. Hemphill, Austin, Texas, 182pp.

Folk, R.L. (1976) Reddening of desert sands: Simpson Desert, Northern Territory, Australia. *Journal of Sedimentary Petrology* 46, 604–615.

Folk, R.L. and Ward, W.C. (1957) Brazos River Bar: a study in the significance of grain size parameters. *Journal of Sedimentary Petrology* 26, 3–32.

Forensic Science Service (1994) The *Scenes of Crime Handbook*, 2nd ed. Forensic Science Service, London.

Fraser, D.G. (1995) The nuclear microprobe — PIXE, PIGE, RBS, NRA and ERDA. In Potts, P.J., Bowles, J.F.W., Reed, S.J.B., and Cave, M.R. (eds.) *Microprobe Techniques in the Earth Sciences*. Chapman and Hall, London, 141–162.

Freestone, I.C. (1982) Applications and potential of electron probe micro-analysis in technological and provenance investigations of ancient ceramics. *Archaeometry* 24, 99–116.

Freile, D., Faulkner, C., Malchow, R., and Ghazi, M. (2001) Determination of trace metal variability in ooid laminae using LA-ICPMS. *Geological Society of America Abstracts With Programs* 33(7), A313.

Friedman, G.M. and Sanders, J.E. (1978) *Principles of Sedimentology*. Wiley, New York, 792pp.

Freile, D., Faulkner, C., Malchow, R., and Ghazi, M. (2001) Determination of trace-metal variability in ooid laminae using LA-ICP-MS. *Geological Society of America Annual Meeting, November 5–8, 2001 (Abstract of Paper No.100-0)*.

Froude, D.O., Irelenad, T.R., Kinny, P.D., Williams, I.S., and Compston, W. (1983) Ion microprobe identification of 4,100-4,200 Myr-old terrestrial zircons. *Nature* 304, 616–618.

Frye, J.C. (1981) *The Encyclopedia of Mineralogy*. Encyclopedia of Earth Sciences Volume IVB. Hutchinson Ross Publishing Company, Stroudsburg, Pennsylvania, 794pp.

Full, W.E. and Ehrlich, R. (1982) Some approaches for location of centroids of quartz grain outlines to increase homology between Fourier amplitude spectra. *Mathematical Geology* 14, 43–55.

Full, W.E., Ehrlich, R., and Kennedy, S.K. (1984) Original configuration and information content of sample data generally displayed as histograms or frequency plots. *Journal of Sedimentary Petrology* 54, 117–126.

Gagosian, R.B., Peltzer, E.T., and Zafiriou, O.C. (1981) Atmospheric transport of continentally derived lipds to the tropical North Pacific. *Nature* 291, 321–324.

Gagosian, R.B., Zafiriou, O.C., Peltzer, E.T., and Alford, J.B. (1982) Lipids in aerosols from the tropical North Pacific; temporal variability. *Journal of Geophysical Research* 87, 11,133–11,144.

Gagosian, R.B., Peltzer, E.T., and Merril, J.T. (1987) Long-range transport of terrestrially derived lipids in aerosols from the South pacific. *Nature* 325, 800–803.

Galehouse, J.S. (1971a) Sedimentation analysis. In Carver, R.E. (ed.) *Procedures in Sedimentary Petrology*. Wiley Interscience, New York, 69–94.

Galehouse, J.S. (1971b) Point counting. In Carver, R.E. (ed.) *Procedures in Sedimentary Petrology*. Wiley Interscience, New York, 385–408.

Gallop, A. and Stockdale, R. (2005) Trace and contact evidence. In White, P.C. (ed.) *Crime Scene to Court. The Essentials of Forensic Science*, 2nd ed. Royal Society of Chemistry, Cambridge, 56–81.

Ganesh, G.J.S., Raghunath, R., and Sekharan, P.C. (1989) Forensic examination of micro-quantity soil samples. *Indian Journal of Forensic Science* 3, 37–43.

Garboczi, E.J. (2002) Three-dimensional mathematical analysis of particle shape using X-ray tomography and spherical harmonics: application to aggregates used in concrete. *Cement and Concrete Research* 32, 1621–1638.

Gee, G.W. and Or, D. (2002) Particle size analysis. In Dane, J.H. and Topp, G.G. (eds.) *Methods of Soil Analysis. Part 4. Physical Methods.* Soil Science Society of America Inc., Madison, Wisconsin, 255–294.

Geiger, C.A. (2004) An introduction to spectroscopic methods in the mineral sciences and geochemistry. In Beran, A. and Libowitzky, E. (eds.) *Spectroscopic Methods in Mineralogy.* Eotvos University Press, Budapest, 1–42.

Gettinby, G. (1991) Application of statistics to particular areas of forensic science. In Aitken, C.G.G. and Stoney, D.A. (eds.) The Use of Statistics in Forensic Science. Ellis Horwood, Chichester, 156–162.

Ghazi, A.M. and Millette, J.R. (204) Environmental forensic application of lead isotope ratio determination: a case study using laser ablation sector ICP-MS. *Environmental Forensics* 5, 97–108.

Gilbert, R.O. and Pulpisher, B.A. (2005) Role of sampling designs in obtaining representative data. *Environmental Forensics* 6, 27–33.

Gillott, J.E. (1980) Use of the scanning electron microscope and Fourier methods in characterization of microfabric and texture of sediments. *Journal of Microscopy* 120, 261–277.

Giodarno, P., Jr., Barrot, D., Mease, P., Garrison, K., Mandayam, S., and Sukumaran, B. (2006) An optical tomography system for characterising 3D shapes of particle aggregates. Paper presented at the Conference of the Institute of Electonic and Electrical Engineers Sensors Application Symposium, Houston, Texas, USA, 7-9 February 2006, 3pp.

Gniadecka, M., Edwards, H.G.M., Hansen, J.P.H., Nielsen, O.F., Christensen, D.H., Gullen, S.E., and Wolf, H.C. (1999) NIT-FT Raman spectroscopy of the mummified skin of the Alpine Iceman, Qilakitosoq Greenland mummies and the Chiribaya mummies from Peru. *Journal of Raman Spectroscopy* 30, 147–154.

Goddard, E.N., Trask, P.D., DeFord, R.K., Rove, O.N., Singlwald, J.T., and Overbeck, R.M. (1948) *Rock Color Chart.* Geological Society of America, Boulder.

Goin, L.J. and Kirk, P.L. (1947) Application of microchemical techniques: identity of soil samples. *Journal of Criminal Law, Criminology and Police Science* 38, 267–281.

Goldstein, J., Newbury, D., Joy, D., Lyman, C., Echlin, P., Lifshin, E., Sawyer, L., and Michael, J. (2003) *Scanning Electron Microscopy and X-ray Microanalysis,* 3rd ed. Kluwer Academic/Plenum Publishers, New York, 689pp.

Goldstein, S.L., Arndt, N.T., and Stallard, R.F. (1997) The history of a continent from U-Pb ages of zircons from Orinoco River sand and Sm-Nd isotopes in Orinoco basin river sediments. *Chemical Geology* 139, 271–286.

Goudie, A.S. and Pye, K. (1983) *Chemical Sediments and Geomorphology.* Academic Press, London, 439pp.

Graham, A. (1997) Forensic palynology and the Ruidoso, New Mexico plane crash — the Pollen Evidence II. *Journal of Forensic Sciences* 42, 391–393.

Graves, W.J. (1979) A mineralogical soil classification technique for the forensic scientist. *Journal of Forensic Sciences* 24, 323–338.

Greaves, P.H. and Saville, B.P. (1995) *Microscopy of Textile Fibres.* Royal Microscopical Society Handbooks 32. BIOS Scientific Publishers, Oxford, 92pp.

Green, O.R. (2001) *A Manual of Practical Laboratory and Field Techniques in Palaeobiology.* Kluwer Academic Publishers, Dordrecht, 538pp.

Greensmith, J.T. (1989) *Petrology of the Sedimentary Rocks*, 7th ed. Unwin Hyman, London. 262pp.

Gregory, P. (2005) *Bayesian Logical Data Analysis for the Physical Sciences.* Cambridge University Press, Cambridge, 488pp.

Gribble, C.D. and Hall, A.J. (1985) *A Practical Introduction to Optical Mineralogy.* George Allen and Unwin, London.

Grigsby, J.D. (1992) Chemical fingerprinting in detrital ilmenite — a viable alternative in provenance research. *Journal of Sedimentary Petrology* 62, 331–337.

Grim, R.E. (1968) *Clay Mineralogy.* McGraw Hill, New York, 596pp.

Grip, W.M., Grip, R.W., and Morrison, R.D. (2000) Application of aerial photography and photogrammetry in environmental forensic investigations. *Journal of Environmental Forensics* 1, 121–129.

Gross, H. (1962) *Criminal Investigation. A Practical Textbook for Magistrates, Police Officers and Lawyers. Adapted by John Adam and J. Collyer Adams from the System Der Kriminalistik of Dr Hans Gross*, 5th ed., by Richard Leofric Jackson. Sweet and Maxwell Ltd., London, 448pp.

Gross, M.G. (1971) Carbon determination. In Carver, R.E. (ed.) *Procedures in Sedimentary Petrology.* Wiley Interscience, New York, 573–596.

Grossman, R.B. and Reinsch, T.G. (2002) Bulk density and linear extensibility. In Dane, J.H. and Topp, G.G. (eds.) *Methods of Soil Analysis. Part 4. Physical Methods.* Soil Science Society of America Inc., Madison, Wisconsin, 201–228.

Grousset, F.E., Biscaye, P.E., Zindler, A., Prospero, J., and Chester, R. (1988) Neodymium isotopes as tracers in marine sediments and aerosols: North Atlantic. *Earth and Planetary Science Letters* 87, 367–378.

Grousset, F.E., Biscaye, P.E., Revel, M., Petit, J.-R., Pye, K., Joussaume, S., and Jouzel, J. (1992) Antarctic (Dome C) ice core dust at 18 kyr BP: isotopic constraints on origins. *Earth and Planetary Science Letters* 111, 175–182.

Haines, J. and Mazzullo, J.M. (1988) The original shapes of quartz silt grains: a test of the validity of the use of quartz grain shape analysis to determine the sources of terrigenous silt in marine sedimentary deposits. *Marine Geology* 78, 227–240.

Hall, A. (1996) *Igneous Petrology*, 2nd ed. Longman Group Limited, Harlow, 551pp.

Hall, D.W. (1997) Forensic botany. In Haglund, W.D. and Song, M.H. (eds.) *Forensic Taphonomy.* CRC Press, Boca Raton, 353–363.

Hall, P.L. (1987) Clays: their significance, properties, origins and uses. In Wilson, M.J. (ed.) *A Handbook of Determinative Methods in Clay Mineralogy.* Blackie, Glasgow and London, 1–25.

Hamblin, W.K. (1971) X-ray photography. In Carver, R.E. (ed.) *Procedures in Sedimentary Petrology.* Wiley Interscience, New York, 251–284.

Hamilton, E.M. and Jarvis, W.D. (1962) *The Identification of Atmospheric Dust by Use of the Microscope.* Central Electricity Generating Board Research and Developing Department. Leatherhead, Surrey, 31pp.

Hanson, I.D. (2004) The importance of stratigraphy in forensic investigation. In Pye, K. and Croft, D.J. (eds.) *Forensic Geoscience: Principles, Techniques and Applications.* Geological Society, London, Special Publications 232, 39–47.

Haq, B.U. and Boersma, A. (1998) *Introduction to Marine Micropaleontology.* Elsevier, Amsterdam, 376pp.

Harben, P.W. and Kuzvart, M. (1996) *Industrial Minerals. A Global Geology.* Industrial Minerals Information Ltd and Metals Bulletin, London, 462pp.

Hardy, R. and Tucker, M.E. (1988) X-ray powder diffraction of sediments. In Tucker, M.E. (ed.) *Techniques in Sedimentology,* Blackwell, Oxford, 191–228.

Harrison, R.M. (1986) Analysis of particulate pollutants. In Harrison, R.M. and Perry, R. (eds.) *Handbook of Air Pollution Analysis,* 2nd ed. Chapman and Hall, New York, 155–214.

Harvey, S.D., Vucelik, M.E., Lee, R.N. and Wright, B.W. (2002) Blind field test evaluation of Raman spectroscopy as a forensic tool. *Forensic Science International* 125, 12–21.

Harwood, G. (1988) Microscopic techniques. II. Principles of sedimentary petrography. In Tucker, M.E. (ed.) *Techniques in Sedimentology,* Blackwell, Oxford, 108–173.

Haslett, S.K. (ed.) (2002) *Quaternary Environmental Micropalaeontology.* Arnold, London, 340pp.

Head, K.H. (1984) *Manual of Soil Laboratory Testing, Volume 1: Soil Classification and Compaction Tests.* Pentech Press Limited, Devon, 339pp.

Helmke, P.A. (2002) Neutron activation analysis. In Sparks, D.L., Page, A.L., Helmke, P.A., Loeppert, R.H., Soltanpour, P.N., Tabatabai, M.A., Johnston, C.T., and Sumner, M.E. (eds.) (1996) *Methods of Soil Analysis. Part 3. Chemical Methods.* Soil Science Society of America Book Series No.5, Soil Science Society of America Inc. and American Society of Agronomy Inc., Madison, Wisconsin, 141–159.

Henderson, J. (2000) *The Science and Archaeology of Materials: An Investigation of Inorganic Materials.* Routledge, London, 334pp.

Henderson, J. (2004) The archaeologist as detective: scientific techniques and the investigation of past societies. In Pye, K. and Croft, D.J. (eds.) *Forensic Geoscience: Principles, Techniques and Applications.* Geological Society, London, Special Publications 232, 147–158.

Henderson, P. (1996) The rare earth elements: introduction and review. In Jones, A.P., Wall, F., and Williams, C.T. (eds.) *Rare Earth Minerals. Chemistry, Origin and Ore Deposits.* Chapman and Hall, London, 1–19.

Herman, B. (1998) *Fluorescence Microscopy,* 2nd ed. BIOS Scientific Publishers, Oxford, 170pp.

Higgs, R. (1979) Quartz grain surface features of Mesozoic–Cenozoic sands from the Labrador and Western Greenland continental margins. *Journal of Sedimentary Petrology* 49, 599–610.

Hinton, R.W. (1995) Ion microprobe analysis in geology. In Potts, P.J., Bowles, J.F.W., Reeds, S.J.B., and Cave, M.R. (eds.) *Microprobe Techniques in Earth Sciences.* Chapman and Hall, London, 235–289.

Hiraoka, Y. (1994) A possible approach to soil discrimination using X-ray fluorescence analysis. *Journal of Forensic Sciences* 39, 1381–1392.

Hiraoka, Y. (1997) Characterization of weathered products from granites around southern Lake Biwa, central Japan — application to forensic geology. *Journal of the Geological Society of Japan*, 103, 36–46.

Hodgson, J.M. (ed.) (1974) *Soil Survey Field Handbook.* Soil Survey Technical Monograph No. 5, Soil Survey of England and Wales, Harpenden, 99pp.

Hoefs, J. (2004) *Stable Isotope Geochemistry*, 5th ed. Springer, Berlin, 244pp.

Hoek, van den, C., Mann, D.G., and Jahns, H.M. (1995) *Algae: An Introduction to Phycology.* Cambridge University Press, Cambridge, 623pp.

Hoffman, C.M., Brunelle, R.I., and Snow, K.B. (1969) Forensic comparisons of soils by neutron activation and atomic absorption analysis. *Journal of Criminal Law, Criminology and Police Science* 60, 395–401.

Hollander, M. and Wolfe, D.A. (1999) *Nonparametric Statistical Methods*, 2nd ed. John Wiley and Sons, New York, 816pp.

Holmes, A. and Chivas, A.R. (2002) *The Ostracoda: Applications in Quaternary Research.* American Geophysical Union Monograph No. 131, American Geophysical Union, Washington DC, 313pp.

Hopen, T.J. (2004) The value of soil evidence. In Houck, M.M. (ed.) *Trace Evidence Analysis.* Elsevier, Amsterdam, 105–122.

Horrocks, J.M. (2004) Sub-sampling and preparing forensic samples for pollen analysis. *Journal of Forensic Sciences* 49, 1–4.

Horrocks, M. and Walsh, K.A.J. (1998) Forensic palynology: assessing the value of the evidence. *Review of Palaeobotany and Palynology* 103, 69–74.

Horrocks, M. and Walsh, K.A.J. (1999) Fine resolution of pollen patterns in limited space: differentiating a crime scene and alibi scene seven metres apart. *Journal of Forensic Sciences* 44, 417–420.

Horrocks, M. and Walsh, K.A.J. (2001) Pollen on grass clippings: putting the suspect at the scene of the crime. *Journal of Forensic Sciences* 46, 947–949.

Horrocks, M., Coulson, S.A., and Walsh, K.A.J. (1998) Forensic palynology: variation in the pollen content of surface soil samples. *Journal of Forensic Sciences* 43, 320–323.

Horrocks, M., Coulson, S.A., and Walsh, K.A.J. (1999) Forensic palynology: variation in the pollen content of soil on shoes and shoeprints in soil. *Journal of Forensic Sciences* 44, 119–122.

Horswell, J. (ed.) (2004) *The Practice of Crime Scene Investigation.* CRC Press, Boca Raton, 418pp.

Horswell, J. and Fowler, C. (2004) Associative evidence — the Locard exchange principle. In Horswell, J.T. (ed.) *The Practice of Crime Scene Investigation.* CRC Press, Boca Raton, 45–55.

Horton, B.P., Boreham, S., and Hillier, C. (2006) The development and application of a diatom-based quantitative reconstruction technique in Forensic Science. *Journal of Forensic Sciences* 51, 643–650.

Houck, M.M. (1999) Statistics and trace evidence: the tyranny of numbers. *Forensic Science Communications* 1(3), 1–8.

Houck, M.M. (ed.) (2001) *Mute Witness: Trace Evidence Analysis.* Academic Press, San Diego, 192pp.

Houck, M.M. (ed.) (2004) Trace *Evidence Analysis: More Cases in Mute Witnesses.* Elsevier, Amsterdam, 259pp.

House, W.A. and Ou, Z. (1992) Determination of pesticides on suspended solids and sediment: investigations on the handling and separation. *Chemosphere* 24, 819–832.

Hubert, J.F. (1971) Analysis of heavy-mineral assemblages. In Carver, R.E. (ed.) *Procedures in Sedimentary Petrology.* Wiley Interscience, New York, 453–478.

Humphries, D.W. (1992) *The Preparation of Thin Sections of Rocks, Minerals and Ceramics.* Royal Microscopical Society Handbooks 24, Oxford Science Publications, Oxford, 83pp.

Huntley, D.J., Godfrey-Smith, D.I., Thewalt, M.L.W., Prescott, J.R., and Hutton, J.T. (1988) Some quartz thermoluminescent spectra relevant to thermoluminescence dating. *Nuclear Tracks and Radiation Measurements* 14, 27–33.

Hunt, J.B. and Hill, P.G. (2001) Tephrological implications of beam size — sample-size effects in electron microprobe analysis of glass shards. *Journal of Quaternary Science* 16, 105–117.

Illenberger, W.K. (1991) Pebble shape (and size!). *Journal of Sedimentary Petrology* 61, 756–767.

Ingham, J.D. and Lawson, D.D. (1973) Thermoluminescence — potential application in forensic science. *Journal of Forensic Sciences* 18, 217–224.

Ingram, B.L. and Lin, J.C. (2002) Geochemical tracers of sediment sources to San Francisco Bay. *Geology* 30, 575–578.

Ingram, R.L. (1971) Sieve analysis. In Carver, R.E. (ed.) *Procedures in Sedimentary Petrology.* Wiley Interscience, New York, 49–68.

Innes, J.B., Blackford, J.J., and Simmons, I.G. (2004) Testing the integrity of fine spatial resolution palaeoecological records: microcharcoal data from near-duplicate peat profiles from the North York Moors, UK. *Palaeogeography, Palaeocolimatology, Palaeoecology* 214, 295–307.

Institute of Geological Sciences (1978a) *Regional Geochemical Atlas: Shetland.* Institute of Geological Sciences, London.

Institute of Geological Sciences (1978b) *Regional Geochemical Atlas: Orkney.* Institute of Geological Sciences, London.

Institute of Geological Sciences (1978c) *Regional Geochemical Atlas: South Orkney and Caithness.* Institute of Geological Sciences, London.

Institute of Geological Sciences (1982) *Regional Geochemical Atlas: Sutherland.* Institute of Geological Sciences, London.

Institute of Geological Sciences (1983) *Regional Geochemical Atlas: The Hebrides.* Institute of Geological Sciences, London.

International Standards Institution (2002a) *Soil quality — Sampling. Part. 1. Guidance on the Design of Sampling Programmes.* ISO 10381-1. International Standards Organization, Geneva, 32pp.

International Standards Institution (2002b) *Soil quality — Sampling. Part. 2. Guidance on Sampling Techniques.* ISO 10381-2. International Standards Organization, Geneva, 23pp.

Ireland, H.A. (1971) The preparation of thin sections. In Carver, R.E. (ed.) *Procedures in Sedimentary Petrology.* Wiley Interscience, New York, 367–383.

ISO (1993-2005) Soil Quality — Sampling. Parts 1–8. International Standards Organization, Geneva.

Isphording, W.C. (2004) The right way and the wrong way of presenting statistical and geological evidence in a court of law (a little knowledge is a dangerous thing!). In Pye, K. and Croft, D.J. (eds.) *Forensic Geoscience: Principles, Techniques and Applications.* Geological Society, London, Special Publications 232, 281–288.

Jackson, M.L. (1958) *Soil Chemical Analysis.* Prentice Hall, Englewood Cliffs, New Jersey, 498pp.

James, S.H. and Nordby, J.J. (eds.) (2003) *Forensic Science: An Introduction to Scientific and Investigative Techniques.* CRC Press, Boca Raton, 689pp.

Janoo, V.C. (1998) *Quantification of Shape, Angularity, and Surface Texture of Base Course Materials.* US Army Corps of Engineers, Cold Regions Research and Engineering Laboratory, Special Report 98-1, 22pp.

Janssen, D.W., Ruhf, W.A., and Richard, W.W. (1983) The use of clay for soil color comparisons. *Journal of Forensic Sciences* 28, 773–776.

Jarvis, K.E. (1997) Inductively coupled plasma-mass spectrometry. In Gill, R.E. (ed.) *Modern Analytical Geochemistry.* Addison Wesley, Harlow, 171–187.

Jarvis, K.E., Gray, A.L., and Houk, R.S. (2003) *Inductively Coupled Plasma Spectrometry.* Viridian Publishing, Surrey, 380pp.

Jarvis, K.E., Wilson, H.E., and James, S.L. (2004) Assessing element variability in small soil samples taken for forensic investigation. In Pye, K. and Croft, D.J. (eds.) *Forensic Geoscience: Principles, Techniques and Applications.* Geological Society, London, Special Publications 232, 171–182.

Jenkins, D.G. and Murray, J.W. (eds.) (1989) *Stratigraphical Atlas of Fossil Foraminifera.* Ellis Horwood, Chichester, 593pp.

Jenny, H. (1941) *Factors of Soil Formation. A System of Quantitative Pedology.* McGraw Hill Book Company, New York, 281pp.

Johannsen, A. (1931) *A Descriptive Petrography of the Igneous Rocks. Volume I.* University of Chicago Press, Chicago, 318pp.

Johannsen, A. (1932) *A Descriptive Petrography of the Igneous Rocks. Volume II.* University of Chicago Press, Chicago, 428pp.

Johannsen, A. (1937) *A Descriptive Petrography of the Igneous Rocks. Volume III.* University of Chicago Press, Chicago, 300pp.

Johannsen, A. (1938) *A Descriptive Petrography of the Igneous Rocks. Volume IV.* University of Chicago Pres, Chicago, 523pp.

Johnson, G.W. and Ehrlich, R. (2002) State of the Art report on multivariate chemometric methods in environmental forensics. *Environmental Forensics* 3, 59–79.

Johnson, W.M. and Maxwell, J.A. (1981) *Rock and Mineral Analysis*, 2nd ed. John Wiley and Sons, New York, 489pp.

Jones, A.P., Wall, F., and Williams, C.T. (eds.) (1996) *Rare Earth Minerals. Chemistry, Origin and Ore Deposits*. Chapman and Hall, London, 372pp.

Jones, K.W., Feng, H., Lindquist, W.B., Adler, P.M., Thovert, J.F., Vekemans, B., Vincze, L., Szaloki, I., Van Grieken, R., Adams, F. and Riekel, C. (2003) Study of the microgeometry of porous materials using synchrotron computed microtomography. In: Mees, F., Swennen, R., Van Geet, M. and Jacobs, P. (eds.) *Applications of X-ray Computed Tomography in the Geosciences*. Geological Society, London, Special Publications 215, 39–50.

Jongerius, A. and Heintzberger, G. (1963) *The Preparation of Mammoth-sized Thin Sections*. Soil Survey Papers No. 1, Netherlands Soil Survey Institute, Wageningen.

Jordan, C., Higgins, A., Hamill, K., and Cruickshank, J.G. (2000) *The Soil Geochemical Atlas of Northern Ireland*. Department of Agriculture and Rural Development, Northern Ireland.

Juckes, L.M. and Pitt, G.J. (1977) The standardization of petrographic analysis of coal. *Journal of Microscopy* 109, 13–21.

Junger, E.P. (1996) Assessing the unique characteristics of close-proximity soil samples: just how useful is soil evidence? *Journal of Forensic Sciences* 41, 27–34.

Kaiser, T.M. and Katterwe, H. (2001) The application of 3D-Microprofilometry as a tool in the surface diagnosis of fossil and sub-fossil vertebrate hard tissue. An example from the Pliocene Upper Laetolil Beds, Tanzania. *International Journal of Osteoarchaeology* 11, 350–356.

Kaplan, I.R. (2003) Age dating of environmental organic residues. *Environmental Forensics* 4, 95–141.

Kaplan, I.R., Galperin, Y., Lu, S.-T., and Lee, R.-P. (1997) Forensic environmental geochemistry: differentiation of fuel types, their sources and release time. *Organic Geochemistry* 27, 289–317.

Karakus, M. (2005) Cathodoluminescence properties of raw materials used in refractory ceramics. *Refractories Applications and News* 10(5), 16–19.

Karathanasis, A.D. and Hajek, B.F. (1996) Elemental analysis by x-ray fluorescence spectroscopy. In Sparks, D.L., Page, A.L., Helmke, P.A., Loeppert, R.H., Soltanpour, P.N., Tabatabai, M.A., Johnston, C.T., and Sumner, M.E. (eds.) *Methods of Soil Analysis. Part 3. Chemical Methods.* Soil Science Society of America Book Series No.5, Soil Science Society of America Inc and American Society of Agronomy Inc., Madison, Wisconsin, 161–223.

Keegan, N. (1999) *Raw Materials for Pigments, Fillers and Extenders.* Industrial Minerals Information Limited, Worcester Park, Surrey, 114pp.

Keegan, N. (ed.) (2000) *Industrial Clays*, 3rd ed. Industrial Minerals Information Limited, Worcester Park, Surrey, 104pp.

Kelley, J.C. (1971) Mathematical analysis of point count data. In Carver, R.E. (ed.) *Procedures in Sedimentary Petrology.* Wiley Interscience, New York, 409–426.

Kemp, R.A. (1985) *Soil Micromorphology and the Quaternary.* Quaternary Research Association Technical Guide No. 2, Quaternary Research Association, Cambridge, 80pp.

Kempe, D.R.C. and Harvey, A.P. (eds.) (1983) *The Petrology of Archaeological Artefacts.* Clarendon Press, Oxford, 374pp.

Kennedy, S.K., Walker, W., and Forslund, B. (2002) Speciation and characterization of heavy metal-contaminated soils using computer-controlled scanning electron microscopy. *Environmental Forensics* 3, 131–143.

Knudsen, J.W. (1972) *Collecting and Preserving Plants and Animals.* Harper and Row Publishers, New York, 320pp.

Knudsen, T.L., Griffith, W.L., Harz, E.H., Andersen, A., and Jackson, S.E. (2001) *In situ* hafnium and lead isotope analyses of detrital zircons from the Devonian sedimentary basin of NE Greenland: a record of repeated crustal reworking. *Contributions to Mineralogy and Petrology* 141, 83–94.

Kirk, R.E. (1999) *Statistics: An Introduction*, 4th ed. Harcourt Brace College Publishers, Fort Worth, 755pp.

Kohn, S.C. (2004) NMR studies of silicate glasses. In Beran, A. and Libowitzky, E. (eds.) *Spectroscopic Methods in Mineralogy.* Eotvos University Press, Budapest, 399–419.

Komar, P.D. and Cui, B. (1984) The analysis of grain-size measurements by sieving and settling tube techniques. *Journal of Sedimentary Petrology* 54, 603–614.

Kondo, R., Childs, C., and Atkinson, I. (1994) *Opal Phytoliths of New Zealand.* Manaaki Whenua Press, New Zealand, 85pp.

Koons, R.D. (1999) Use of statistics in trace evidence. Paper presented at the *International Symposium on Setting Quality Standards for the Forensic Community. Federal Bureau of Investigation,* San Antonio, Texas, 3-7 May 1999, 2pp.

Koons, R.D. and Buscaglia, J. (2002) Interpretation of glass composition measurements: The effects of match criteria on discrimination capability. *Journal of Forensic Sciences* 47, 505–512.

Kovda, V. (ed.) (1977) *The Soil Map of the World Scale 1: 10 000 000.* USSR Academy of Sciences Publishing House, Moscow.

Korpelainen, H. and Virtanen, V. (2003) Case report. DNA fingerprinting of mosses. *Journal of Forensic Sciences* 48, 1–4.

Krinsley, D.H. and Donahue, J. (1968) Environmental interpretation of sand grain surface textures by electron microscopy. *Bulletin of the Geological Society of America* 79, 743–748.

Krinsley, D.H. and Doornkamp, J.C. (1973) *Atlas of Quartz sand Surface Textures.* Cambridge University Press, Cambridge, 91pp.

Krinsley, D.H. and Margolis, S. (1969) A study of quartz sand grain surfaces with the scanning electron microscope. *Transactions of the New York Academy of Sciences* 31, 457–477.

Krinsley, D.H. and Takahashi, T. (1962) The surface texture of sand grains, an application of electron microscopy. *Science* 135, 923–925.

Krinsley, D.H., Pye, K., Boggs, Jr., and Tovey, N.K. (1998) *Backscattered Scanning Electron Microscopy and Image Analysis of Sediments and Sedimentary Rocks.* Cambridge University Press, Cambridge, 193pp.

Krstic, S., Duma, A., Janevska, B., Levkov, Z., Nikolova, K., and Noveska, M. (2002) Diatoms in forensic expertise of drowning: a Macedonian experience. *Forensic Science International* 127, 198–203.

Krumbein, W.C. (1941) Measurement and geologic significance of shape and roundness of sedimentary particles. *Journal of Sedimentary Petrology* 11, 64–72.

Krumbein, W.C. and Pettijohn, F.J. (1938) *Manual of Sedimentary Petrography.* Appleton Century Crofts Inc, New York, 549pp. Classic Facsimile Edition published in 1988 as SEPM Reprint Series Number 13 by the Society of Economic Paleontologists and Mineralogists, Tulsa, Oklahoma, 549pp.

Kubic, T. and Petraco, N. (2002) Microanalysis and examination of trace evidence. In James, S.H. and Nordby, J.J. (eds.) *Forensic Science: An Introduction to Scientific Investigative Techniques.* CRC Press, Boca Raton, 251–296.

Kubiena, W.L. (1938) *Micropedology.* Collegiate Press Inc., Ames, Iowa.

Kuenen, P.H. and Perdok, W.G. (1962) Experimental abrasion. 5. Frosting and defrosting of quartz grains. *Journal of Geology* 70, 648–658.

Kuisma-Kursula, P. (2000) Accuracy, precision and detection limits of SEM-WDS, SEM-EDS and PIXE in the multi-elemental analysis of Medieval glass. *X-ray Spectrometry* 29, 111–118.

Kuo, C.Y., Frost, J.D., Lai, J.S., and Wang, L.B. (1996) Three-dimensional image analysis of aggregate particles from orthogonal projections. *Transport Research Record* 1526, 98–103.

Ladd, C. and Lee, H.C. (2004) The use of biological and botanical evidence in criminal investigations. In: Coyle, H.M. (ed.) *Forensic Botany: Principles and Applications to Criminal Casework.* CRC Press, Boca Raton, 97–115.

Langmuir, D. (1971) Eh-pH determination, In Carver, R.E. (ed.) *Procedures in Sedimentary Petrology.* Wiley Interscience, New York, 597–635.

Larsen, G. (1981) Tephrochronology by microprobe glass analysis. In Self, S. and Sparks, R.S.J. (eds.) *Tephra Studies*. Reidel, Dordrecht, 95–102.

Lazzarini, L. and Lombardi, G. (1995) Provenance and authenticity of Roman sculptures by petrographic techniques. *Journal of Forensic Sciences* 40, 1090–1096.

Leary, E. (1983) The *Building Limestones of the British Isles*. HMSO, London, 91pp.

Leary, E. (1986) *The Building Sandstones of the British Isles*. Department of Environment, Building Research Station, Garston, Watford, 115pp.

Lee, B.D., Williamson, T.N., and Graham, R.C. (2002) Identification of stolen palm trees by soil morphological and mineralogical properties. *Journal of Forensic Sciences* 47, 190–194.

Lee, P.K., Touray, J.C., Baillif, P., and Ildefonse, J.P. (1997) Heavy metal contamination of settling particles in a retention pond along the A-71 motorway in Sologne, France. *The Science of the Total Environment* 201, 1–15.

Lees, G. (1964) The measurement of particle shape and its influence in engineering materials. *Journal of the British Granite and Whinstone Federation* 4(2), 1–22.

Lees, G. (1965) A new method for determining the angularity of particles. *Sedimentology* 3, 2–21.

Lewis, R.J., Sr. (2000a) *Sax's Dangerous Properties of Industrial Materials*, 10th ed. Wiley Interscience, New York, 3 volumes.

Lewis, R.J., Sr. (2000b) *Rapid Guide to Hazardous Chemicals in the Workplace*, 4th ed. Wiley Interscience, New York, 260pp.

Lewis, W.H. (1997) Pollen composition in crashed plane's engine. *Journal of Forensic Sciences* 42, 387–390.

Li, X.H., Liang, X.R., Sun, M., Guan, H., and Malpas, J.G. (2001) Precise Pb-206/U-238 age determination on zircons by laser ablation microprobe-inductively coupled plasma-mass spectrometry using continuous linear ablation. *Chemical Geology* 175, 209–219.

Lindemann, J.W. (2000) Forensic geology. *The Professional Geologist* 37, 4–7.

Livens, F.R. and Baxter, M.S. (1988a) Particle size and radionuclide levels in some West Cumbria soils. *Science of the Total Environment* 70, 1–17.

Livens, F.R. and Baxter, M.S. (1988b) Chemical associations of artifical radionuclides in Cumbrian soils. *Journal or Environmental Research* 7, 75–86.

Locard, E. (1920) *L'Enquete Criminelle et les Methodes Scientifiiques*. Flammarion, Paris, 300pp. Translated in English as "The Criminal Investigation and Scientific Methods" by C. Lennard.

Locard, E. (1928) Dust and its analysis. An aid to criminal investigation. *Police Journal* 1, 177–192.

Locard, E. (1930a) The analysis of dust traces. Part I. *American Journal of Police Science* 1, 276–298.

Locard, E. (1930b) The analysis of dust traces. Second Part. *American Journal of Police Science* 1, 401–418.

Locard, E. (1930c) The analysis of dust traces. Part III (Conclusion). *American Journal of Police Science* 1, 496–514.

Locard, E. (1931–1932) *Traite de Criminalistique*. Johannes Desvigne et Fils, Lyon, 4 volumes.

Loeblich, A.R., Jr. and Tappan, H. (1988) *Foraminiferal Genera and Their Classification*. Van Nostrand Reinhold, New York, 970pp.

Lombardi, G. (1999) The contribution of forensic geology and other trace evidence analysis to the investigation of the killing of Italian Prime Minister Aldo Moro. *Journal of Forensic Sciences* 44, 634–642.

Long, D.A. (2002) *Raman Spectroscopy*. John Wiley and Sons, Chichester, xxx pp.

Long, J.V.P. (1977) Electron microprobe analysis. In Zussman, J. (ed.) *Physical Methods in Determinative Mineralogy*, 2nd ed. Academic Press, London, 273–341.

Long, J.V.P. and Agrell, S.O. (1965) The cathodoluminescence of minerals in thin section. *Mineralogical Magazine* 34, 318–326.

Lowenthal, G.C. and Airey, P.L. (2001) *Practical Applications of Radioactivity and Nuclear Radiations*. Cambridge University Press, Cambridge, 337pp.

Lucy, D. (2005) *Introduction to Statistics for Forensic Scientists*. Wiley, Chichester, 251pp.

Machado, N. and Simonetti, A. (2001) I-Pb dating and Hf isotopic composition of zircon by laser-ablation MC-ICP-MS. In Sylvester, P. (ed.) *Laser Ablation-ICPMS in the Earth Sciences, Principles and Applications*. Mineralogical Association of Canada, Short Course Volume 29, 121–146.

MacKenzie, A.B. and Scott, R.D. (1982) Radiocaesium and plutonium in intertidal sediments from southern Scotland. *Nature* 299, 613–616.

MacKenzie, W.S. and Adams, A.E. (1994) *A Colour Atlas of Rocks and Minerals Under the Microscope*. Manson Publishing, London, 192pp.

MacKenzie, W.S. and Guilford, C. (1980) *Atlas of Rock-Forming Minerals in Thin Section*. Longman, Harlow, 98pp.

MacKenzie, W.S., Donaldson, C.H., and Guilford, C. (1982) *Atlas of Igneous Rocks and Their Textures*. Longman Scientific and Technical, Harlow, 148pp.

Madigan, M.T., Martinko, J.M., and Parker, J. (2003) *Brock Biology of Microorganisms*, 10th ed. Pearson Education Inc., Upper Saddle River, New Jersey, 1019pp plus appendices.

Mahaney, W.C. (2002) *Atlas of Sand Grain Surface Textures and Applications*. Oxford University Press, Oxford, 237pp.

Mahaney, W.C., Stewart, A., and Kalm, V. (2001) Quantification of SEM microtextures useful in sedimentary environmental discrimination. *Boreas* 30, 165–171.

Mange, M.A. and Mauer, H.F.W. (1992) *Heavy Minerals in Colour*. Chapman & Hall, London, 147pp.

Mark, D.M. and Church, M. (1977) On the misuse of regression in earth science. *Journal of the International Association of Mathematical Geology* 9, 63–77.

Marokowitz, A. and Milliken, K.L. (2003) Quantification of brittle deformation in burial compaction, Frio and Mount Simon Formation sandstones. *Journal of Sedimentary Research* 73, 1007–1021.

Marques, M.J., Salvador, A., Morales-Rubio, A.E., and de la Guardia, M. (2000) Trace element determination in sediments: a comparative study between neutron activation analysis (NAA) and inductively coupled plasma mass-spectrometry (ICP-MS). *Microchemical Journal* 65, 177–187.

Marshall, D.J. (1988) *Cathodoluminescence of Geological Materials.* Unwin Hyman, London and New York.

Marumo, Y. and Yanai, H. (1986) Morphological analysis of opal phytoliths for soil discrimination in forensic science investigations. *Journal of Forensic Sciences* 31, 1039–1049.

Marumo, Y. and Sugita, R. (1998) Forensic examination of soil evidence. *Paper presented at The 12th INTERPOL Forensic Science Conference, Lyon, France,* 20-23 October 1998, 8pp.

Marumo, Y. and Sugita, R. (2001) Forensic examination of soil evidence. *Proceedings of the 13th INTERPOL Forensic Science Symposium, Lyon, France,* 16–19 October 2001, D1-175 - D1-191.

Marumo, Y., Nagatsuka, S., and Oba, Y. (1986) Clay mineralogical analysis using the <0.05 mm fraction for forensic science investigation — its application to volcanic ash soils and yellow-brown forest soils. *Journal of Forensic Sciences* 31, 92–105.

Marumo, Y., Nagatsuka, S., and Oba, Y. (1988) Rapid clay mineralogical analysis for forensic science investigation — clay mineralogy over short distances. *Journal of Forensic Sciences* 33, 1360–1368.

Marumo, Y., Sugita, R., and Seta, S. (1999) Soil as evidence in crime investigation. *International Criminal Police Review* 474–475, 75–84.

Mason, B.J. (1992) *Preparation of Soil Sampling Protocols: Sampling Techniques and Strategies.* Environmental Protection Agency Report EPA/600/R-92/128. Environmental Monitoring Systems Laboratory, Office of Research and Development, U.S. Environmental Protection Agency, Las Vegas.

Mathers, S.J., Harrison, D.J., Mitchell, C.J., and Evans, E.J. (2000) *Exploration, Evaluation and Testing of Volcanic Raw Materials for Use in Construction.* British Geological Survey, Keyworth, Nottingham, 116pp.

Mattey, D.P. (1977) Gas source mass spectrometry: isotopic composition of lighter elements. In Gill, R. (ed.) *Modern Analytical Geochemistry.* Addison Wesley Longman, Harlow, 154–170.

Mayr, E. (1982) *The Growth of Biological Thought*: Diversity, Evolution and Inheritance. Belknap Press, Cambridge, Massachusetts, 974pp.

Mazzullo, J.M. and Magenheimer, S. (1987) The original shapes of quartz grains. *Journal of Sedimentary Petrology* 57, 479–487.

Mazzullo, J.M. and Ritter, C. (1991) Influence of sediment source on the shapes and surface features of glacial quartz sand grains. *Geology* 19, 384–388.

Mazzullo, J.M., Ehrlich, R., and Hemming, M.A. (1983) Provenance and area distribution of Late Pleistocene and Holocene quartz sand on the southern New England continental shelf. *Journal of Sedimentary Petrology* 54, 1335–1348.

Mazzullo, J.M., Leschak, P., and Prusak, D. (1988) Sources and distribution of Late Quaternary silt in the surficial sediment of the northeastern continental shelf of the United States. *Marine Geology* 78, 241–254.

McCrone Research Institute (1992) *The Particle Atlas. Electronic Edition.* McCrone Research Institute, Chicago (CD Rom and Installation Manual, revised 1993 and 1997).

McCrone, W.C. (1980) *The Asbestos Particle Atlas.* Ann Arbor Science Publishers, Ann Arbor, Michigan, 122pp.

McCrone, W.C. (1982) Soil comparison and identification of constituents. *The Microscope* 30, 17–25.

McCrone, W.C. (1992) Forensic soil examination. *The Microscope* 40, 109–121.

McCrone, W.C. and Bayard, M. (1967) Microprobe analysis of small particles. *The Microscope* 15, 26–42.

McCrone, W.C. and Delly, J.G. (1973a) *The Particle Atlas. Edition Two. Volume I* McCrone Research Institute, Chicago.

McCrone, W.C. and Delly, J.G. (1973b) *The Particle Atlas. Edition Two. Volume II* McCrone Research Institute, Chicago.

McCrone, W.C. and Delly, J.G. (1973c) *The Particle Atlas. Edition Two. Volume III* McCrone Research Institute, Chicago.

McCrone, W.C. and Delly, J.G. (1973d) *The Particle Atlas. Edition Two. Volume IV* McCrone Research Institute, Chicago.

McCrone, W.C., Draftz, R. and Delly, J.G. (1967) *The Particle Atlas.* Ann Arbor Science Publishers, Ann Arbor, 421pp.

McCrone, W.C., Delly, J.G., and Palenik (1979) *The Particle Atlas. Edition Two. Volume V.* Ann Arbor Science Publishers, Ann Arbor, Michigan.

McCrone, W.C., Brown, J.A., and Stewart, I.M. (1980) *The Particle Atlas. Edition Two. Volume VI.* Ann Arbor Science Publishers, Ann Arbor, Michigan.

McGrath, S.P. and Loveland, P.J. (1992) *The Soil Geochemical Atlas of England and Wales.* Blackie Academic and Professional, Glasgow, 101pp.

McHardy, W.J. and Birnie, A.C. (1987) Scanning electron microscopy. In Wilson, M.J. (ed.) *Quantitative Methods in Clay Mineralogy.* Blackie, Glasgow, 174, 208.

McKeever, S.W.S., Moscovitch, M., and Townsend, P.D. (eds.) (1995) *Thermoluminescence Dosimetry Materials: Properties and Uses.* Ramstrans Publishing, Ashford, Kent, 212pp.

McLaren, A.C. (1991) *Transmission Electron Microscopy of Minerals and Rocks.* Cambridge University Press, Cambridge, 387pp.

McManus, J. (1988) Grain size determination and interpretation. In Tucker, M.E. (ed.) *Techniques in Sedimentology,* Blackwell, Oxford, 63–85.

McPhee, J. (1996) Grounds for Murder. *The New Yorker* January, 44–69.

McVicar, M.J. and Graves, W.J. (1997) The forensic comparison of soils by automated scanning electron microscopy. *Canadian Society of Forensic Science Journal* 30, 241–261.

Mees, F., Swennen, R., Van Geet, M., and Jacobs, P. (eds.) (2003) *Applications of X-ray Computed Tomography in the Geosciences*. Geological Society, London, Special Publications 215, 243pp.

Mehta, P.K. and Montero, P.J.M. (2005) *Concrete: Micro-structure, Properties and Materials*, 3rd ed. Prentice Hall, Englewood Cliffs, New Jersey, 659pp.

Melville, M.D. and Atkinson, G. (1985) Soil colour: its measurement and its designation in models of uniform colour space. *Journal of Soil Science* 36, 495–512.

Merefield, J.R., Stone, I.M., Roberts, J., Jones, J., Barron, J., and Dean, A. (2000) Fingerprinting airborne particles for identifying provenance. In Foster, I.D.L. (ed.) *Tracers in Geomorphology*, Wiley, Chichester, 85–100.

Mildenhall, D.C. (1990) Forensic palynology in New Zealand. *Review of Palaeobotany and Palynology* 64, 227–234.

Mildenhall, D.C. (2004) An example of the use of forensic palynology in assessing an alibi. *Journal of Forensic Sciences* 49, 1–5.

Mildenhall, D.C. (2006) An unusual appearance of a common pollen type indicates the scene of crime. *Forensic Science International* 163, 236–240.

Mildenhall, D.C., Wiltshire, P.E.J. and Bryant, V.M. (2006) Forensic palynology: why do it and how it works. *Forensic Science International* 163, 163–172.

Miller, J. (1988a) Microscopical techniques. I. Slices, slides, stains and peels. In Tucker, M.E. (ed.) *Techniques in Sedimentology*, Blackwell, Oxford, 86–107.

Miller, J. (1988b) Cathodoluminescenc microscopy. In Tucker, M.E. (ed.) *Techniques in Sedimentology*, Blackwell, Oxford, 174–190.

Miller, J.N. and Miller, J.C. (2005) *Statistics and Chemometrics for Analytical Chemistry*, 5th ed. Pearson Education Limited, Harlow, 268pp.

Miller, P.S. (1996) Disturbances in soil: finding buried bodies and other evidence using ground penetrating radar. *Journal of Forensic Sciences* 41, 648–652.

Milliken, K.L. (1994) Cathodoluminescence textures and the origin of quartz silt in Oligocene mudrocks, south Texas. *Journal of Sedimentary Research* A64, 567–571.

Milne, L.A., Bryant, V.M. Jr., and Mildenhall, D.C. (2004) Forensic palynology. In Coyle, H.M. (ed.) *Forensic Botany: Principles and Applications to Criminal Casework*. CRC Press, Boca Raton, 217–252.

Milner, H.B. (1962a) *Sedimentary Petrography. Volume 1. Principles and Applications*, 4th ed. Allen and Unwin, London.

Milner, H.B. (1962b) *Sedimentary Petrography. Volume 2. Methods in Sedimentary Petrography*, 4th ed. Allen and Unwin, London.

Milsom, J. (1989) *Field Geophysics*, 2nd ed. John Wiley & Sons, Chichester, 187pp.

Minnis, M.M. (1984) An automatic point-counting method for mineralogical assessment. *American Association of Petroleum Geologists Bulletin* 68, 744–752.

Mohr, E.C.J., van Baren F.A. and van Schulenborgh, J. (1972) *Tropical soils. A Comprehensive Study of Their Genesis*, 3rd ed. Mouton-Ichtiar Baru-Van Hoeve, The Hague, 481pp.

Montero, S., Hobbs, A.L., French, B.S., and Almirall, J.R. (2003) Elemental analysis of glass fragments by ICP-MS as evidence of association of a case. *Journal of Forensic Sciences* 48, 1–5.

Moore, C.A. and Donaldson, C.F. (1995) Quantifying soil microstructure using fractals. *Geotechnique* 45, 105–116.

Moore, D.M. and Reynolds, R.C. Jr. (1997) *X-ray Diffraction and the Identification and Analysis of Clay Minerals*, 2nd ed. Oxford University Press, Oxford, 378pp.

Moore, P.D., Webb, J.A., and Collinson, M.E. (1991) *Pollen Analysis*, 2nd ed. Blackwell Science, Oxford, 216pp.

Moorehead, W. (2004) Dr. Walter McCrone: contributions to criminalistics. *The Microscope* 52(1), 49–55.

Morgan, R.M., Wiltshire, P.J.E., Parker, A. and Bull, P.A. (2006) The role of forensic geoscience in wildlife crime detection. *Forensic Science International* 162, 152–162.

Morrison, R.D. (2000) *Environmental Forensics. Principles and Applications*. CRC Press, Boca Raton, 351pp.

Morton, A.C. (1985) A new approach to provenance studies — electron microprobe analysis of detrital garnets from Middle Jurassic sandstones of the Northern North Sea. *Sedimentology* 32, 553–566.

Moss, A.J. (1966) Origin, shaping and significance of quartz sand grains. *Journal of the Geological Society of Australia* 13, 97–136.

Mottana, A. (2004) X-ray absopption spectroscopy in mineralogy: theory and experiment in the XANES region. In: Beran, A. and Libowitzky (eds.) *Spectroscopic Methods in Mineralogy*. EMU Notes in Mineralogy No. 6, Eotvos University Press, Budapest, 465–528.

Mudroch, A. and Azcue, J.M. (1995) *Manual of Aquatic Sediment Sampling*. Lewis Publishers, Boca Raton, 219pp.

Muir, I.D. (1977) Microscopy: transmitted light. In Zussman, J. (ed.) *Physical Methods in Determinative Mineralogy*, 2nd ed. Academic Press, London, 35–108.

Munroe, R. (1995) Forensic geology. *Royal Canadian Mounted Police Gazette* 57(3), 10–17.

Munsell Color (1994) *Munsell Soil Color Charts*. Gretag Macbeth, New Windsor, New York, Revised edition, 1–10 plus charts.

Murphy, B.L. and Morrison, R.D. (eds.) (2002) *Introduction to Environmental Forensics*. Academic Press, San Diego, 560pp.

Murphy, C.P. (1982) A comparative study of three methods of water removal prior to resin impregnation of two soils. *Journal of Soil Science* 33, 719–735.

Murphy, C.P. (1985) Faster methods of liquid-phase acetone replacement of water from soils and sediments prior to resin impregnation. *Geoderma* 35, 39–45.

Murphy, C.P. (1986) *Thin Section Preparation of Soils and Sediments.* AB Academic Publishers, Berkhampstead, 149pp.

Murray, J.W. (1991) *Ecology and Palaeoecology of Benthic Foraminifera.* Longman, Harlow, 397pp.

Murray, R.C. (1976) Soil and rocks as physical evidence. *Law and Order* July, 36–40.

Murray, R.C. (1982) Forensic examination of soil. In R. Saferstein (ed.) *Forensic Science Handbook.* Prentice Hall, Englewood Cliffs, New Jersey, 653–671.

Murray, J.C. (1988) Forensic geology — 100 years. *The Microscope* 36, 303–308.

Murray, R.C. (2000) Devil in the details: the science of forensic geology. *Geotimes* 45, 14–17.

Murray, R.C. (2004) *Evidence from the Earth: Forensic Geology and Criminal Investigation.* Mountain Press Publishing Company, Missoula, Montana, 226pp.

Murray, R.C. (2005) Collecting crime evidence from Earth. *Geotimes* January 2005, 1–7 (www.geotimes.org/jan05/feature_evidence.html).

Murray, R.C. and Murray, R. (1980) Soil evidence. *Law and Order* July, 26–28.

Murray, R.C. and Tedrow, J.C.F. (1975) *Forensic Geology.* Rutgers University Press, New Brunswick, New Jersey, 217pp.

Murray, R.C. and Tedrow, J.C.F. (1992) *Forensic Geology,* 2nd ed. Prentice Hall Inc., Englewood Cliffs, New Jersey, 203pp.

Nachtergaele, F.O. (2000) From the soil map of the world to the digital global soils and terrain database: 1960–2002. In Sumner, M.E. (ed.) *Handbook of Soil Science.* CRC Press, Boca Raton, H-5 to II-17.

Nakayama, Y., Fujita, Y., Kanbara, K., Nakayama, N., Mitsuo, H., Matsumoto, H., and Satoh, T. (1992) Forensic chemical study on soils, (1) discrimination of area by pyrolysis of products of soil. *Japanese Journal of Toxicolgy and Environmental Health* 38, 38044.

Nasdala, L., Gotze, J., Hanchar, J.M., Gaft, M., and Krbetschek, M.R. (2004a) Luminescence techniques in Earth Sciences. In Beran, A. and Libowitzky, E. (eds.) *Spectroscopic Methods in Mineralogy.* Eotvos University Press, Budapest, 43–91.

Nasdala, L., Smith, D.C., Kaindl, R., and Ziemann, M.A. (2004b) Raman spectroscopy: analytical perspectives in mineralogical research. In Beran, A. and Libowitzky, E. (eds.) *Spectroscopic Methods in Mineralogy.* Eotvos University Press, Budapest, 281–343.

Nelms, S.M. (ed.) (2005) *ICP Mass Spectrometry Handbook.* Blackwell Publishing Ltd., Oxford, 485pp.

Nelson, D.W. and Sommers, L.E. (1996) Total carbon, organic carbon, and organic matter. In Sparks, D.L., Page, A.L., Helmke, P.A., Loppert, R.H., Soltanpour, P.N., Tabatabal, M.A., Johnson, C.T., and Sumner, M.E. (eds.) *Methods of Soil Analysis. Part 3. Chemical Methods.* Soil Science Society of America Book

Series No. 5, Soil Science Society of America Inc. and American Society of Agronomy Inc., Madison, Wisconsin, 961–1010.

Neville, A.M. and Brooks, J.J. (1987) *Concrete Technology.* Longman Scientific and Technical, Harlow, 438pp.

Nic Daeid, N. (2001) Statistical interpretation of glass evidence. In: Caddy, B. (ed.), *Forensic Examination of Glass and Paint.* Taylor and Francis, London and New York, 85–95.

Nicholls, J. and Stout, M.Z. (1986) Electron beam analytical instruments and the determination of modes, spatial variations in minerals and textural features of rocks in polished section. *Contributions to Mineralogy and Petrology* 94, 395–404.

Nickolls, L.C. (1956) *The Scientific Investigation of Crime.* Butterworth & Co (Publishers) Ltd., London, 398pp.

Nickolls, L.C. (1962) Forensic problems involving crime. In Neil, M.W. and Warren, F.L. (eds.) *Soil.* British Academy of Forensic Sciences Teaching Symposium No.1. Sweet and Maxwell Ltd., London, 23–26.

Nobes, D.C. (2000) The search for "Yvonne": a case example of the delineation of a grave using near-surface geophysical methods. *Journal of Forensic Sciences* 45, 715–721.

Nocerino, J.M., Schumacher, B.A., and Dary, C.C. (2005) Role of laboratory sampling devices and laboratory subsampling methods in representative sampling strategies. *Environmental Forensics* 6, 35–44.

Norrish, K. and Chappell, B.W. (1977) X-ray fluorescence spectrometry. In Zussman, J. (ed.) *Physical methods of Determinative Mineralogy,* 2nd ed. Academic Press, London, 201–272.

Norrish, K. and Hutton, J.T. (1969) An accurate X-ray spectrographic method for the analysis of a wide range of geological samples. *Geochemica et Cosmochimica Acta* 33, 431–453.

Norton, L.D., Bigham, J.M., Hall, G.F., and Smeck, N.E. (1983) Etched thin sections for coupled optical and electron microscopy and microanalysis. *Geoderma* 30, 55–64.

Nute, H.D. (1975) An improved density gradient system for forensic science soil study. *Journal of Forensic Sciences* 20, 668–673.

O'Brien, N.R. (1981) SEM study of shale fabric — a review. *Scanning Electron Microscopy* 1, 569–575.

O'Brien, N.R. and Slatt, R.M. (1990) *Argillaceous Rock Atlas.* Springer Verlag, Berlin, 141pp.

Ogle, R.R., Jr. and Fox, M.J. (1999) *Atlas of Human Hair.* Microscopic Characteristics. CRC Press, Boca Raton, 83pp.

Oldfield, F. (1999) Environmental magnetism; the range of applications. In Walden, J., Oldfield, F., and Smith, J. (eds.) *Environmental Magnetism: A Practical Guide.* Quaternary Research Association Technical Guide, No. 6, Quaternary Research Association, London, 212–239.

Oliver, M.A., Loveland, P.J., Frogbrook, Z.L., Webster, R., and McGrath, S.P. (2002) *Statistical and Geostatistical Analysis of The National Soil Inventory of England*

and Wales. The Technical Report. Project No SP0124, UK Department of Environment, Food and Rural Affairs, London.

Orford, J.D. (1990) Particle form. In Goudie, A.S. (ed.) *Geomorphological Techniques*, 2nd ed. Unwin Hyman, London, 122–126.

Orford, J.D. and Whalley, W.B. (1983) The use of fractal dimensions to quantify the morphology of irregular-shaped particles. *Sedimentology* 30, 655–668.

Orford, J.D. and Whalley, W.B. (1987) The quantitative description of highly irregular sedimentary particles: the use of the fractal dimension. In Marshall, J.R. (ed.) *Clastic Particles: Scanning Electron Microscopy and Shape Analysis of Sedimentary and Volcanic Clasts*. Von Nostrand Reinhold, New York, 267–280.

Orford, J.D. and Whalley, W.B. (1991) Quantitative grain form analysis. In Syvitski, J. (ed.) *Principles, Methods and Application of Particle Size Analysis*. Cambridge University Press, Cambridge, 88–108.

Orton, C. (2000) *Sampling in Archaeology: Cambridge Manuals in Archaeology*. Cambridge University Press, Cambridge, 261pp.

Ovianki, S.M. (1996) Forensic geology: geologic investigation as a tool for enforcement of environmental regulations. *Mississippi Geology* 17, 45–50.

Palenik, S.J. (1979) The determination of geographical origin of dust samples. In McCrone, W.C., Delly, J.G., and Palenik, S.J. (eds.) *The Particle Atlas, Volume V*, 2nd ed. Ann Arbor Science Publishers, Ann Arbor, Michigan, 1347–1368.

Palenik, S.J. (1982) Microscopic trace evidence — the overlooked clue. *The Microscope* 30, 163–169.

Palenik, S.J. (1988) Microscopy and microchemistry of physical evidence. In Saferstein, R. (ed.) *Forensic Science Handbook, Volume II*. Prentice Hall, Englewood Cliffs, New Jersey, 161–208.

Palenik, S.J. (1997) Forensic microscopy. *Microscopy and Analysis* November, 7–9.

Palenik, S.J. (2000a) Dust. In Siegel, J., Knupfer, G., and Saukko, P (eds.) *Encyclopedia of Forensic Science. Volume 2*. Academic Press, New York, 662–667.

Palenik, S.J. (2000b) Microscopy. In Siegel, J., Knupfer, G., and Saukko, P (eds.) *Encyclopedia of Forensic Science. Volume 1*. Academic Press, New York, 161–166.

Parry, S.J. (2003) *Handbook of Neutron Activation Analysis*. Viridian Publishing, Woking, Surrey, 243pp.

Paterson, E. and Swaffield, R. (1987) Thermal analysis. In Wilson, M.J. (ed.) *A Handbook of Determinative Methods in Clay Mineralogy*. Blackie, Glasgow and London, 99–132.

Patrick, W.H., Jr., Gambrell, R.P., and Faulkner, S.P. (1996) Redox measurements of soils. In Sparks, D.L., Page, A.L., Helmke, P.A., Loppert, R.H., Soltanpour, P.N., Tabatabal, M.A., Johnson, C.T., and Sumner, M.E. (eds.) *Methods of Soil Analysis. Part 3. Chemical Methods*. Soil Science Society of America Book Series No. 5, Soil Science Society of America Inc. and American Society of Agronomy Inc., Madison, Wisconsin, 1255–1274.

Peabody, A.J. (1977) Diatoms in forensic science. *Journal of the Forensic Science Society* 17, 81–87.

Peabody, A.J. (1999) Forensic science and diatoms. In Stoermer, E.F. and Smol, J.P. (eds.) *The Diatoms: Applications for the Environmental and Earth Sciences.* Cambridge University Press, Cambridge, 413–418.

Pearce, N.J.G., Perkins, W.T., and Fuge, R. (1992) Developments in the quantitative and semi-quantitative determination of trace elements in carbonates by laser ablation inductively coupled plasma mass-spectrometry. *Journal of Analytical Atomic Spectrometry* 7, 595–598.

Pell, S.D., Williams, I.S., and Chivas, A.R. (1997) The use of protolith zircon-age fingerprints in determining the protosource areas for some Australian dune sands. *Sedimentary Geology* 109, 233–260.

Pell, S.D., Chivas, A.R., and Williams, I.S. (2000) The Simpson, Strzelecki and Tirari Deserts: development and sand provenance. *Sedimentary Geology* 130, 107–130.

Pella, E. (1990a) Elemental organic analysis. Part 1. Historical development. *American Laboratory* 22, 116–125.

Pella, E. (1990b) Elemental organic analysis. Part 2. State of the art. *American Laboratory* 22, 28–32.

Pennell, K.D. (2002) Specific surface area. In Dane, J.H. and Topp, G.G. (eds.) *Methods of Soil Analysis. Part 4. Physical Methods.* Soil Science Society of America Inc., Madison, Wisconsin, 295–316.

Perkins, W.T., Fuge, R., and Pearce, N.J.G. (1991) Quantitative analysis of trace elements in carbonates using laser-ablation inductively coupled plasma mass-spectrometry. *Journal of Analytical Atomic Spectrometry* 6, 445–449.

Perrin, R.M.S. (1971) *The Clay Mineralogy of British Sediments.* Mineralogical Society (Clay Minerals Group), London, 247pp.

Peters, D.W.A. (1962) The examination of soil with particular reference to the density gradient test. In Neil, M.W. and Warren, F.L. (eds.) *Soil.* British Academy of Forensic Sciences Teaching Symposium No. 1. Sweet and Maxwell Ltd., London, 27–36.

Petraco, N. (1994a) Microscopic examination of mineral grains in forensic soil analysis. Part 1. *American Laboratory* 26(6), 35–40.

Petraco, N. (1994b) Microscopic examination of mineral grains in forensic soil analysis. Part 2. *American Laboratory* 26(14), 33–35.

Petraco, N. and DeForrest, P.R. (1993) A guide to the analysis of forensic dust samples. In Saferstein, R. (ed.) *Forensic Handbook. Volume III.* Prentice Hall, Englewood Cliffs, New Jersey, 24–69.

Petraco, N. and Gale, F. (1984) A rapid method for cross-sectioning of multilayered paint chips. *Journal of Forensic Sciences* 29, 597–600.

Petraco, N. and Kubic, T. (2000) A density gradient technique for use in forensic soil analysis. *Journal of Forensic Sciences* 45, 872–873.

Petraco, N. and Kubic, T.A. (2004) *Color Atlas and Manual of Microscopy for Criminalists, Chemists and Conservators*. CRC Press, Boca Raton, 313pp.

Petrisor, I.G. (2005) Fingerprinting in environmental forensics. *Environmental Forensics* 6, 101–102.

Pettijohn, F. (1949) *Sedimentary Rocks*. Harper and Brothers, New York.

Pettijohn, F.J., Potter, P.E., and Siever, R. (1972) *Sand and Sandstone*. Springer-Verlag, Berlin, 618pp.

Pettijohn, F.J., Potter, P.E., and Siever, R. (1987) *Sand and Sandstone*, 2nd ed. Springer Verlag, Berlin.

Pilcher, J.R. (1991) Radiocarbon dating. In Smart, P.L. and Frances, P.D (eds.) *Quaternary Dating Methods — A User's Guide*. Quaternary Research Association Technical Guide No. 4, Quaternary Research Association, Cambridge, 16–36.

Pirrie, D., Butcher, A.R., Power, M.R., Gottlieb, P., and Miller, G.L. (2004) Rapid quantitative mineral and phase analysis using automated scanning electron microscopy (QemSCAN): potential applications in forensic geoscience. In Pye, K. and Croft, D.J. (eds.) *Forensic Geoscience: Principles, Techniques and Applications*. Geological Society, London, Special Publications 232, 123–136.

Pollanen, M.S (1997) The diagnostic value of the diatom test for drowning. II Validity: analysis of diatoms in bone marrow and drowning medium. *Journal of Forensic Sciences* 42, 286–290.

Pollanen, M.S. (1998a) *Forensic Diatomology and Drowning*. Elsevier, Amsterdam, 159pp.

Pollanen, M.S. (1998b) Diatoms and homicide. *Forensic Science International* 91, 29–34.

Pollanen, M.S., Cheung, C., and Chiasson, D.A. (1997) The diagnostic value of the diatom test for drowning. I Utility: a retrospective analysis of 771 cases of drowning in Ontario, Canada. *Journal of Forensic Sciences* 42, 281–285.

Porter, J.J. (1962) Electron microscopy of sand surface textures. *Journal of Sedimentary Petrology* 32, 124–135.

Post, D.R., Bryant, R.B., Batchily, A.K., Levine, S.J., Mays, M.D., and Escadafal, R. (1993) Correlations between field and laboratory measurements of soil color. In Bigham, J.M. and Ciolkhoz, E.J. (eds.) *Soil Colour*. Soil Science Society of America Special Publication 31, 35–49.

Potter, P.E. (1986) South America and a few grains of sand. I. Beach sands. *Journal of Geology* 94, 301–319.

Potter, P.E., Maynard, J.B., and Pryor, W.A. (1980) *Sedimentology of Shale. Study Guide and Reference Source*. Springer Verlag, New York, 306pp.

Potts, P.J. (1987) *A Handbook of Silicate Rock Analysis*. Blackie, Glasgow, 622pp.

Powers, M.C. (1953) A new roundness scale for sedimentary particle. *Journal of Sedimentary Petrology* 23, 117–119.

Powers, L.S., Bruceckner, H.K., and Krinsley, D.H. (1979) Rb-Sr provenance ages from weathered and stream transported quartz grains from the Harvey Peak Granite, Black Hills, South Dakota. *Geochimica et Cosmochimca Acta* 43, 137–146.

Pownceby, M. (2005) Compositional and textural variation in detrital chrome-spinels from the Murray basin, southeastern Australia. *Mineralogical Magazine* 69, 191–204.

Prentice, J.E. (1990) *Geology of Construction Materials.* Chapman and Hall, London, 202pp.

Pryor, W.A. (1971) Grain shape. In Carver, R.E. (ed.) *Procedures in Sedimentary Petrology.* Wiley Interscience, New York, 131–150.

Putnis, A. (1992) *Introduction to the Mineral Sciences.* Cambridge University Press, Cambridge, 457pp.

Pye, K. (1987) *Aeolian Dust and Dust Deposits.* Academic Press, London, 344pp.

Pye, K. (2004a) Forensic geology. In Selley, R.C., Cocks, R.M., and Plimer, I.R. (eds.) *The Encyclopedia of Geology, Volume 2.* Elsevier, Amsterdam, 261–273.

Pye, K. (2004b) Forensic examination of rock, sediments, soils and dusts using scanning electron microscopy and X-ray chemical microanalysis. In Pye, K. and Croft, D.J. (eds.) *Forensic Geoscience: Principles, Techniques and Applications.* Geological Society, London, Special Publications 232, 103–122.

Pye, K. (2004c) Isotope and trace element analysis of human teeth and bones for forensic purposes. In Pye, K. and Croft, D.J. (eds.) *Forensic Geoscience: Principles, Techniques and Applications.* Geological Society, London, Special Publications 232, 215–236.

Pye, K. (2007) Sediment fingerprints: a forensic technique using quartz and grains — a comment. *Science and Justice* (in press).

Pye, K. and Blott, S.J. (2004a) Particle size analysis of sediments, soils and related particulate materials for forensic purposes using laser granulometry. *Forensic Science International* 144, 19–27.

Pye, K. and Blott, S.J. (2004b) Comparison of soils and sediments using major and trace element data. In Pye, K. and Croft, D.J. (eds.) *Forensic Geoscience: Principles, Techniques and Applications.* Geological Society, London, Special Publications 232, 183–196.

Pye, K. and Blott, S.J. (2007) Development of a searchable major and trace element database for use in forensic soil comparisons. *Forensic Science International* (in press).

Pye, K. and Croft, D.J. (eds.) (2004a) *Forensic Geoscience: Principles, Techniques and Applications.* Geological Society, London, Special Publication 232, 318pp.

Pye, K. and Croft, D.J. (2004b) Forensic geoscience: introduction and overview. In Pye, K. and Croft, D.J. (eds.) *Forensic Geoscience: Principles, Techniques and Applications.* Geological Society, London, Special Publication 232, 1–4.

Pye, K. and Croft, D.J. (2007) Forensic analysis of soil and sediment traces by scanning electron microscopy and energy-dispersive X-ray analysis: an experimental investigation. *Forensic Science International* 165, 52–63.

Pye, K. and Mazzullo, J. (1994) Effects of tropical weathering on quartz grain shape: an example from northern Australia. *Journal of Sedimentary Research (A)* 64, 500–507.

Pye, K. and Schiavon, N. (1989) Cause of sulphate attack on concrete, render and stone indicated by sulphur isotope ratios. *Nature* 342, 663–664.

Pye, K. and Tsoar, H. (1990) *Aeolian Sand and Sand Dunes*. Unwin Hyman, London, 396pp.

Pye, K., Blott, S.J., and Wray, D.S. (2006a) Elemental analysis of soil samples for forensic purposes by inductively coupled plasma spectrometry — precision considerations. *Forensic Science International* 160, 178–192.

Pye, K., Blott, S.J., Croft, D.J., and Carter, J.F. (2006b) Forensic comparison of soil samples: assessment of small-scale spatial variability in elemental composition, carbon and nitrogen isotope ratios, colour, and particle size distribution. *Forensic Science International*, 163, 59–80.

Pye, K., Blott, S.J., Croft, D.J., and Witton, S.J. (2007) Discrimination between sediment and soil samples for forensic purposes using elemental data: an investigation of particle size effects. *Forensic Science International* (in press).

Pye, K., Coleman, M.L., and Duan, W.M. (1997) Microbial activity and diagenesis in saltmarsh sediments, North Norfolk, England. In Jickells, T.D. and Rae, J.E. (eds.) *Biogeochemistry of Intertidal Sediments*. Cambridge University Press, Cambridge, 119–151.

Pye, W. and Pye, M. (1943) Sphericity determinations of pebbles and sand grains. *Journal of Sedimentary Petrology* 13, 28–34.

Ramsay, M.H. (1997) Sampling and sample preparation. In Gill, R. (ed.) *Modern Analytical Geochemistry*. Addison Wesley, Harlow, 12–28.

Ramsey, C.A. and Hewitt, A.D. (2005) A methodology for assessing sample representativeness. *Environmental Forensics* 6, 71–75.

Rapp, J.S. (1987) Forensic geology and a Colusa County murder. *California Geology* July, 147–153.

Rawlins, B.G. and Cave, M. (2004) Investigating multi-element soil geochemical signatures and their potential for use in forensic studies. In Pye, K. and Croft, D.J. (eds.) *Forensic Geoscience: Principles, Techniques and Applications*. Geological Society, London, Special Publications 232, 197–206.

Rawlins, B.G., Lister, T.R., and MacKenzie, A.C. (2002) Trace-metal pollution of soils in northern England. *Environmental Geology* 42, 612–620.

Rawlins, B.G., Webster, R., and Lister, T.R. (2003) The influence of parent material on topsoil geochemistry in eastern England. *Earth Surface Processes and Landforms* 28, 1389–1409.

Rawlins, B.G., Kemp, S.J., Hodgkinson, E.H., Riding, J.B., Vane, C.H., Poulton, C. & Freeborough, K. (2006) Potential and pitfalls in establishing the provenance of Earth-related samples in forensic investigations. *Journal of Forensic Sciences* 51, 832–845.

Reed, S.J.B. (2005) *Electron Microprobe Analysis and Scanning Electron Microscopy in Geology*, 2nd ed. Cambridge University Press, Cambridge, 192pp.

Reidy, L.J., Meier-Augenstein, W., and Kalin, R.M. (2005) ^{13}C isotope ratio mass spectrometry as a potential tool for the forensic analysis of white architectural paint: a preliminary study. *Rapid Communications in Mass Spectrometry* 19, 1899–1905.

Reimann, C., Siewers, U., Tarvainen, T., Bityukova, L., Eriksson, J., Gilucis, A., Gregorauskiene, V., Lukashev, V.K., Matinian, N.N., and Pasieczna, A. (2003) *Agricultural Soils in Northern Europe: A Geochemical Atlas*. Geologisches Jahrbuch Sonderhefte, Reihe D Heft SD5, 279pp.

Rencher, A.C. (2002) *Methods of Multivariate Analysis*, 2nd ed. John Wiley & Sons, Hoboken, New Jersey, 708pp.

Rendell, H.M., Steer, D.C., Townsend, P.D., and Wintle, A.G. (1985) The emission spectra of TL from obsidian. *Nuclear Tracks and Radiation Measurements* 23, 441–449.

Reuland, D.J. and Trinler, W.A. (1981) An investigation of the potential of high performance liquid chromatography for the comparison of soil samples. *Forensic Science International* 18, 201–208.

Reuland, D.J., Trinler, W.A., and Farmer, M.D. (1992) Comparison of soil samples by high performance liquid chromatography augmented by absorbance ratioing. *Forensic Science International* 52, 131–142.

Rhoades, J.D. (1996) Salinity: electrical conductivity and total dissolved solids. In Sparks, D.L., Page, A.L., Helmke, P.A., Loppert, R.H., Soltanpour, P.N., Tabatabal, M.A., Johnson, C.T., and Sumner, M.E. (eds.) *Methods of Soil Analysis. Part 3. Chemical Methods*. Soil Science Society of America Book Series No. 5, Soil Science Society of America Inc. and American Society of Agronomy Inc., Madison, Wisconsin, 417–435.

Righi, D. and Elsass, F. (1996) Characterization of soil clay minerals: decomposition of X-ray diffraction diagrams and high resolution electron microscopy. *Clays and Clay Minerals* 44, 791–800.

Riley, M.C. (1941) Projection sphericity. *Journal of Sedimentary Petrology* 11, 94–97.

Rittenhouse, G. (1943) A visual method of estimating two-dimensional sphercity. *Journal of Sedimentary Petrology* 13, 79–81.

Robertson, B. and Vignaux, G.X. (1995) *Interpreting Evidence: Evaluating Forensic Evidence in the Courtroom*. John Wiley & Sons, Chichester, 240pp.

Robertson, J. (2004) Botanical and soil evidence at the crime scene. In Horswell, J. (ed.) *The Practice of Crime Scene Investigation*. CRC Press, Boca Raton, 317–346.

Robertson, J. and Grieve, M. (eds.) (2001) *Forensic Examination of Fibres*. Taylor and Francis, London, 447pp.

Robertson, J., Thomas, C.J., Caddy, B., and Lewis, M. (1984) Particle size analysis of soils: a comparison of dry and wet sieving techniques. *Forensic Science International* 24, 209–217.

Rogasik, H., Onasch, I., Brunotte, J., Jegou, D., and Wendroth, O. (2003) Assessment of soil structure using X-ray computed tomography. In Mees, F., Swennen, R., Van Geet, M., and Jacobs, P. (eds.) (2003) *Applications of X-ray Computed Tomography in the Geosciences*. Geological Society, London, Special Publications 215, 151–165.

Rogers, R.S. (1999) Forensic geology and mineral exploration projects. *Exploration and Mining Geology* 7, 25–27.

Rollinson, H. (1993) *Using Geochemical Data*. Longman, Harlow, 352pp.

Rong, Y. (2000) Statistical methods and pitfalls in environmental data analysis. *Environmental Forensics* 1, 213–220.

Rose, A.W., Hawkes, H.E., and Webb, J.S. (1979) *Geochemistry in Mineral Exploration*, 2nd ed. Academic Press, London, 657pp.

Round, F.E., Crawford, R.M., and Mann, D.G. (1990) *The Diatoms: Biology and Morphology of the Genera*. Cambridge University Press, Cambridge, 757pp.

Rowan, J.S., Goodwill, P., and Franks, S.W. (1999) Uncertainty estimation in fingerprinting suspended sediment sources. In Foster, I.D.L. (ed.) *Tracers in Geomorphology*. Wiley, Chichester, 279–290.

Rowan, J.S., Goodwill, P., and Franks, S.W. (2000) Uncertainty estimation in fingerprinting suspended sediment sources. In Foster, I.D.L. (ed.) *Tracers in Geomorphology*, Wiley, Chichester, 279–290.

Rowe, T., Ketcham, R., Denison, C., Colbert, M., Xu, X., and Currie, P.J. (2001) The *Archaeoraptor* forgery. *Nature* 410, 539–540.

Ruffell, A. (2002) Remote detection and identification of organic remains: an assessment of archaeological potential. *Archaeological Prospecting* 9, 115–122.

Ruffell, A. (2005) Burial location using cheap and reliable quantitative probe measurements. *Forensic Science International* 151, 207–211.

Ruffell, A. (2006) Forensic geoscience. *Geology Today* 22, 68–70.

Ruffell, A. and McKinley, J. (2005) Forensic geoscience: applications of geology, geomorphology and geophysics to criminal investigations. *Earth Science Reviews* 69, 235–247.

Ruffell, A. and Wiltshire, P. (2004) Conjunctive use of quantitative and qualitative X-ray diffraction analysis of soils and rocks for forensic analysis. *Forensic Science International* 145, 13–23.

Russell, R.D. and Taylor, R.E. (1937) Roudness and shape of Mississipii River sands. *The Journal of Geology* 45, 225–267.

Russell, R.J. (1987) Infrared methods. In Wilson, M.J. (ed.) *A Handbook of Determinative Methods in Clay Mineralogy.* Blackie, Glasgow and London, 133–173.

Saferstein, R. (2001) *Criminalistics: An Introduction to Forensic Science.* Prentice Hall, Englewood Cliffs, New Jersey, 576pp.

Salminen, R., Tarvainen, T., Demetriades, A., Duris, M., Fordyce, F.M., Gregorauskiene, V., Kahelin, H., Kivisilla, J., Klaver, G., Klein, H., Larson, J.O., Lis, J., Locutura, K., Marsina, H., Mjartanova, C., Mouvet, P., O'Connor, L., Odor, G., Ottonello, T., Paukola, J.A., Plant, C., Reimann, O., Schermann, U., Siewers, A., Sluys, V.D., DeVivo, B., and Williams, L. (1998) *FOREGS Geochemical Mapping Field Manual.* Geological Survey of Finland Guide 47, Espoo, 36pp.

Sasieni, P.D. and Royston, P. (1996) Dotplots. *Applied Statistics* 45, 219–234.

Saye, S.E. and Pye, K. (2004) Development of a coastal dune sediment database for England and Wales: forensic applications. In Pye, K. and Croft, D.J. (eds.) *Forensic Geoscience: Principles, Techniques and Applications.* Geological Society, London, Special Publications 232, 75–96.

Saye, S.E. and Pye, K. (2006) Variations in chemical composition and particle size of dune sediments along the west coast of Jutland, Denmark. *Sedimentary Geology* 183, 217–242.

Saye, S.E., Pye, K., and Clemmensen, L.B. (2006) Development of a cliff-top dune indicated by particle size and geochemical characteristics: Rubjerg Knude, Denmark. *Sedimentology* 53, 1–22.

Schatz, W., Halle, A., and Saale, D. (1930) Dirt scraped from shoes, as a means of identification. *The American Journal of Police Science* 31, 55–59.

Scholle, P.A. (1978) *A Color Illustrated Guide to Carbonate Rock Constituents, Textures, Cements and Porosities.* American Association of Petroleum Geologists Memoir 27, American Association of Petroleum Geologists, Tulsa, Oklahoma, 241pp.

Scholle, P.A. (1979) *A Color Illustrated Guide Constituents, Textures, Cements and Porosities of Sandstones and Associated Rocks.* American Association of Petroleum Geologists Memoir 28, American Association of Petroleum Geologists, Tulsa, Oklahoma, 201pp.

Schwarz, H.B. and Shane, K.C. (1969) Measurement of particle shape by Fourier analysis. *Sedimentology* 13, 213–231.

Schwarzenbach, R.P., Gschwend, P.M., and Imboden, D.M. (1993) *Environmental Organic Chemistry.* John Wiley & Sons, New York, 681pp.

Scott, A.C. and Jones, T.P. (1991) Microscopical observations of recent and fossil charcoal. *Microscopy and Analysis* July, 13–15.

Scott, A.C., Moore, J., and Brayshaw, B. (2000a) Introduction to fire and the palaeoenviroment. *Palaeogeography, Palaeoclimatology, Palaeoecology* 164, vii–xi.

Scott, A.C., Cripps, J.A., Collinson, M.E., and Nichols, G.J. (2000b) The taphonomy of charcoal following a recent heathland fire and some implications for the interpretation of fossil charcoal deposits. *Palaeogeography, Palaeoclimatology, Palaeoecology* 164, 1–31.

Scott, J. and Hunter, J.R. (2004) Environmental influences on resistivity mapping for the location of clandestine graves. In Pye, K. and Croft, D.J. (eds.) *Forensic Geoscience: Principles, Techniques and Applications*. Geological Society, London, Special Publications 232, 33–38.

Selley, R.C. (1988) *Applied Sedimentology*. Academic Press, London, 446pp.

Sen Gupta, B.K. (ed.) (1999) *Modern Foraminifera*. Kluwer Academic Publishers, Dordrecht, 371pp.

Sergeyev, Y.M., Grabowska-Olszewska, B., Osipov, V.I., Sokolov, V.N., and Kolomenski, Y.N. (1980) The classification of microstructures of clay soils. *Journal of Microscopy* 120, 237–260.

Sever, M. (2005) Murder and mud in the Shenandoah. *Geotimes* January, 1–7.

Sheorei, P.R., Barat, D., Das, M.N., Muckerjee, K.P., and Singh, B. (1984) Schmidt Hammer rebound data for estimation of large scale in situ coal strength. *International Journal of Rock Mechanics, Mining Sciences and Geomechanics Abstracts* 21, 39–42.

Shuirman, G. and Slosson, J.E. (1992) *Forensic Engineering. Environmental Case Histories for Civil Engineers and Geologists*. Academic Press, San Diego, 296pp.

Siegel, J.A. and Precord, C. (1985) Analysis of soil samples by reverse phase high performance liquid chromatography using wavelength ratioing. *Journal of Forensic Sciences* 30, 511–525.

Siegel, J.A., Saukko, P.J., and Knupfer, G.C. (eds.) (2000) *Encyclopedia of Forensic Sciences. 3 Volumes*. Academic Press, London, 1440pp plus glossary and index.

Sinclair, D.J., Kinsley, L.P.J., and McCulloch, M.T. (1998) High resolution analysis of trace elements in corals by laser ablation-ICP-MS. *Geochimica et Cosmochimica Acta* 62, 1889–1901.

Sippel, R.F. (1965) Simple device for luminescence petrography. *Review of Scientific Instruments* 36, 1556–1558.

Sippel, R.F. (1968) Sandstone petrology, evidence from luminescence petrography. *Journal of Sedimentary Petrology* 38, 530–554.

Sippel, R.F. and Glover, E.D. (1965) Structures in carbonate rocks made visible by luminescent petrography. *Science* 150, 1283–1287.

Siver, P.A., Lord, W.D., and McCarthy, D.J. (1993) Forensic limnology: the use of freshwater algal community ecology to link suspects to an aquatic crime scene in southern New England. *Journal of Forensic Sciences* 67, 847–853.

Skinner, D.N.B. (1988) *Sampling Procedures in Forensic Earth Science*. New Zealand Geological Survey Report G129, 18pp.

Skinner, H.C.W., Ross, M., and Frondel, C. (1988) *Asbestos and Other Fibrous Materials. Mineralogy, Crystal Chemistry, and Health Effects*. Oxford University Press, Oxford, 204pp.

Slater, G.F. (2003) Stable isotope forensics — when isotopes work. *Environmental Forensics* 4, 13–23.

Slattery, M.C., Walden, J., and Burt, T.P. (2000) Use of mineral magnetic measurements to fingerprint suspended sediment sources: results from a linear mixing model. In Foster, I.D.L. (ed.) *Tracers in Geomorphology*. Wiley, Chichester, 309–322.

Smale, D. (1973) The examination of paint flakes, glass and soils for forensic purposes, with special reference to electron probe microanalysis. *Journal of the Forensic Science Society* 13, 5–15.

Smale, D. and Trueman, N.A. (1969) Heavy mineral studies as evidence in a murder case in Outback Australia. *Journal of the Forensic Science Society* 9, 123–128.

Small, I.F., Rowan, J.S., Franks, S.W., Wyatt, A., and Duck, R.W. (2004) Bayesian sediment fingerprinting provides a robust tool for environmental forensic geoscience applications. In Pye, K. and Croft, D.J. (eds.) *Forensic Geoscience: Principles, Techniques and Applications*. Geological Society, London, Special Publications 232, 207–214.

Smart, P. and Tovey, N.K. (1981) *Electron Microscopy of Soils and Sediments: Examples*. Oxford University Press, Oxford, 177pp.

Smart, P. and Tovey, N.K. (1982) *Electron Microscopy of Soils and Sediments: Techniques*. Oxford University Press, Oxford, 264pp.

Smith, A.P. and Prentice, D.A. (1993) Exploratory data analysis. In Keren, G. and Lewis, C. (eds.) *A Handbook of Data Analysis in the Behavioural Sciences: Statistical Issues*. Lawrence Erlbaum Associates, Publishers, Hillsdale, New Jersey, 349–390.

Smith, M.R. and Collis, L. (eds.) (2001) *Aggregates. Sand, Gravel and Crushed Rock Aggregates For Construction Purposes*. Third edition revised by Fookes, P.G., Lay, J., Sims, I., Smith, M.R., and West, G. Geological Society Engineering Geology Special Publication 17, The Geological Society, London, 360pp.

Smith, A.S. (2003) A brick's a brick, isn't it? (Michael Sams — Case Study). In Pye, K. and Croft, D.J. (eds.) *Forensic Geoscience, Principles, Techniques and Applications, Programme and Abstracts, 3 & 4 March 2003, the Geological Society, London, UK*, p.44.

Sneed, E.D. and Folk, R.L. (1958) Pebbles in the lower Colorado River, Texas: a study in particle morphogenesis. *Journal of Geology* 66, 114–150.

Soil Survey of England and Wales (1983) *Soil Map of England and Wales Scale1:250,000. Six Sheets plus Legend. Soil Survey of England and Wales*. Rothamsted Experimental Station, Harpenden.

Soil Survey Staff (1990) *Keys to Soil Taxonomy*. Fourth Printing. SMSS Technical Monograph No. 19, Blacksburg, Virginia.

Soltanpour, P.N., Johnson, G.W., Workman, S.M., Jones, J.B., Jr., and Miller, R.O. (2002) Inductively coupled plasma emission spectrometry and inductively coupled plasma spectrometry. In Sparks, D.L., Page, A.L., Helmke, P.A., Loeppert, R.H., Soltanpour, P.N., Tabatabai, M.A., Johnston, C.T., and Sumner, M.E. (eds.) (1996) *Methods of Soil Analysis. Part 3. Chemical Methods*. Soil

Science Society of America Book Series No.5, Soil Science Society of America Inc and American Society of Agronomy Inc., Madison, Wisconsin, 91–139.

SOTER (1990) *Procedures Manual for Small-scale Map and Database Compilation.* International Soil Reference Information Centre, Wageningen.

Sparks, D.L., Page, A.L., Helmke, P.A., Loeppert, R.H., Soltanpour, P.N., Tabatabai, M.A., Johnston, C.T., and Sumner, M.E. (eds.) (1996) *Methods of Soil Analysis. Part 3. Chemical Methods.* Soil Science Society of America Book Series No.5, Soil Science Society of America Inc and American Society of Agronomy Inc., Madison, Wisconsin, 1390pp.

Sperazza, M., Moore, J.N., and Hendrix, M.S. (2004) High resolution particle size analysis of naturally occurring very fine-grained sediment through laser diffractometry. *Journal of Sedimentary Research* 74, 736–743.

Stam, M. (2004) Soil as significant evidence in a sexual assault/attempted homicide case. In Pye, K. and Croft, D.J. (eds.) *Forensic Geoscience: Principles, Techniques and Applications.* Geological Society, London, Special Publications 232, 295–300.

Stanley, E.A. (1992) Application of palynology to establish the provenance and travel history of illicit drugs. *The Microscope* 40, 149–152.

Stoermer, E.F. and Smol, J.P. (eds.) (1999) *The Diatoms: Applications for the Environmental and Earth Sciences.* Cambridge University Press, Cambridge, 469pp.

Stokes, S. (1999) Luminescence dating applications in geomorphological research. *Geomorphology* 29, 153–171.

Strangeways, I. (2000) *Measuring the Natural Environment.* Cambridge University Press, Cambridge, 365pp.

Streickeisen, A.L. (1976) To each plutonic rock its proper name. *Earth Science Reviews* 12, 1–34.

Sugita, R. and Marumo, Y. (1996) Validity of colour examination for forensic soil identification. *Forensic Science International* 83, 201–210.

Sugita, R. and Marumo, Y. (2001) Screening of soil evidence by a combination of simple techniques: validity of particle size distribution. *Forensic Science International* 122, 155–158.

Sugita, R. and Marumo, Y. (2004) "Unique" particles in soil evidence. In Pye, K. and Croft, D.J. (eds.) *Forensic Geoscience: Principles, Techniques and Applications.* Geological Society, London, Special Publications 232, 97–102.

Sugita, R., Suzuki, S., and Marumo, Y. (2004) Validity of micro algae for forensic soil identification. *Japanese Journal of Forensic Identification* 9, 113–122 (in Japanese, with English abstract).

Sukumaran, B. and Ashmawy, A.K. (2001) Quantitative characterization of the geometry of discrete particles. *Geotechnique* 51, 619–627.

Sumner, M.E. (ed.) (2000) *Handbook of Soil Science.* CRC Press, Boca Raton.

Sumner, M.E. and Miller, W.P. (1996) Cation exchange capacity and exchange coefficients. In Sparks, D.L., Page, A.L., Helmke, P.A., Loppert, R.H., Soltanpour, P.N.,

Tabatabal, M.A., Johnson, C.T., and Sumner, M.E. (eds.) *Methods of Soil Analysis. Part 3. Chemical Methods*. Soil Science Society of America Book Series No.5, Soil Science Society of America Inc. and American Society of Agronomy Inc., Madison, Wisconsin, 1201–1230.

Sumner, M. and Wilding, L.P. (2000) What is soil? In Sumner, M. (ed.) *Handbook of Soil Science*. CRC Press, Boca Raton, introductory un-numbered pages.

Sutherland, R.A., Analysis and commentary on "Statistical methods and pitfalls in environmental data analysis" by Yue Rong. *Environmental Forensics* 2, 265–274.

Swan, A.R.H. and Sandilands, M. (1995) *Introduction to Geological Data Analysis*. Blackwell Science Ltd, Oxford, 446pp.

Swift, D.J.P. (1971) Grain mounts. In Carver, R.E. (ed.) *Procedures in Sedimentary Petrology*. Wiley Interscience, New York, 499–511.

Swift, R.S. (1996) Organic matter characterization. In Sparks, D.L., Page, A.L., Helmke, P.A., Loeppert, R.H., Soltanpour, P.N., Tabatabai, M.A., Johnston, C.T., and Sumner, M.E. (eds.) *Methods of Soil Analysis. Part 3. Chemical Methods*. Soil Science Society of America Book Series No.5, Soil Science Society of America Inc. and American Society of Agronomy Inc., Madison, Wisconsin, 1011–1069.

Sylvester, P. (ed.) (2001) *Laser-Ablation-ICPMS in the Earth Sciences: Principles and Applications*. Mineralogical Association of Canada Short Course Series Volume 29, Mineralogical Association of Canada, Ottawa, 243pp.

Syvitski, J.P.M. (ed.) (1991) *Principles, Methods and Applications of Particle Size Analysis*. Cambridge University Press, Cambridge, 368pp.

Taggart, J.E. (1977) Polishing technique for geologic samples. *American Mineralogist* 62, 824–827.

Tan, K.H. (2000) *Environmental Soil Science*, 2nd ed. Marcel Dekker Inc., New York, 452pp.

Taroni, F. and Aitken, C.G.G. (1998) Probabilistic reasoning in the law. Part 2: assessment of probabilities and explanation of the value of trace evidence other than DNA. *Science and Justice* 38, 179–188.

Taylor, S.T. and McLennan, S.M. (1985) *The Continental Crust: its Composition and Evolution. An Examination of the Geochemical Record Preserved in Sedimentary Rocks*. Blackwell Scientific Publications, Oxford, 312pp.

Teichmuller, M. and Wolf, M. (1977) Application of fluorescence microscopy in coal petrology and oil exploration. *Journal of Microscopy* 109, 49–73.

Teixeira, S.R., Dixon, J.B., White, G.N., and Newsom, L.A. (2002) Charcoal in soil: a preliminary view. In Dixon, J.B. and Schulze, D.G. (eds.) *Soil Mineralogy with Environmental Applications*. Soil Science Society of America, Madison, Wisconsin, 819–830.

Teller, J.T. (1976) Equantcy versus sphericity. *Sedimentology* 23, 427–428.

Textoris, D.A. (1971) Grain-size measurement in thin section. In Carver, R.E. (ed.) *Procedures in Sedimentary Petrology*. Wiley Interscience, New York, 95–127.

Thomas, G.W. (1996) Soil pH and soil acidity. In Sparks, D.L., Page, A.L., Helmke, P.A., Loeppert, R.H., Soltanpour, P.N., Tabatabai, M.A., Johnston, C.T., and Sumner, M.E. (eds.) *Methods of Soil Analysis. Part 3. Chemical Methods.* Soil Science Society of America Book Series No.5, Soil Science Society of America Inc. and American Society of Agronomy Inc., Madison, Wisconsin, 475–490.

Thomas, M.C., Wiltshire, R.J., and Williams, A.T. (1995) The use of Fourier descriptors in the classification of particle shape. *Sedimentology* 42, 635–645.

Thompson, M. and Walsh, J.N. (2003) *Inductively Coupled Plasma Atomic Emission Spectrometry.* Viridian Publishing, Woking, Surrey. 316pp.

Thompson, R. and Oldfield, F. (1986) *Environmental Magnetism.* Allen and Unwin, London, 227pp.

Thompson, R., Battarbee, R.W., O'Sullivan, P.E., and Oldfield, F. (1975) Magnetic susceptibility of lake sediments. *Limnology and Oceanography.* 20, 687–698.

Thorndycraft, V.R., Pirrie, D., and Brown, A.G. (1999) Tracing the record of early alluvial tin mining on Dartmoor, UK. In Pollard, A.M. (ed.) *Geoarchaeology: Exploration, Environments, Resources.* Geological Society, London, Special Publications 165, 91–102.

Thornton, J.I. (1974) *Forensic Biochemical Characterization of Soils.* PhD Thesis, University of California, Berkeley.

Thornton, J.I. and Crim, D. (1986) Forensic soil characterization. In Maehly, A. and Williams, R.L. (eds.) *Forensic Science Progress Volume I.* Springer Verlag, Berlin, 1–35.

Thornton, J.I. (1997) Visual color comparisons in forensic science. *Forensic Science Review* 9, 37–57.

Thornton, J.I. and Fitzpatrick, F. (1978) Forensic science characterization of sand. *Journal of Forensic Sciences* 20, 460–475.

Thornton, J.I., Crim, D., and McLaren, A.D. (1975) Enzymatic characterization of soil evidence. *Journal of Forensic Sciences* 20, 674–692.

Tickell, F.G. (1931) *The Examination of Fragmental Rocks.* Stanford University Press, Stanford.

Tickell, F.G. (1965) *The Techniques of Sedimentary Mineralogy.* Developments in Sedimentology 4. Elsevier, Amsterdam, 220pp.

Tilley, R. (2000) *Colour and the Optical Properties of Materials.* John Wiley & Sons Ltd, Chichester, 335pp.

Tillman, S.E. (1973) The effect of grain orientation on Fourier shape analysis. *Journal of Sedimentary Petrology* 43, 867–869.

Tolonen, K. (1986) Rhizopod analysis. In Berglund, B.E. (ed.) *Handbook of Holocene Palaeoecology and Palaeohydrology.* John Wiley, Chichester, 645–666.

Torrent, J. and Barron, B. (1993) Laboratory measurement of soil colour: theory and practice. In Bigham, J.M. and Ciolkhoz, E.J. (eds.) *Soil Colour.* Soil Science Society of America Special Publication 31, 51–69.

Tovey, N.K. and Krinsley, D.H. (1980) A cathodoluminescent study of quartz sand grains. *Journal of Microscopy* 120, 279–289.

Townley, L. and Ede, R. (2004) *Forensic Practice in Criminal Cases*. The Law Society, London, 471pp.

Trejos, T., Montero, S., and Almirall, J.R. (2003) Analysis and comparison of glass fragments by laser ablation inductively coupled plasma mass spectrometry (LA-ICP-MS) and ICP-MS. *Analytical and Bioanalytical Chemistry* 376, 1255–1264.

Trewin, N. (1988) Use of the scanning electon microscope in sedimentology. In Tucker, M.E. (ed.) *Techniques in Sedimentology*. Blackwell, Oxford, 229–273.

Tucker, M.E. (ed.) (1988) *Techniques in Sedimentology*. Blackwell Scientific Publications, Oxford, 394pp.

Tucker, M.E. (1991) *Sedimentary Petrology: An Introduction to the Origin of Sedimentary Rocks*, 2nd ed. Blackwell Scientific Publications, Oxford, 260pp.

Tucker, M.E. and Wright, V.P. (1990) *Carbonate Sedimentology*. Blackwell Scientific Publications, Oxford, 482pp.

Tukey, J.W. (1977) *Exploratory Data Analysis*. Addison Wesley, Reading, Massachusetts.

Tunlid, A. and White, D.C. (1992) Biochemical analysis of biomass, community structure, nutritional status, and metabolic activity of the microbial community in soil. In Stotzky, G. and Bollag, J.M. (eds.) *Soil Biochemistry, Volume 7*. Marcel Dekker, New York, 229–262.

Udden, J.A. (1914) Mechanical composition of clastic sediments. *Bulletin of the Geological Society of America* 25, 655–744.

Ugolini, F.C., Corti, G., Agnelli, A., and Piccarli, F. (1996) Mineralogical, physical and chemical properties of rock fragments in soil. *Soil Science* 161, 521–542.

Ulery, A.L. and Graham, R.C. (1993) Forest fire effects on soil colour and texture. *Soil Science Society of America Journal* 57, 135–140.

U.S. Department of Agriculture (1972) *Soil Survey Laboratory Methods and Procedures for Collecting Soil Samples*. Soil Survey Investigations Report No. 2. U.S. Government Printing Office, Washington D.C.

U.S. Environmental Protection Agency (2002) *Guidance for Choosing a Sampling Design for Environmental Data Collection for Use in Developing a Quality Assurance Plan, EPA QA/G-58*. EPA/240/R-02/0005, Final report. Office of Environmental Information, Washington D.C.

Vallejo, L.E. (1995) Fractal analysis of granular materials. *Geotechnique* 45, 159–163.

Van Es, H.M. (2002) Soil variability. In Dane, J.H. and Topp, G.G. (eds.) *Methods of Soil Analysis. Part 4. Physical Methods*. Soil Science Society of America, Madison, Wisconsin, 1–13.

Vennemann, T.W., Kesler, S.E., and O'Neil, J.R. (1992) Stable isotope composition of quartz pebbles and their fluid inclusions as tracers of sediment provenance: implications for gold- and uranium-bearing quartz pebble conglomerates. *Geology* 20, 837–840.

Vermeij, E. (2003) The characterization of building materials. In Pye, K. and Croft, D.J. (eds.) *Forensic Geoscience: Principles, Techniques and Applications, Programme and Abstracts, 3 & 4 March 2003, The Geological Society, London, UK*, p.49.

Vernon, R.H. (2004) *A Practical Guide to Rock Microstructure*. Cambridge University Press, Cambridge, 606pp.

Vogt, C. and Latkoczy, C. (2005) Laser ablation ICP-MS. In Nelms, S.M. (ed.) *ICP Mass Spectrometry Handbook*. Blackwell Publishing, Oxford, and CRC Press, Boca Raton, Florida, 228–258.

Wade, C. (ed.) (2003) *Handbook of Forensic Services*. Federal Bureau of Investigation Laboratory Publication, Virginia, 180pp.

Wadell, H. (1932) Volume, shape and roundness of rock particles. *The Journal of Geology* 40, 443–451.

Wadell, H. (1933) Sphericity and roundness of rock particles. *The Journal of Geology* 41, 310–331.

Wadell, H. (1935) Volume, shape and roundness of quartz particles. *The Journal of Geology* 43, 250–280.

Walden, J., (1999) Remanence measurements. In Walden, J., Oldfield, F., and Smith, J. (eds.) *Environmental Magnetism: A Practical Guide*. Quaternary Research Association Technical Guide, No. 6, Quaternary Research Association, London, 63–88.

Walkley, A. and Black, I.A. (1934) An examination of the Degtjareff method for determining soil organic matter and a proposed modification of the chromic acid titration method. *Soil Science* 63, 251–263.

Wallace, T.P. and Wintz, P.A. (1980) An efficient three dimensional aircraft recognition algorithm using normalised Fourier descriptors. *Computer Graphics and Image Processing* 13, 96–126.

Walls, H.J. (1974) *Forensic Science: An Introduction to Scientific Crime Detection*, 2nd ed. Sweet and Maxwell, London, 257pp.

Walsh, J.N. (1997) Inductively-coupled plasma-atomic emission spectrometry (ICP-AES). In Gill, R. (ed.) *Modern Analytical Geochemistry*. Addison Wesley, Harlow, 41–66.

Walsh, J.N., Gill, R., and Thirlwall, M.F. (1997) Dissolution procedures for geological and environmental samples. In Gill, R. (ed.) *Modern Analytical Geochemistry*. Addison-Wesley, Harlow, 29–40.

Wanogho, S. (1985) *The Forensic Comparison of Soils with Particular Reference to Particle Size Distribution Analysis*. PhD Thesis, University of Strathclyde.

Wanogho, S, Gettinby, B., Caddy, B., and Robertson, J. (1985) A statistical method for assessing soil comparisons. *Journal of Forensic Sciences* 30, 864–872.

Wanogho, S., Gettinby, B., and Caddy, B. (1987a) Particle size distribution analysis of soils using laser diffraction. *Forensic Science International* 33, 117–128.

Wanogho, S., Gettinby, B., Caddy, B., and Robertson, J. (1987b) A statistical procedure for the forensic evaluation of soil particles. In Lloyd, P.J. (ed.) *Particle Size Analysis.* John Wiley, Chichester, 105–113.

Wanogho, S., Gettinby, B., Caddy, B., and Robertson, J. (1987c) Some factors affecting soil sieve analysis in forensic science. 1. Dry sieving. *Forensic Science International* 33, 129–137.

Wanogho, S., Gettinby, B., Caddy, B., and Robertson, J. (1987d) Some factors affecting soil sieve analysis in forensic science. 2. Wet sieving. *Forensic Science International* 33, 139–147.

Wanogho, S., Gettinby, B., Caddy, B., and Robertson, J. (1989) Determination of particle size distribution of soils in forensic science using classical and modern instrumental methods. *Journal of Forensic Sciences* 34, 823–835.

Warren, J. (2005) Representativeness of environmental samples. *Environmental Forensics* 6, 21–25.

Watling, R.J., Lynch, B.F., and Herring, D. (1997) Use of laser ablation inductively coupled plasma mass spectrometry for fingerprinting scene of crime evidence. *Journal of Analytical and Atomic Spectrometry* 12, 195–203.

Watson, J. (1911) *British and Foreign Building Stones. A Descriptive Catalogue of the Specimens in the Sedgwick Museum, Cambridge.* Cambridge University Press, Cambridge, 483pp.

Watt, J. (1988) Automated characterization of individual carbonaceous fly ash particles by computer controlled scanning electron microscopy: analytical techniques. *Water Air and Soil Pollution*, 106, 309–327.

Watters, M. and Hunter, J.R. (2004) Geophysics and burials: field experience and software development. In Pye, K. and Croft, D.J. (eds.) *Forensic Geoscience: Principles, Techniques and Applications.* Geological Society, London, Special Publications 232, 21–31.

Weaver, C.E. and Pollard, L.D. (1973) *The Chemistry of Clay Minerals.* Developments in Sedimentology 15. Elsevier, Amsterdam, 213pp.

Weaver, R.W. (2003) The life work of Walter C. McCrone Jr. — bibliography and selected historical facts. *The Microscope* 51, 31–44.

Webster, R. (1989) Is regression what you really want. *Soil Use and Management* 5, 47–53.

Webster, R. (1997) Regression and functional relations. *European Journal of Soil Science* 48, 557–566.

Webster, R. (2001) Statistics to support soil research and their presentation. *European Journal of Soil Science* 52, 331–340.

Webster, R. and Oliver, M.A. (1990) *Statistical Methods in Soil and Land Resource Survey.* Oxford University Press, Oxford.

Webster, R. and Oliver, M.A. (2001) *Geostatistics for Environmental Scientists: Statistics in Practice.* John Wiley & Sons, Chichester, 271pp.

Weltje, G.J. (2002) Quantitative analysis of detrital modes: statistically rigorous confidence regions in ternary diagrams and their use in sedimentary petrology. *Earth Science Reviews* 57, 211–253.

Weltje, G.J. (2004) A quantitative approach to capturing the compositional variability of modern sands. *Sedimentary Geology* 171, 59–78.

Weltje, G.J. and von Eynatten, H. (2004) Quantitative provenance analysis of sediments: review and outlook. *Sedimentary Geology* 171, 1–11.

Welton, J.E. (1984) *SEM Petrology Atlas*. American Association of Petroleum Geologists, Tulsa, 237pp.

Wenk, H.-R. and Bulakh, A. (2004) *Minerals: Their Constitution and Origin*. Cambridge University Press, Cambridge, 645pp.

Wentworth, C.K. (1922a) A scale of grade and class terms for classic sediments. *The Journal of Geology* 30, 377–392.

Wentworth, C.K. (1922b) The shapes of pebbles. *Bulletin of the United States Geological Survey* 730-C, 91–114.

Wentworth, C.K. (1922c) The shapes of beach pebbles. *United States Geological Survey Professional Paper* 131-C, 75–83.

Wentworth, C.K. (1933) The shapes of rock particles: a discussion. *The Journal of Geology* 41, 306–309.

Westgate, J.A. and Gorton, M.P. (1981) Correlation techniques in tephra studies. In Self, S. and Sparks, R.S.J. (eds.) *Tephra Studies*. Reidel, Dordrecht, 73–94.

Westgate, J.A., Perkins, W.T., and Fuge, R. (1995) Trace element analysis of volcanic glass by laser ablation inductively coupled plasma mass spectrometry: application to tephrochronological studies. *Applied Geochemistry* 9, 323–335.

Whalley, W.B. (1972) The description and measurement of sedimentary particles and the concept of form. *Journal of Sedimentary Petrology* 42, 961–965.

Wheals, B.B. and Noble, W. (1972) Forensic applications of pyrolysis gas chromatography. *Chromatography* 5, 553–557.

White, D.C. (1993) In situ measurement of microbial biomass, community structure and nutritional status. *Philosophical Transactions of the Royal Society of London A*, 344, 59–67.

White, P.C. (ed.) (2004) *Crime Scene to Court. The Essentials of Forensic Science*, 2nd ed. Royal Society of Chemistry, Cambridge, 448pp.

Wilding, L.P. and Drees, L.R. (1971) Biogenic opal in Ohio soils. *Soil Science Society of America Proceedings* 35, 1004–1010.

Wilding, L.P. and Drees, L.R. (1973) Scanning electron microscopy of opaque opaline forms isolated from forest soils in Ohio. *Proceedings of the Soil Science Society of America* 37, 647–650.

Wilding, L.P. and Drees, L.R. (1974) Contributions of forest opal and associated cystal-line phases to fine silt and clay fractions of soils. *Clays and Clay Minerals* 22, 295–306.

Wilding, L.P., Smeck, N.E., and Drees, L.R. (1977) Silica in soils: quartz, cristobalite, tridymite, and opal. In Dixon, J.B. and Weed, S.B. (eds.) *Minerals in Soil Environments*. Soil Science Society of America, Madison, Wisconsin, 471–552.

Willard, H.H., Merritt, L.L., Jr., Dean, J.A., and Settle, F.A. Jr. (1988) *Instrumental Methods of Analysis*, 7th ed. Wadsworth Publishing Company, Belmont, California, 895pp.

Willetts, B.B. and Rice, M.A. (1983) Practical representation of characteristic grain shape of sands: a comparison of methods. *Sedimentology* 30, 557–566.

Williams, A.C., Edwards, H.G.M., and Barry, B.W. (1995) The iceman: molecular structure of 5200-year old skin characterised by Raman spectroscopy and electron microscopy. *Biochimica et Biophysica Acta* 1246, 98–104.

Williams, E.M. (1965) A method of indicating pebble shape with one parameter. *Journal of Sedimentary Petrology* 35, 993–996.

Willis, S., McCullough, J., and McDermott, S. (2001) The interpretation of paint evidence. In Caddy, B. (ed.) *Forensic Examination of Glass and Paint*. Forensic Science Series. Taylor & Francis, London, 273–287.

Wilson, M.J. (1987) X-ray powder diffraction methods. In Wilson, M.J. (ed.) *Quantitative Methods in Clay Mineralogy*. Blackie, Glasgow, 26–98.

Wiltshire, P.J. (2006) Consideration of some taphonomic variables of relevance to forensic palynological investigation in the United Kingdom. *Forensic Science International* 163, 173–182.

Winspear, N.R. and Pye, K. (1995a) Textural, geochemical and mineralogical evidence for the origin of Peoria Loess in Central and Southern Nebraska. *Earth Surface Processes and Landforms* 20, 735–745.

Winspear, N.R. and Pye, K. (1995b) Textural, geochemical and mineralogical evidence for the sources of aeolian sand in central and southwestern Nebraska, USA. *Sedimentary Geology* 101, 85–98.

Winter, A. and Siesser, W.G. (eds.) (1994) *Coccolithophores*. Cambridge University Press, Cambridge, 242pp.

Wintle, A.G. (1996) Archaeologically-relevant dating techniques for the next century. *Journal of Archaeological Science* 23, 123–138.

Wintle, A.G. (1997) Luminescent dating: laboratory procedures and protocols. *Radiation Measurements* 27, 769–818.

Wintle, A.G. and Murray, A.S. (1997) The relationship between quartz thermoluminescence, photo-transferred luminescence, and optically-stimulated luminescence. *Radiation Measurements* 27, 611–624.

Wray, D. (2005) Geological analysis. In Nelms, S.M. (ed.) *ICP Mass Spectrometry Handbook*. Blackwell Publishing, Oxford, and CRC Press, Boca Raton, 432–451.

Wright, R.J. (2002) Atomic absoption and flame emission spectrometry. In Sparks, D.L., Page, A.L., Helmke, P.A., Loeppert, R.H., Soltanpour, P.N., Tabatabai, M.A.,

Johnston, C.T., and Sumner, M.E. (eds.) (1996) *Methods of Soil Analysis. Part 3. Chemical Methods*. Soil Science Society of America Book Series No.5, Soil Science Society of America Inc and American Society of Agronomy Inc., Madison, Wisconsin, 65–90.

Yardley, B.W. D. (1989) *An Introduction to Metamorphic Petrology*. Longman, Harlow, 248pp.

Yardley, B.W.D., MacKenzie, W.S., and Guilford, C. (1990) *Atlas of Metamorphic Rocks and Their Textures*. Longman, Harlow, 120pp.

Yates, S.R. and Warwick, A.W. (2002) Geostatistics. In De Gruijter, J.J. *Designing a sampling scheme*. In Dane, J.H. and Topp, G.G. (eds.) *Methods of Soil Analysis. Part 4. Physical Methods*. Soil Science Society of America Book Series No.5. Soil Science Society of America Inc., Madison, Wisconsin, 81–118.

Yinon, J. (ed.) (1995) *Forensic Applications of Mass Spectrometry*. CRC, Press Boca Raton, 296pp.

Yinon, J. (ed.) (2004) *Advances in Forensic Applications of Mass Spectrometry*. CRC Press, Boca Raton, 279pp.

Zingg, T. (1935) Beitrage zue Schotteranlyse. *Schweizerische Mineralogische und Petrographische Mitteilungen* 15, 39–140.

Zussman, J. (ed.) (1977) *Physical Methods in Determinative Mineralogy*, 2nd ed. Academic Press, London, 720pp.

Index

A

AAS, *see* atomic absorption spectrometry (AAS)
absent features, 152
abundant features, 152
acid humas fractions, 112
adsorption, 149
affirmative, numerical category scales, 266
Alabama (United States), 38, 106
Alaska (United States), 258
albite, *37*
algae
    trace evidence acceptability, 46–47, *47–48*
    water body particulates, 33
algebraic reconstruction technique (ART), 160
allochthonous sediments, 17
alpha radiation, 102
alternative analytical techniques, 167
aluminosilicate gels, 41
aluminum, 38
amorphous materials, 34, 41
amphiboles, 35
analysis of variance (ANOVA), 231
anatase, *36*
andalusite, *36*
angularity, 141, *144,* 144–145
angularity factor, 145
anhydrite, *36*
anions, 88
anorthite, *36*
anthropogenic materials, 51–57
apatite, *36*
appropriateness, samples, 265
aragonite, 35
Arborfield Bridge site, 79, 251
Archaeoraptor, 160
area method, 105
argon/argon dating method, 104
ART, *see* algebraic reconstruction technique (ART)
artificial muds, 28
asymmetry, 159
atomic absorption spectrometry (AAS), 89

Australia
    automated mineralogical analysis, 107
    radioactive isotopes, 104
    shape frequency distributions, 82
autochthonous sediments, 17
automated mineralogical analysis, 107–109, *109–110*
Avicennia-Rhizophora, 114
Ayers Rock, 63, *64–65*

B

Baby Lara, 52
backscattered electron mode (BSE)
    electron probe analysis, 167–168
    microfabrics, 180
    surface texture, 157
backscattering, 148
bacteria, 31
Baldwin County, Alabama, 38, 106
Barepot, Cumbria, 52
Bayesian statistical approach, 247, *267*
Bentham, Jeremy, 266
benthic foraminifera, 48
Berkshire, England
    correlations, 235
    data comparison, 251
    hypothesis testing, 233
    light elements stable isotopes, 99
    particle size distributions, 77
best fit, 235
beta radiation, 102
bicarbonates, 88
binocular microscope, 153
biogenic carbonate gravel, 146
biological influences, 30
biotite, *37*
bivalves, 42
bivariate scatter plots, 231
bladed particles, 127
blades, 129
Blu-Tack, 136

bodywork, 213
bones, 31, 98
boulders, 21
box and whisker plot, 231
brachiopods, 42
Bradford-on-Avon, Wiltshire, 264
brick, *48, 50,* 52–53
Brisbane, Australia, 107
bromide, 88
bromoform, 35
BSE, *see* backscattered electron mode (BSE)
bulk density, 68
bulk properties
    anions, 88
    automated mineralogical analysis, 107–109,
        109–110
    bulk luminescence properties, 111
    cation exchange capacity, 87
    chemical characteristics, 86–104
    clay mineral assemblage, 109–110
    color, 60–63, *62, 64–68,* 68
    computerized mineralogical analysis, 107–109,
        109–110
    concentrations, light elements, 96–97, *97*
    density, 68–69
    diatom assemblages, 117–118
    Eh, 86–87
    electrical conductivity, 87, *88*
    hardness, 69
    heavy elements, 100–102, *103*
    heavy mineral analysis, 106
    light elements, *97–102*
    magnetic characteristics, 73, 75
    major elements, 89, *90,* 91, *92–95,* 93
    microbial populations, 112–113
    microfabrics, 69, *70–72,* 71
    mineralogical characteristics, 104–111
    modal mineralogy, 104–106, *105*
    organic carbon content, 111
    organic compounds, 111–112
    organic matter characteristics, 111–118
    particle size distributions, 75–77, *78–83,* 79, 81
    permeability, 71–72
    pH, 86
    physical characteristics, 59–85, *60*
    pollen assemblages, 113–117, *115–116*
    porosity, 71–72
    radioactive isotopes, 102, 104
    rare earth elements, 93, *95–96,* 96
    shape frequency distributions, 82, *84–85,* 85
    specific surface, 72–73, *73–74*
    stable isotopes, 98–102, *98–103*
    trace elements, 89, *90,* 91, *92–95,* 93
bumps, 145
burial metamorphism, 13

C

calcareous nannoplankton, 49
calcite, 35, 178
calcium, 38
California (United States), 43
Canada, 258
Canada Balsam, 106
canonical analysis, 242
carbon, 98
carbonates, 37
cardboard, 56
Carlo Erba instrument, 97
carpets, 213
cathodoluminescence, 111, 163–164
cation exchange capacity (CEC), 87
CEC, *see* cation exchange capacity (CEC)
cement, 32
cenospheres, 55
ceramics
    cathodoluminescence, 163
    thermoluminescence, 164
    trace evidence acceptability, 53
cerium, 37, 96
CF-IRMS, *see* continuous helium flow isotope
        ratio mass spectrometry (CF-IRMS)
C-GC-IRMS, *see* combustion-gas
        chromatography-isotope ratio mass
        spectrometry (C-GC-IRMS)
Chalk of Upper Cretaceous ago, 49
chalks and chalky marls
    calcareous nannoplankton, 49
    chert pebbles, 145
    foraminifera, 48
chance coincidence probabilities, 246–247
charcoal, 46, *see also* macro-charcoal;
        micro-charcoal
chemical analysis, 52
chemical characteristics, 86–104
chemical sediments, 17
chert pebbles, 145
chloride, 88
chlorite, *37*
CHN elemental analyzer, 111
Citronelle Formation, 38–39, 106
classical hypothesis testing, 231–233, *232*
classic optical petrography, 177
clast lithological analysis, 22
clays and clay minerals
    fundamentals, 17
    mineralogical characteristics, 109–110
    mud, 28
    trace evidence acceptability, 38–39, *40–41*
Clerici's solution, 35
climate, 30

clinker, 53
clothing, 210, 219, 223
cobbles, 21
coccolithophores, 49
coccoliths, 49
coefficient of determination, 242
coincidence probabilities, 246–247
color
    bricks, 52
    particle properties, 161, *161–162*
    physical characteristics, 60–63, *62, 64–68,* 68
    soil constituents, 31
colorimetric method, 111
combined approaches, 245–246
combustion-gas chromatography-isotope ratio
        mass spectrometry (C-GC-IRMS), 98
common features, 152
comparison samples, 183
composite characterizations
    gravel-size particles and rocks, 164–165
    sand-size particles, 165, *166,* 167
computer image analysis, 75
computerized mineralogical analysis, 107–109,
        109–110
computer tomography, MRI, 72
concentrations, light elements, 96–97, *97*
conchoidal fractures, 150
concrete, 51–52
constant head permeameters, 72
contact exchange principle, 4–5
contact metamorphism, 15
containers, samples, 205–208, *206–207*
context, 226–227
continuous helium flow isotope ratio mass
        spectrometry (CF-IRMS), 98
control samples, 183
coralline, 146
corner angularity, 144–145
correlations, *228,* 233–235, *236–241*
costs, 195, 223
Coulter-Counter, 75–76
Coulter Rapid Vue instrument, 82, 85
Cretaceous age, 49
crime scenes, 186–200, *187–200*
cross-sectional area, 122, *123,* 124
crystalline overgrowths, 150
CSIRO studies, 31
cumulative frequency curve, 231

D

database and database interrogation, 256–264,
        260–263
dating methods, *see specific type*
defense investigation stages, *264,* 264–265

degree of circularity, 134
degree of roundness, 139
deionized water, 87
density, 52, 68–69, *see also* particle density
descriptive statistics, 227
detrital sedimentary rocks, 16
detrital sediments, 21, 37
devitrifiction, 41
Devon dirt track, 251
diagenesis, 15, 38
diatomite, 47
diatoms
    microfossils, 46
    organic matter characteristics, 117–118
    trace evidence acceptability, 46–47, *47–48*
    water body particulates, 33
differences between samples, 248–251, 266
differential thermal analysis (DTA), 39
direct data comparison, 248–251, *250, 252–254*
discoasters, 49
discoid particles, 127
discriminant analysis, 242
disc-rod index, 128
dish-shaped depressions, 150
distinguishability, 248
distortion angle, 145
distributions, 229
DNA techniques and evidence
    macrofossils, 44
    popularity, 6–7
    trace evidence comparison, 226
dolomite, 35
domestic tap water, 33
dotplots, 231
Doyle, Sir Arthur Conan, 3–4
drowning victims, 34, 47
dry sieving, 75–77
DTA, *see* differential thermal analysis (DTA)
dusts and particulates, 4–5, 32–34

E

Eh, 31, 86–87
electrical conductivity, 87, *88*
electron diffraction, 177
electron microprobe analysis (EMPA)
    elemental composition, 167–168, *168,*
        170–174
    heavy mineral analysis, 106
    major and trace elements, 89
    microfabrics, 180
    modal mineralogy, 106
electron microscopy, 110
electron spin resonance (ESR), 110, 177
elemental analyzers, 96

elemental composition
    alternative analytical techniques, 167
    electron microprobe analysis, 167–168, *168,*
        170–174
    laser ablation ICP-MS analysis, 169, *176*
    SEM-EDXRA analysis, 168–169, *175*
elongated particles, 127
elongation, 126, 159
EMPA, *see* electron microprobe analysis (EMPA)
Encyclopedia of Forensic Sciences, 8
energy-dispersive (ED) electron microprobe
        analysis (EMPA), 167
England
    brick, *50*
    correlations, 235
    data comparison, 251
    electrical conductivity, 87
    hypothesis testing, 233
    light elements stable isotopes, 99
    multivariate analysis, 245
    particle size distributions, 77, 79
    soil databases, 257–259
environmental forensics, 2, 8
environmental settings, 10
epdiote, *36*
equant particles, 127
equivalent circular diameter, 122
equivalent spherical diameter, 122
ESR, *see* electron spin resonance (ESR)
estuaries, 9
Eucalyptus woodland, 113–114
Euclidean distance, 244, 261
europium, 96
evaluation of significance
    Bayesian statistical approach, 247, *267*
    chance coincidence probabilities, 246–247
    classical hypothesis testing, 231–233, *232*
    coincidence probabilities, 246–247
    combined approaches, 245–246
    correlations, *228,* 233–235, *236–241*
    database and database interrogation, 256–264,
        260–263
    differences between samples, 248–251
    direct data comparison, 248–251, *250, 252–254*
    exclusion, 255–256
    exploratory data analysis, 227–231, *228–230*
    future directions, 269
    guidelines, 225–227
    hierarchical cluster analysis, 243–244,
        244–245
    inclusion, 255–256
    indistinguishability of samples, 248–251
    likelihood ratios, 246–247
    multi-technique comparison data, 255–256
    multivariate analysis, 242–244

    principle component analysis, 243, *244*
    procedures, 225–227
    regression analysis, 233–235, 242
    similarity of samples, 248–251
    strength of evidence, *264–265,* 264–269, *267*
"every contact leaves a trace," 4–5
evidential significance and value, 227, 268
exclusion, 248, 255–256
exotic pollens, 46
exploratory data analysis, 227–231, *228–230*
extremely strong evidence, 247
extrusive rocks, 13

F

factor analysis, 242
falling head permeameters, 72
FBI *Handbook of Forensic Sciences,* 7
feldspar
    cathodoluminescence, 163
    light minerals, 35
    modal mineralogy, 106
    noncrystalline materials, 178
    sand-size particles, 165
    thermoluminescence, 164
Feret diameter, 121–122
ferrihydrite, 41
fibers
    household dusts, 33
    organic particles, 181
    soil constituents, 32
    trace evidence acceptability, 56–57
fingernail clippings, 213
firearms, 213
fish scales, 33
flatness, 126
flattened and elongated particles, 127
flattened particles, 127
Fleet method, 105
fluorescence microscopy, 180
fluoride, 88
foot pedal rubbers, 213
footwear
    crime scenes, 186, 196–197
    data comparison, 249
    requested samples, 210, 223
footwell mats, 213
foraminifera, 46, 48
forensic geology, nature and development
    fundamentals, 1–3, *2*
    geomorphological data, 10
    geophysical survey data, 8–9
    historical developments, 3–8
    lake level data, 9
    meterological data, 9

river flow, 9
tidal data, 10
forensic pedology, 1
forensic soil science, 1
form, 126–130, *129–133*
forward scattering, 149
fossils
    algae, 46–47, *47–48*
    brick, *48*
    calcareous nannoplankton, 49
    coccolithophores, 49
    diatoms, 46–47, *47–48*
    discoasters, 49
    foraminifera, 48
    fossils, 42
    macrofossils, 42–43, *43–44*
    macroscopic-charcoal, 44–45, *45*
    micro-charcoal, 51
    microfossils, 46–51
    opal phytoliths, 50
    ostracods, 50, *50*
    pollen, 46
    spores, 46
    testate amoebae, 49
    trace evidence acceptability, *40–41*, 42
Fourier techniques
    organic compounds, 112
    organic particles, 181
    particle morphology characterization,
        157–159, *158*
    shape frequency distributions, 82
fractal analysis, 157–159, *158*
Frantz isomagnetic separator, 35
freeze-drying vegetation, 210
frequency distributions
    shape, 82, *84–85*, 85
    size, 75–81, *78–83*
freshwaters, 87
frosted appearance, 150
frosted texture, 152
frustule, algae, 46
F-test, 231–232
FTIS, *see* Fourier techniques
fungi, 31
future directions, 269

G

gamma radiation, 102
garnet, 35, *36*
gas pycnometry, 71
gastropods, 42
Gatwick Airport, Sussex, 251
gels, 41
geological evidence, 2

geological parent material, 30
geomorphological data, 10
geophysical survey data, 8–9
gladolinium, 96
Glasgow, 233
glass containers, 208
glasses, 32, 41
glassy texture, 152
gloss, 148
goethite, 41
granules, 21
graptolites, 42
gravel, 17, 21–22, *23–26*
gravimetric methods, 71
gregite, 41
grid-based strategies, 195, *197–199*
grooves, 150
Gross, Hans, 3
guidelines, 184–186, 225–227
gypsum, *36*

H

hair, 33, 98, 181
halides, 37
halite, *36*
Hanbuch fur Untersuchungsrichter, 3
Handbook for Examining Magistrates, 3
Handbook of Forensic Sciences (FBI), 7
hardness, 52, 69
harmonics, 157
HCA, *see* hierarchical cluster analysis (HCA)
heavy elements, 100–102, *103*
heavy minerals
    cathodoluminescence, 163
    mineralogical characteristics, 106
    sand-size particles, 165
    trace evidence acceptability, 35, *36–37*, 37
Heinrich, Edward Oscar, 6
hematite, 35
hierarchical cluster analysis (HCA), 242–244,
    244–245
highly-elongated particles, 128–129
high performance liquid chromatography
    (HPLC), 112
high pressure liquid chromatography (HPLC), 111
high resolution XCT (HR-XCT) systems, 159–160
histograms, 231
historical developments, 3–8
hollows, 145
Home Office Central Research Establishment at
    Aldermaston, 6–7
Horiba CAP A-300 particle size analyzer, 76
horizons, 31
Hotelling's T^2 test, 232

household dusts, 33, *34*
HPLC, *see* high performance liquid
  chromatography (HPLC); high pressure
  liquid chromatography (HPLC)
human origin of particles
  fibers, 56
  glass particles, 55, *55*
  metallic fragments, 53–54
  soil constituents, 32
humidity, 9
hydraulic conductivity, 72
hydrogen, 38, 98
hydrometer method, 75
hypothesis testing, 231–233, *232*

I

ice, *37*
ICP, *see* inductively coupled plasma spectrometry
  (ICP)
ICP-atomic emission spectrometry (ICP-AES)
  elemental composition, 167
  major and trace elements, 89, 91
  soil databases, 259
ICP-mass spectrometry (ICP-MS)
  elemental composition, 167
  major and trace elements, 89, 91
  soil databases, 259
igneous rocks, 13, *15*
Ilfracombe, Devon, 79
illite, *37*
ilmenite, 35, *36*
IMMA, *see* ion microprobe mass analysis (IMMA)
Imperial Valley, California, 43
inclusion, 255–256
index of best fit, 235
indistinguishability of samples, 225, 248–251, 266
inductively coupled plasma spectrometry (ICP),
  89, 259
industrial dusts, 33
inferential statistics, 231
information, 208–210, *209*
infrared analysis
  automated mineralogical analysis, 108
  clay minerals, 38–39, 110
infrared (IR) spectroscopy, 177
inorganic particles, 11
inscribed circle sphericity, 135
insect carapaces, 31
intercept sphericity parameter, 136
intrusive rocks, 13
invertebrate animals, 31
iodine tincture, 208
ion microprobe mass analysis (IMMA), 167
ions, 38

Ireland, 259
IRMS, *see* isotope ratio mass spectrometry (IRMS)
iron, 38
iron mono-sulfides, 41
iron oxy-hydrides, 41
irregularity, 126, 145–146, *147–148*
isotope ratio mass spectrometry (IRMS), 98
isotopes
  chemical characteristics, 98–102, *98–103*
  composition, 177
  dating, 177
  stable isotope ratios, 177

J

jigsaw match, 225

K

kaolinite, *37*
K-feldspar, *36*, *see also* feldspar
knives, 206, 213
knobbly, 145
Kruskal-Wallis test, 231
Ku Klux Klan, 38, 106

L

labeling samples, 208–210, *209*
labradorite, *36*
L*a*b* system, 61, 63, 161
Lake Cahuilla, 43
lake level data, 9, *see also* water
Lakenheath, Suffolk, 49
landscape features, 10
lanthanide group, 93
lanthanum, 37, 96
Lara (Baby), 52
large pits, 150
laser ablation ICP-MS analysis
  elemental composition, 167, 169, *176*
  SEM-EDXRA analysis, 169
  stable isotopes, 177
laser diffraction, 75, 77
laser granulometry, 77
laser profilometry, *160*
lead/lead dating method, 104
level of overall scientific support, 266
light elements, *97–102*, 98–100
light minerals, 35, *36–37*
light scattering, 148
likelihood ratios, 246–247
limitations, samples and sample handling,
  183–184

Lincolnshire, 77
linear dimensions, 120–122, *121*
lithic grains, 24
lithium metaborate fusion beads, 89
Locard, Edmond, 3–5
Locard's Exchange Principle, 4
London, England, *50*, 99
Lugol's iodine solution, 208
luminescence
    mineralogical characteristics, 111
    particle properties, *163*, 163–164
Lycopodium, 113

M

Mackereth-type gravity corer, 208
macro-charcoal, 45
macrofossils, 42–43, *43–44*
macroscopic-charcoal, 44–45, *45*
magnesium, 38
magnetic characteristics, 73, 75
magnetic resonance imaging (MRI), 72, 180
magnetite, 35, *36*, 108
major elements, 89, *90*, 91, *92–95*, 93
man-made glass particles, 55, *55*, *see also* human
    origin of particles
Mann-Whitney U-test, 231
marker compounds, 112
mass spectrometry (MS), 98
material density, 68
maximum caliper diameter, 121
maximum entropy method, 157
maximum Feret diameter, 121
maximum projection sphericity, 136
McCrone Particle Atlas, 6
mercury injection, 71
metal detectors, 213
metallic fragments
    soil constituents, 32
    trace evidence acceptability, 53–54, *54*
metamorphic rocks, 13, 15–16, *16*
metamorphism, 38
meterological data, 9
methylene iodide, 35
methyl tertiary butyl ether (MTBE), 112
Metropolitan Police Laboratory, 6
mica, 165
microbial populations, 112–113
micro-charcoal, 45, 51
microcline, *36*
microfabrics
    physical characteristics, 69, *70–72*, 71
    rocks and soil, 178–180, *179*
microfossils, *40–41*, 42, 46–51
micromorphology, 69

micro-photometry, 180
microprofilometry, 160–161
microscopic-charcoal, 45
microscopic methods, 106
mineralogical characteristics
    automated mineralogical analysis, 107–109,
        109–110
    bulk luminescence properties, 111
    clay mineral assemblage, 109–110
    computerized mineralogical analysis, 107–109,
        109–110
    heavy mineral analysis, 106
    mineralogical characteristics, 104–111
    modal mineralogy, 104–106, *105*
    organic carbon content, 111
    organic compounds, 111–112
    organic matter characteristics, 104–118
mineraloids, 34
minerals
    clay minerals, 38–39, *40–41*
    heavy type, 35, *36–37*, 37
    light type, 35, *36–37*
    nature of, 34–35
    rare earth type, 37–38, *39*
mites, 33
Mobile Bay, Alabama, 39
modal analysis, 105
modal mineralogy, 104–106, *105*, 106
moderately flattened particles, 129
Moh's scale of hardness, 69
monazite, 37–38
monocrystalline grains, 24
monomineralic sand grains, 23
Moro, Aldo (Italian Prime Minister), 76, 167
morphology, 126
Mossbauer spectroscopy, 177
MRI, *see* magnetic resonance imaging (MRI)
MS, *see* mass spectrometry (MS)
MTBE, *see* methyl tertiary butyl ether (MTBE)
mud
    artificial, 28
    evidence, 2
    Locard's definition, 5
    pollen, 117
    trace evidence acceptability, 27–28, *29–30*
mud flaps, trace evidence, 213
mud fraction, 17, 27
multiple correlation method, 242
multiple regression method, 242
multiple stepped surfaces, 150
multiple working hypotheses, 265
multi-technique comparison data, 255–256
multivariate analysis, 242–244
multivariate analysis of variance (MANOVA), 232,
    242

Munsell color charts, 61–63
muscovite, *37*

N

NAA, *see* neutron activation analysis (NAA)
nannoconids, 49
National Geographic magazine, 160
natural dusts, 32–33
natural glasses, 41, *see also* glasses
negative, numerical category scales, 266
neutron activation analysis (NAA), 89
New Scotland Yard, 6
nitrate, 88
nitrogen, 98
nitrogen gas injection, 71
NMR, *see* nuclear magnetic resonance (NMR)
noncrystalline materials, 177–178
non-elongated particles, 128
non-flattened particles, 128
normal distributions, 229
Northern Ireland, 259
nuclear magnetic resonance (NMR), 177
numerical category scales, 266, 268

O

obsidian glass, 41
oceans, 9
olivine, *36*
OLLS, *see* ordinary linear least squares regression
    (OLLS)
opal phytoliths, 46, 50
optically stimulated luminescence (OSL), 164
optical petrography, 177
ordinary linear least squares regression (OLLS),
    235
ore deposits, 35
organic carbon content, 111
organic compounds, 111–112
organic matter characteristics
    diatom assemblages, 117–118
    microbial populations, 112–113
    mineralogical characteristics, 104–111
    organic carbon content, 111
    organic compounds, 111–112
    pollen assemblages, 113–117, *115–116*
organic particles, 11, 180–181
orthoclase, *36*
OSL, *see* optically stimulated luminescence
    (OSL)
ostracods, 50, *50*
outline circularity parameter, 136
oxides, 35, 37
oxygen, 38, 98

P

paint
    soil constituents, 32
    trace evidence acceptability, 55, *55*–56
palygorskite, *37*
paper, 32, 56, *56*
paper bags, 207
particle density, 68, 120
particle morphology, 126
particle properties
    alternative analytical techniques, 167
    angularity, 141, *144,* 144–145
    characterizations, 157–159, *158*
    color, 161, *161–162*
    composite characterization, 164–167
    cross-sectional area, 122, *123,* 124
    dating, isotropic, 177
    electron microprobe analysis, 167–168, *168,*
        170–174
    elemental composition, 167–169
    form, 127–130, *129–133*
    Fourier techniques, 157–159, *158*
    fractal analysis, 157–159
    fundamentals, 119
    gravel-size particles and rocks, 164–165
    irregularity, 145–146, *147–148*
    isotropic composition, 177
    laser ablation ICP-MS analysis, 169, *176*
    laser profilometry, 159–161
    linear dimensions, 120–122, *121*
    luminescence properties, *163,* 163–164
    microfabrics, 178–180, *179*
    noncrystalline materials, 177–178
    organic particles, 180–181
    perimeter, 122, *123,* 124
    roundness, 138–141, *139–140, 142–143*
    sand-size particles, 165, *166,* 167
    SEM-EDXRA analysis, 168–169, *175*
    shape, 126–146, *127–128*
    size, 119–124
    sphericity, 133–136, *135, 137–138*
    stable isotope ratios, 177
    surface area, 124
    surface texture, 148–157, *149, 151–152, 154–156*
    three-dimensional shape, 159–161, *160*
    volume, 124, *126*
    weight, 120
    x-ray tomography, 159–161, *160*
particle size distributions, 75–77, *78–83,* 79, 81
particle typology analysis, 165
particulates, *see* dusts and particulates
PCA, *see* principal component analysis (PCA)
PCBs, *see* polychlorinated biphenyls (PCBs)
PCDD/Fs, *see* polychlorinated dibenzo dioxins
        and furans (PCDD/Fs)

Pearson's product-moment correlation coefficient, 234–235
pea shingle, 21
perimeter, 122, 124
permeability, 71–72
pesticides, 112
petrographic analysis, 52
pH, 31, 86
phosphates, 37, 88
phospholipid fatty acids (PLFAs), 112–113
photographic images, 216
photoluminescence, 111
physical characteristics
    color, 60–63, *62, 64–68,* 68
    density, 68–69
    fundamentals, 59, *60*
    hardness, 69
    magnetic characteristics, 73, 75
    microfabrics, 69, *70–72,* 71
    particle size distributions, 75–77, *78–83,* 79, 81
    permeability, 71–72
    porosity, 71–72
    shape frequency distributions, 82, *84–85,* 85
    specific surface, 72–73, *73–74*
phytoliths, 46, 50
pipes, 53
pitted texture, 152
PIXE, *see* proton-induced x-ray emission (PIXE)
placers, 35
plagioclase feldspar, *36, see also* feldspar
planktonic foraminifera, 48
planting evidence, 7
plaster, 32
plastic, 32, 57
plastic containers, 205–208
plastic tools, 205
plates, 129
PLFAs, *see* phospholipid fatty acids (PLFAs)
plutonic rocks, 13
Poaceae, 117
point counting, 105, 108
polish, 148, 153
pollen
    microfossils, 46
    organic matter characteristics, 113–117, *115–116*
    trace evidence acceptability, 46
polychlorinated biphenyls (PCBs), 112
polychlorinated dibenzo dioxins and furans (PCDD/Fs), 112
polycrystalline grains, 24
polyethylene bottles, 208
polymineralic sand grains, 23
polythene bags, 207–208

ponds, *see* water
Popp, Georg, 3
porosity, 52, 71–72
posterior probabilities, 247
potassium, 38
potassium/argon dating method, 104
pottery, 53, 164
praseodymium, 96
prediction error, 242
pressed powder pellets, 89
principal component analysis (PCA)
    multivariate analysis, 242
    significance evaluation, 243, *244*
prior probabilities, 247
procedures, 225–227
projected area, 122
projection sphericity, 134
prosecution investigation stages, *264,* 264–265
proton-induced x-ray emission (PIXE), 167
purposeful sampling, *197–199,* 198
PyGC, *see* pyrolysis gas chromatography (PyGC)
pyrite, 41
pyrolysis gas chromatography (PyGC), 112
pyroxenes, 35

Q

QEMSCAN, 107
QUANTIMET image analyzer, 145
quantitative methods, 226
quartz
    cathodoluminescence, 163
    light minerals, 35, *37*
    noncrystalline materials, 178
    quartz-feldspar-rock fragment diagrams, 106
    sand-size particles, 165
    surface texture, 150
    thermoluminescence, 164
Quaternary deposits, 39, 145
Queensland, Australia, 82
questioned soil samples
    fundamentals, 183
    handling, 210–223, *211–223*

R

radioactive isotopes, 102, 104
radiocarbon accelerator mass spectrometry (^{14}C MAS) dating, 45–46
radiocarbon dating, 42–43, 46
radiogenic isotopes, 102
radionuclides, 102
rainfall, 9
Raman spectroscopy, 169, 177–178
randomness, 226

random sampling, 184–185
rare earth elements (REE), 93, *95–96, 96*
rare earth minerals, 37–38, *39*
rare features, 152
rarity, 226
redox potential, 86
REE, *see* rare earth elements (REE)
reference samples, 183
reflectance, 148
reflectance microscopy, 180
reflected light microscopy, 149
refractory ceramics, 163
Regent's Canal, London, 50
regression analysis, 233–235, 242
reniform shape, 127
representativeness, 226
residual, 242
Rhizopoda amoebae, 49
rhyolite glass, 41
ribbon method, 105–106
ridges, 150
River Avon, England, 87
rivers, 9
robust ANOVA, 246
Rock Color Chart, 61
rocks
    character, 12–13, *14*
    composite characterization, 164–165
    exclusion/inclusion, 255
    fragments, 24
    igneous, 13, *15*
    metamorphic, 13, 15–16, *16*
    microfabric, 178–180, *179*
    microfabrics, 180
    samples, 210
    sedimentary, 16–17, *17–18*
rod-like particles, 127
rods, 129
roots, higher plants, 31
rope, 57, 213, *see also* fibers
roughness, 148
roughness ratio, 152
rough texture, 152
rounded particles, 141
roundness, 126, 138–141, *139–140, 142–143*
roundness parameter, 134, 136
rubber, 57
rubidium/strontium dating method, 104
rutile, 35, *36*

S

sacking, 57, *see also* fibers
samarium, 96
samarium/neodymium dating method, 104

samples and sample handling
    associated information, 208–210, *209*
    containers, 205–208, *206–207*
    crime scenes, 186–200, *187–200*
    differences between samples, 248–251
    fundamentals, 183–184
    guidance, 184–186
    indistinguishability of samples, 248–251
    labeling, 208–210, *209*
    limitations, 183–184
    questioned soil samples, 210–223, *211–223*
    similarity of samples, 248–251
    size of samples, 201–202, *203–205*
    storage, 210
    strategies, 186–201
    tools, 205–208, *206–207*
    type of samples, 201–202, *203–205*
    wider areas, 201
sand, 17, 22–25, *26–29*
scanning electron microscope (SEM), *see also*
        SEM-EDXRA analysis
    cathodoluminescence, 163
    composite characterization, particles, 165
    micro-charcoal, 51
    microfabric, 69, 178
    organic particles, 180
    size and shape, 122
    sphericity, 136
    surface texture, 149–150, 153, 157
Schmidt rebound hammer, 69
SCMT, *see* synchrotron computed
        micro-tomography (SCMT)
sea, *see* water
secondary ion mass spectrometry (SIMS), 167
secondary transfer, 268
sedimentary rocks
    fundamentals, 17, *17–18,* 21
    gravel, 21–22, *23–26*
    mud, 27–28, *29–30*
    sand, 22–25, *26–29*
sedimentation/pipette method, 75
sediments, 17, *19–21,* 21–28
SEM-EDXRA analysis, *see also* Scanning electron
        microscope (SEM)
    automated mineralogical analysis, 107–108
    composite characterization, particles, 165
    elemental composition, 167–169, *175*
    major and trace elements, 91, 93
    modal mineralogy, 106
semiquantitative grain surface textural analysis, 153
settling tube, 75
shape
    angularity, 141, *144,* 144–145
    form, 127–130, *129–133*
    frequency distributions, 82, *84–85,* 85

fundamentals, 126
gravel, 22
  irregularity, 145–146, *147–148*
  roundness, 138–141, *139–140, 142–143*
  sphericity, 133–136, *135, 137–138*
  three-dimensional analysis, 159–161, *160*
shape factor, 145
shell materials, 31, 48
shingle, 21
shock metamorphism, 15
shoes, *see* footwear
siderite, *36*
sieving, 75–77
signature lipid biomarker (SLB) technique, 112
significance evaluation
  Bayesian statistical approach, 247, *267*
  chance coincidence probabilities, 246–247
  classical hypothesis testing, 231–233, *232*
  coincidence probabilities, 246–247
  combined approaches, 245–246
  correlations, *228*, 233–235, *236–241*
  database and database interrogation, 256–264,
    260–263
  differences between samples, 248–251
  direct data comparison, 248–251, *250,
    252–254*
  exclusion, 255–256
  exploratory data analysis, 227–231, *228–230*
  future directions, 269
  guidelines, 225–227
  hierarchical cluster analysis, 243–244, *244–245*
  inclusion, 255–256
  indistinguishability of samples, 248–251
  likelihood ratios, 246–247
  multi-technique comparison data, 255–256
  multivariate analysis, 242–244
  principal component analysis, 243, *244*
  procedures, 225–227
  regression analysis, 233–235, 242
  similarity of samples, 248–251
  strength of evidence, *264–265*, 264–269, *267*
silicates, 37
silicon, 38
sillimanite, *36*
silts, 17, 28
similarity of samples, 225, 248–251, 266
Similarity Probability Model, 233
Simon's Wood site, 79, 235
Simpson Desert, 104
SIMS, *see* secondary ion mass spectrometry
    (SIMS)
single crystal x-ray diffraction, 177
size, *see also* particle size distributions
  aspects, 119–120
  cross-sectional area, 122, *123*, 124

distributions, 75–81, *78–83*
  gravel-size particles, 164–165
  linear dimensions, 120–122, *121*
  perimeter, 122, 124
  samples, 201–202, *203–205*
  sand-size particles, 165, *166*, 167
  surface area, 124
  volume, 124, *126*
  weight, 120
Skegness, Lincolnshire, 77
skin cells, 33
slag, 53
SLB, *see* signature lipid biomarker (SLB)
    technique
small pits, 150
smectites, *37*, 41
smoothness, 148
smooth texture, 152
soaps, 41
sodium, 38
sodium polytungstate, 35
soils, 2, 30–32, 178–180
spades, 213
sparse features, 152
Spearman's rank correlation coefficient,
    234–235
specific surface, 72–73, *73–74*
spectacles, 213
spherical particles, 127
sphericity, 126, 133–136, *135, 137–138*
spores, 46
squareness, 159
stable isotopes
  chemical characteristics, 98–102, *98–103*
  composition, 177
  light elements, 97
  ratios, 177
stainless steel tools, 205–206
staurolite, *36*
storage, 210
strategies, crime scenes, 186–201, *187–200*
streams, *see* water
street dusts, 33, *34*
strength of evidence, *264–265*, 264–269, *267*
striations, 150
Strzelecki Desert, 104
Student's *t* test, 231–232
sub-angular particles, 141
sub-aqueous sediment, 208
sub-rounded particles, 141
sulfates, 37, 88
sulfur, 98
surface, specific, 72–73, *73–74*
surface area, 124
surface markings, 148

surface texture
  gravel, 22
  particle properties, 148–157, *149, 151–152,*
    154–156
  shape, 126
surface transport processes, 10
suspension, 213
Sussex, 251
Sutton-on-Sea, Lincolnshire, 77
synchrotron computed micro-tomography
    (SCMT), 160

T

talc, *37*
tap water, 33, *see also* water
teeth, 31, 98
TEM, *see* transmission electron microscope (TEM)
temperatures, 9
ternary diagrams, 106
Tertiary epoch, 49
test, 48
testate amoebae, 49
tetra-kaikahedron, 138
textiles, *see* fibers
thermal analysis, 110
thermoluminescence, 111, 164
thorium, 37
Thoulet's solution, 35
three-dimensional shape analysis, 159–161, *160*
tidal data, 10
tiles, 53
time, 30
Tirari Desert, 104
tire tracks, crime scenes, 196
tools, 205–208, *206–207*
topography, 30
total dissolved solids, 87
touch test, 152
touramline, *36*
trace elements, 89, *90,* 91, *92–95,* 93
trace evidence acceptability
  algae, 46–47, *47–48*
  amorphous materials, 41
  anthropogenic materials, 51–57
  brick, *48,* 52–53
  calcareous nannoplankton, 49
  ceramics, 53
  clay minerals, 38–39, *40–41*
  clinker, 53
  coccolithophores, 49
  concrete, 51–52
  diatoms, 46–47, *47–48*
  discoasters, 49
  dusts, 32–34

  fibers, 56–57
  foraminifera, 48
  fossils, *40–41,* 42
  fundamentals, 11–12
  gels, 41
  glasses, 41
  gravel, 21–22, *23–26*
  heavy minerals, 35, *36–37,* 37
  household dusts, 33, *34*
  igneous rocks, 13, *15*
  industrial dusts, 33
  light minerals, 35, *36–37*
  macrofossils, 42–43, *43–44*
  macroscopic-charcoal, 44–45, *45*
  man-made glass particles, 55, *55*
  metallic fragments, 53–54, *54*
  metamorphic rocks, 13, 15–16, *16*
  micro-charcoal, 51
  microfossils, *40–41,* 46–51
  minerals, 34–35
  mud, 27–28, *29–30*
  natural dusts, 32–33
  natural glasses, 41
  opal phytoliths, 50
  ostracods, 50, *50*
  paint, *55,* 55–56
  paper, 56, *56*
  particulates, 32–34
  pipes, 53
  pollen, 46
  pottery, 53
  rare earth minerals, 37–38, *39*
  rocks, 12–17
  sand, 22–25, *26–29*
  sedimentary rocks, 16–17, *17–18*
  sediments, 17, *19–21,* 21–28
  slag, 53
  soaps, 41
  soils, 30–32
  spores, 46
  street dusts, 33, *34*
  testate amoebae, 49
  tiles, 53
  unusual particle types, 57
  water bodies, particulates in, 33–34
transmission electron microscope (TEM), 150, 180
triangularity, 159
triboluminescence, 111
Tryonia protea, 43
tyranny of numbers, 226
Tyrrhenian Sea coast, 167

U

Udden-Wentworth scale, *19,* 38

ultra-high XCT (UHR-XCT) systems, 159
ultraviolet (UV) spectroscopy, 177
uniqueness, 226
United Kingdom, 258, *see also* England
United States, 257–258, *see also specific state*
unusual matter, 32
unusual particle types, 32, 57
uranium/lead dating method, 104
uranium-series dating, 43
uranium/thorium dating method, 104

V

valves, algae, 46
Van-Veen-type grab sampler, 208
variable pressure SEM (VP-SEM), 165
vegetation storage, 210
vermiculite, *37*
vermiform shape, 127
vertebrate animals, 31
very angular category, 140
very rough texture, 152
very smooth texture, 152
Vickers Hardness tester, 69
volcanic glasses, 41
volcanic rocks, 13
volume, 124, *126*
VP-SEM, *see* variable pressure SEM (VP-SEM)

W

Wales, 257–259
water
    body transport, 9
    electrical conductivity, 87
    lake level data, 9
    particulates, 33
    rivers, 9

samples, 208
tap water, 33
Testate amoebae, 49
trace evidence acceptability, 33–34
wavelength-dispersive (WD) electron microprobe
    analysis (EMPA), 167
waxy texture, 150, 152
weathering phenomena, 10, 38, 150, 153
weight, 120
West Yorkshire, 100
wet chemical methods, 89, 96
wet-chemical potassium dichromate oxidation
    method, 111
wet sieving, 75–76
wheel arches, 213
wider areas, 201
Wiltshire, 264
wind speed and direction, 9
wustite, 108

X

x-radiography, 71, 180
x-ray computed tomography (X-ray CT), 159
x-ray computer aided tomography (X-ray CT),
    180
x-ray fluorescence spectrometry (XRF), 89, 167
x-ray micro-tomography, 71, 159
x-ray powder diffraction (XRD)
    automated mineralogical analysis, 108
    clay minerals, 38, 110
x-ray sedigraph method, 75
x-ray spectroscopy, 178
x-ray tomography, 159–161, *160*

Z

zircon, *36*